Sustaining Climate Action With AI

Mohieddine Rahmouni
King Faisal University, Saudi Arabia

IGI Global
Scientific Publishing
Publishing Tomorrow's Research Today

Vice President of Editorial — Melissa Wagner
Director of Acquisitions — Mikaela Felty
Director of Book Development — Jocelynn Hessler
Production Manager — Mike Brehm
Cover Design — Jose Rosado

Published in the United States of America by
IGI Global Scientific Publishing
701 East Chocolate Avenue
Hershey, PA, 17033, USA
Tel: 717-533-8845 | Fax: 717-533-7115
Website: https://www.igi-global.com E-mail: cust@igi-global.com

Library of Congress Cataloging-in-Publication Data

Names: Rahmouni, Mohieddine, 1977- editor
Title: Sustaining climate action with AI / edited by Mohieddine Rahmouni.
Description: Hershey, PA : IGI Global Scientific Publishing, [2026] | Includes bibliographical references and index. | Summary: "This book advances scholarly and practical understanding of how artificial intelligence can be effectively and responsibly leveraged to support long-term climate action and sustainable development"-- Provided by publisher.
Identifiers: LCCN 2026012228 (print) | LCCN 2026012229 (ebook) | ISBN 9798337399782 h/c | ISBN 9798337399799 s/c | ISBN 9798337399805 ebook
Subjects: LCSH: Climate change mitigation--Technological innovations | Artificial intelligence--Environmental aspects | Carbon dioxide mitigation | Renewable energy sources--Data processing | Climatic changes--Prevention--Technological innovations | Sustainability
Classification: LCC TD171.75 .S94 2027 (print) | LCC TD171.75 (ebook)
LC record available at https://lccn.loc.gov/2026012228
LC ebook record available at https://lccn.loc.gov/2026012229
British Cataloguing in Publication Data
A Cataloguing in Publication record for this book is available from the British Library.

Table of Contents

Detailed Table of Contents

Wildfires have turned into one of the most devastating and fast spreading forms of climate changes, posing a threat to ecosystems and human habitats, as well as carbon regimes around the world. Conventional approaches to early warning of wildfires, which are limited to a limited number of sensors and are designed reactive responses to risk conditions, are becoming less sufficient in the context of the growing fire seasons and new fire behavior. Artificial intelligence (AI) can be used to increase wildfire disaster risk reduction (DRR) with transformative capabilities through the fusion of real-time data, modeling of predictive fire behavior, and the automatic dissemination of early warning. In this chapter, the author focuses on the intersections of AI technologies, such as machine learning (ML), deep learning (DL), remote sensing analytics, and Internet of Things (IoT) sensor integration, to construct next-generation early warning systems (EWS) of wildfires.

Climate change is increasingly threatening agricultural productivity, food security, and sustainability of global food systems. Artificial intelligence (AI) and digital agriculture technologies offer new opportunities to enhance climate resilience and optimize resource use. This chapter proposes a five layer AI enabled smart agriculture

The chapter explores the way the Artificial Intelligence can go beyond carbon mitigation to ecological regeneration and restoration-oriented governance of climate action. It states that AI can be the most useful in combination with the ecological objectives, adaptive planning, governance safeguards, and long-term resilience. To reinforce this stand, the chapter presents the AERSI framework, which relates environmental sensing, analytical intelligence, adaptive decision support, integration to governance, and sustainability responsibility. The targeted literature review and analytical evidence reveal that the use of AI can reinforce the monitoring, predictive planning, and restoration decisions; yet, it is conditional to the quality of data, institutional capacities, local involvement, and ecological legitimacy. The chapter concludes that AI can only be used as restoration infrastructure through a combination of technological ability, governance responsibility, and regenerative outcomes.

The transition toward a circular economy requires intelligent systems capable of redesigning traditional linear supply chains into regenerative value networks. This chapter explores AI-enabled circular economy models that integrate advanced analytics, machine learning, IoT, and predictive optimization to enhance resource efficiency, waste minimization, and sustainable value creation. By embedding artificial intelligence across procurement, production, distribution, consumption, and reverse logistics, organizations can achieve real-time visibility, predictive maintenance, demand forecasting, and automated material recovery. The study develops a conceptual framework linking AI capabilities with circular business strategies,

imagery, Internet-of-Things (IoT) sensor feeds, and historical climatological archives. Using a qualitative, secondary-data-based approach, it proposes an Integrated AI-Sustainability Framework with three hierarchical levels: Data Layer, AI Processing Layer, and Decision Layer, along with a Real-Time Adaptive Climate Prediction System. The chapter introduces an AI-Based Sustainability Impact Index to assess environmental, economic, and social outcomes. Emphasis on Explainable AI ensures transparency, accountability, and stakeholder trust while addressing data constraints, biases, and energy demands. Findings indicate AI improves predictive accuracy, resource optimization, and policy formulation for adaptive, transparent, and scalable climate solutions overall.

This paper explores how green innovation systems can be improved with the help of artificial intelligence (AI) to push towards decarbonisation, particularly when it comes to wind-tree-based systems of renewable energy. Based on the Technology-Organisation-Environment (TOE) paradigm, the study elaborates and estimates a structural equation model (SEM) in order to investigate the connections between AI, GIS, and decarbonisation results in Tanzanian universities. The results specify that AI has a strong aptitude to enhance the performance of green innovation systems and is a mediator in the enhancement of the performance of decarbonisation. AI enhances efficient working and sustainable institutional practices by maximising energy consumption and promoting renewable energy sources. This paper offers empirical results in the case of a developing-country setting and illuminates the manner in which AI-based systems can facilitate climate action at the HEI level. Those findings are applicable as practical and policy-relevant information for sustainability in higher education.

Climate change poses disproportionate risks to vulnerable populations in the Global South, where social inequality, limited adaptive capacity, and fragile institutions

Preface

The accelerating climate crisis has emerged as one of the defining challenges of the twenty-first century, demanding transformative, interdisciplinary, and globally coordinated solutions. From intensifying wildfires and biodiversity degradation to food insecurity, water scarcity, and environmental displacement, the consequences of climate change are increasingly visible across all regions of the world. At the same time, rapid advancements in Artificial Intelligence (AI) are reshaping how societies collect data, generate insights, make decisions, and respond to complex environmental challenges. The intersection of climate action and intelligent technologies therefore represents not only a technological frontier, but also a critical opportunity to reimagine sustainability, resilience, and ecological stewardship on a global scale.

This edited volume, *Sustaining Climate Action With AI*, was developed with the objective of exploring how AI-driven systems can contribute to meaningful and sustainable climate solutions across diverse sectors and geographies. The chapters included in this book collectively demonstrate that AI is no longer limited to theoretical applications or isolated experiments; rather, it is increasingly becoming a practical instrument for climate adaptation, environmental governance, resource optimization, ecological restoration, and resilience planning. From predictive climate analytics and disaster preparedness to sustainable agriculture and circular economy systems, the contributors present innovative frameworks and evidence-based discussions that illustrate the growing relevance of AI in supporting long-term sustainability goals.

Importantly, this volume also recognizes that technological advancement alone cannot resolve climate-related challenges. The deployment of AI within climate action must be guided by ethical governance, inclusivity, transparency, accessibility, and ecological responsibility. Throughout the chapters, authors critically examine issues such as data inequality, infrastructure limitations, digital literacy, governance gaps, and social vulnerability. These discussions reinforce the understanding that climate technologies must remain human-centered and context-sensitive, particularly when implemented in developing regions and among vulnerable populations. By bringing together perspectives from scholars, researchers, and practitioners across multiple disciplines and countries, this book contributes to a broader conversation about how AI can support equitable and sustainable climate futures.

The contributions in this volume reflect the growing importance of interdisciplinary collaboration in addressing climate complexity. Climate action today requires expertise that extends beyond environmental science alone, incorporating information systems, engineering, governance, economics, agriculture, sustainability studies, and social sciences. The chapters included in this book showcase how integrated thinking can strengthen resilience strategies and improve decision-making processes at institutional, governmental, and community levels. Collectively, the contributors highlight the importance of balancing technological innovation with social responsibility and environmental integrity.

This book is intended for researchers, policymakers, academics, environmental professionals, sustainability practitioners, graduate students, and technology developers seeking deeper insight into the emerging relationship between AI and climate action. The volume provides both conceptual and applied perspectives, making it valuable to readers interested in climate adaptation, environmental management, sustainable development, and intelligent systems. It is our hope that this book will encourage further research, policy dialogue, and collaborative innovation in support of sustainable climate solutions worldwide.

CHAPTER OVERVIEW

Chapter 1, "AI Driven Early Warning Systems for Wildfire Disaster Risk Reduction in a Changing Climate," examines the growing threat of wildfires and explores how AI-powered technologies such as machine learning, deep learning, remote sensing, and IoT integration can strengthen early warning systems and disaster risk reduction strategies.

Chapter 2, "AI Enabled Climate Resilient Smart Agriculture for Sustainable Food Systems," introduces a five-layer AI-enabled smart agriculture framework designed to improve climate resilience, optimize resource efficiency, and support sustainable food systems through precision agriculture and predictive analytics.

Chapter 3, "AI-Based Biodigesters and University Engagement for Biodiversity Conservation in East Africa," investigates the role of AI and predictive analytics in biodiversity conservation within the Tanzanian context, emphasizing university collaboration, big data integration, and sustainable environmental resource management.

Chapter 4, "AI-Driven Regenerative Intelligence for Ecological Restoration and Climate Action," presents the AERSI framework and explores how AI can contribute to ecological restoration, adaptive governance, and regenerative climate action while addressing governance and ecological legitimacy concerns.

Chapter 5, "AI-Enabled Circular Economy Models and Sustainable Supply Chains," discusses how AI technologies can transform traditional supply chains into circular and regenerative systems through predictive optimization, resource efficiency, and intelligent waste reduction mechanisms.

Chapter 6, "Artificial Intelligence for Climate Smart Agriculture in Developing Countries – Pathways for Resilience, Sustainability, and Inclusive Climate Action," evaluates AI applications in climate-smart agriculture, particularly within developing economies, while examining barriers such as infrastructure gaps and digital inequality.

Chapter 7, "Artificial Intelligence for Integrated Environmental Resilience: Agriculture, Water, and Biodiversity," explores how AI can support integrated environmental resilience by connecting food systems, water governance, and biodiversity conservation through intelligent environmental informatics and geospatial analytics.

Chapter 8, "Artificial Intelligence in Climate Change Prediction and Sustainability Solutions," focuses on AI-enhanced climate forecasting systems and sustainability frameworks that integrate satellite imagery, IoT data, and explainable AI to improve transparency, predictive accuracy, and climate governance.

Chapter 9, "Wind-Tree-Based Green Innovation Systems for Decarbonization through Artificial Intelligence in Tanzanian Universities," investigates the application of AI within renewable energy innovation systems in higher education institutions, particularly for decarbonization and sustainable institutional practices.

Chapter 10, "AI-Enabled Climate Adaptation and Resilience for Vulnerable Populations in the Global South," critically examines the opportunities and governance challenges associated with deploying AI-driven adaptation systems within vulnerable communities and developing-country contexts.

Chapter 11, "An Integrated SOR–TAM Framework using PLS-SEM Evidence for AI-Driven Climate Action in India: From Prediction to Preservation," analyzes the behavioral and technological dimensions of AI adoption for climate action using structural equation modeling and sustainability-focused technology acceptance frameworks.

CONCLUSION

As the chapters in this volume collectively demonstrate, Artificial Intelligence possesses transformative potential in advancing climate action across environmental, social, institutional, and economic dimensions. AI technologies are increasingly enabling more accurate forecasting, real-time monitoring, predictive planning, intelligent resource management, and evidence-based policymaking. Whether applied to wildfire prediction, ecological restoration, biodiversity conservation, renewable energy systems, or climate-smart agriculture, AI is emerging as a significant enabler of sustainable development and climate resilience.

At the same time, this book emphasizes that technological innovation must not be viewed as an isolated solution to climate challenges. The effectiveness of AI-driven climate initiatives depends greatly on governance quality, ethical implementation, accessibility, transparency, and local contextualization. Several chapters highlight concerns related to data inequity, infrastructural limitations, algorithmic bias, institutional readiness, and digital divides, particularly within developing regions. These issues reinforce the need for responsible AI frameworks that prioritize inclusivity, human well-being, and ecological sustainability alongside technological advancement.

Another important insight emerging from this volume is the value of interdisciplinary and cross-sector collaboration. Climate action requires coordinated engagement among governments, academic institutions, industries, local communities, and technology developers. AI systems become most effective when integrated into broader sustainability ecosystems that include environmental policy, public participation, educational institutions, and community-driven resilience strategies. The contributors to this volume demonstrate how collaborative innovation can strengthen adaptive capacity and promote long-term environmental stewardship.

The editors hope that *Sustaining Climate Action With AI* will serve as a meaningful contribution to ongoing global discussions surrounding sustainability, intelligent technologies, and climate governance. Beyond its academic value, this volume seeks to inspire practical action, future research, and informed dialogue regarding the responsible use of AI for environmental resilience and climate adaptation. The climate crisis requires bold thinking, ethical leadership, and innovative collaboration, and AI, when responsibly developed and inclusively implemented, has the potential to become a powerful catalyst for positive environmental transformation.

Finally, we extend our sincere appreciation to all contributing authors, reviewers, researchers, and institutions whose efforts made this volume possible. Their scholarship and commitment to sustainability have enriched this book with diverse perspectives and valuable insights from across the globe. We also thank the academic and professional communities dedicated to advancing climate action and sustainability research. It is our hope that this book will encourage continued exploration of intelligent, equitable, and regenerative pathways toward a more sustainable future for generations to come.

Acknowledgment

Funding: This work was supported by the Deanship of Scientific Research, Vice Presidency for Graduate Studies and Scientific Research, King Faisal University, Saudi Arabia [Grant No. KFU262272].

Chapter 1

AI-Driven Early Warning Systems for Wildfire Disaster Risk Reduction in a Changing Climate

Neeta Raj Sharma
Lovely Professional University, India

Preet Kanwal
http://orcid.org/0009-0006-5114-8381
Lovely Professional University, India

ABSTRACT

Wildfires have turned into one of the most devastating and fast spreading forms of climate changes, posing a threat to ecosystems and human habitats, as well as carbon regimes around the world. Conventional approaches to early warning of wildfires, which are limited to a limited number of sensors and are designed reactive responses to risk conditions, are becoming less sufficient in the context of the growing fire seasons and new fire behavior. Artificial intelligence (AI) can be used to increase wildfire disaster risk reduction (DRR) with transformative capabilities through the fusion of real-time data, modeling of predictive fire behavior, and the automatic dissemination of early warning. In this chapter, the author focuses on the intersections of AI technologies, such as machine learning (ML), deep learning (DL), remote sensing analytics, and Internet of Things (IoT) sensor integration, to construct next-generation early warning systems (EWS) of wildfires.

DOI: 10.4018/979-8-3373-9978-2.ch001

1. INTRODUCTION

The 21st century has seen a staggering escalation of wildfires in almost all fire-prone landscapes worldwide. Disastrous events, such as the 2019-2020 "Black Summer" fires in Australia that affected more than 18 million hectares (Boer et al., 2020), the record-breaking fires in California in 2020-2022, and fires in the Amazon Basin, Mediterranean Europe and Siberia, are not aberrations. Climate change multiplies threats via rising temperatures, prolonged droughts, lower live fuel moisture, and more extreme fire weather days, resulting in large, intense wildfires that exceed firefighting capacities (Abatzoglou & Williams, 2016).

Human and economic losses are profound. California's 2018 Camp Fire alone resulted in 85 lives lost and USD 16.5 billion in damages (Nauslar et al, 2018), while wildfire smoke is thought to result in 339,000 premature deaths annually (Johnston et al, 2012). In addition to human lives lost, wildfires cause population displacement, air and water pollution, loss of biodiversity, and release billions of tonnes of carbon dioxide, exacerbating the climate change factors that increase the risk of future fires (van der Werf et al., 2017).

Early warning figure as a priority area in the Sendai Framework for Disaster Risk Reduction 2015-2030, which states that EWS should include knowledge of risk, monitoring, warning dissemination and capacity for response by the community (UNDRR, 2017). EWS tailored to wildfires are faced with a particularly complex set of data challenges: they must integrate weather data, satellite and vegetation indices, fuel moisture, topographic features, and human activity patterns across timescales from real-time detection to probabilistic forecasts up to weeks in advance. This challenge is beyond the capabilities of classical statistical models and human analysts unaided by computers, driving a clear and pressing need for AI.

1.1 Research Problem and Identified Gap

Over the past decade, rapid advancements in AI-based wildfire research have been made, yet there remains a significant and important gap between the performance of experimental models and operational EWS. Machine learning and deep learning models consistently report impressive detection accuracy in experimental settings, but adoption by fire agencies has been limited by challenges in ingesting real-time data, maintaining models, meeting interpretability demands and institutional constraints (Jain et al., 2020). Crucially, while there are extensive literature reviews that address aspects of AI-based wildfire management (detection algorithms; fuel moisture models; social media analysis), there is no systematic analytical review that critically examines the whole EWS pipeline from wildfire detection to last-mile communications to communities, while also considering the complex challenges of climate resilience, equity, and operational feasibility.

Other reviews also report model performance (accuracy, F1-score, AUC) without critically examining their constraints for new climate conditions, difficulty in model transferability, or false alarm rates that govern trust in operational deployment. The sociotechnical aspects (governance, accountability, risk of AI-EWS perpetuating the digital divide) remain understudied compared with the technical literature.

1.2 Chapter Contribution and Structure

This chapter speaks to these gaps in an integrated critical assessment of AI-based wildfire EWS. This chapter makes three novel contributions. First, it offers a comparative assessment of AI approaches across the entire EWS process, explicitly examining trade-offs, limitations, and transferability issues alongside successes. Second, it translates global case studies into an analytical framework that highlights both successes and challenges in operationalizing EWS. Finally, it contextualises technological capacity within governance, equity and climate-policy frameworks that must be in place for AI-EWS to deliver their societal benefits, particularly to marginalised and data-poor communities.

The chapter is organized as follows: Section 2 discusses the changing wildfire threat and the inadequacy of traditional EWS. Section 3 surveys AI architectures for wildfire EWS with explicit consideration of associated advantages and disadvantages. Section 4 outlines integrated EWS system design for components. Section 5 reviews worldwide operational case studies. Section 6 explores data, algorithmic, equity, governance, validation and cost issues. Section 7 provides recommendations for the future and Section 8 concludes with key recommendations.

2. CLIMATE CHANGE AND WILDFIRE RISK: AN EXPANDING THREAT

2.1 Wildfire and the Climate Change Connection

Wildfire risk is exacerbated by climate change in a number of ways. Higher temperatures decrease live fuel moisture, making vegetation more vulnerable to ignite and burn quickly (Flannigan et al., 2013). Shifted rainfall patterns and increased evaporation and transpiration lengthen fire seasons, as observed in the western United States, Australia, Portugal and sub-Saharan Africa (Jolly et al., 2015). Increasing winter temperatures remove snowpack and enhance spring drying and fuel buildup. Further, shifts in wind patterns associated with warming of the Arctic region are driving more frequent extreme fire weather conditions, including the downslope winds that can cause fires to spread rapidly (Abatzoglou & Williams, 2016).

Fire weather hazard is projected to increase globally, with some regions experiencing an extension of fire seasons by 30-50% by mid-century for both the RCP 4.5 and 8.5 scenarios (Moritz et al., 2012). Fire frequency is increasing in the boreal forests of Canada and Russia, which store large carbon pools, outpacing their ability to suppress fires (Kasischke & Turetsky, 2006; Yang et al., 2014). Such projections highlight a critical weakness of classical risk models based on historical records: their mathematical structure is not suitable to describe future fire regimes that increasingly push beyond the bounds of historical experience.

The ecological facet of wildfire risk is coupled with increasing socioeconomic vulnerability. The wildland-urban interface (WUI) - where human settlements coexist with flammable vegetation - has grown significantly as urban development encroaches into fire-prone areas (Bowman et al., 2017). California has over 11 million people living in very high or high fire hazard severity zones. Socially vulnerable groups, such as low-income households, the elderly and those without personal vehicles, are the least able to evacuate timely and recover from fire, and are overrepresented in WUI communities (Zhang et al., 2023). This geospatial and sociodemographic complexity requires EWS that consider fire behavior models in conjunction with dynamic population exposure estimates, evacuation planning and communication, and targeted outreach to affected communities, which can only be achieved through AI-powered data integration at scale.

2.2 Limitations of Conventional EWS

Traditional EWS for wildfire rely on meteorological indices (primarily the Canadian Forest Fire Weather Index (FWI) System, the McArthur Forest Fire Danger Index (FFDI) and the Haines Index), complemented by ground-based monitoring networks, aerial surveillance, and fire lookouts (de Groot et al., 2015). While these systems offer critical operational support, they are hampered by inherent limitations in today's risk landscape.

The most important constraint is spatial resolution: many fire-prone areas lack enough meteorological stations to capture topographic microclimate effects on fire weather. Fuel maps are generally produced on a seasonal basis and do not account for dynamic changes in vegetation moisture and composition. In remote settings, detection times are long and fires can burn for hours or days before being detected. Most significantly, traditional models lack the capacity to leverage the variety of data streams that are currently available - satellite detected thermal anomalies, social media-based fire reports, atmospheric models, and near real-time sensor networks - to generate operationally relevant products at the necessary frequency (Jain et al., 2020).

These deficiencies directly translate to lives and property lost. The 2018 Camp Fire, which started in steep terrain in the early hours of the morning, highlighted the role of bottlenecks in detection-to-alert systems (lag in ignition detection, lack of effective communication, and lack of lead time for evacuations) in contributing to catastrophic fire outcomes (Nauslar et al., 2018). The 2018 Attica fires in Greece (102 dead) and 2021 Bootleg Fire in Oregon also exhibited similar patterns due to delayed official warnings and the pyroconvective effect, respectively, that confounded conventional fire spread models. AI-based technologies offer tangible ways to counter these potential shortcomings by reducing detection-to-alert lead times, improving the spatial accuracy of risk and enabling dynamic communication in rapidly evolving fire conditions.

3. ARCHITECTURES, APPLICATIONS, AND LIMITATIONS OF AI FOR WILDFIRE EARLY WARNING

3.1 Machine Learning Foundations

Machine learning involves algorithms that learn to predict patterns in data without explicit programming of rules. Supervised machine learning methods, such as random forests, gradient boosting machines (e.g. XGBoost) and support vector machines, have been shown to be effective in fire danger rating, fire probability mapping and post-fire severity mapping in wildfire management (Jain et al., 2020). These tree-based and kernel-based methods are known for their interpretability and computational speed, suitable for operational use with limited computational resources.

Random forests have been commonly used to generate fire probability maps by incorporating environmental information such as temperature, relative humidity, wind speed, fuel type, topographic slope and aspect, and past fire records (Vilar et al., 2016). Gradient boosting has been shown to perform better in terms of prediction accuracy, especially for datasets with low fire occurrence (compared to non-fire events) (Jaafari et al., 2019). Stacking methods that combine various base learners are increasingly used for enhanced operational reliability.

Yet, there are significant challenges in developing machine learning models for wildfire EWS. Class imbalance, where the number of fires is highly disproportionate in most spatial-temporal data, can lead to models that achieve high accuracy but low sensitivity to the fire-positive class (a condition that is unacceptable for life safety). Shift in data distribution is more fundamental: models trained on historical fire weather patterns may struggle to perform under new climate conditions that increasingly lie beyond the data distribution (Huot et al., 2022). Spatial generalizability is also limited, as models trained on regions with rich fire history (e.g., the western

United States, south Europe) often fail to transfer to regions with scant fire history and distinct ecological and climatic conditions. Unsupervised and semi-supervised approaches are increasingly common as a consequence of data poverty: clustering approaches reveal pyroregional typologies that allow risk models to be calibrated at a regional scale, and anomaly detection approaches detect high risk atmospheric states without relying on annotated historical fire events (Jain et al., 2020). Active learning approaches, which carefully select unlabeled data for expert-labeling, offer an efficient way to increase the amount of high-quality training data in data-scarce regions (Prapas et al., 2021).

3.2 Fire Detection and Prediction with Deep Learning

Deep learning approaches have improved on wildfire monitoring and forecasting by learning directly from raw data without feature extraction. Convolutional neural networks (CNNs), designed for image classification, have been applied to active fire detection in multispectral satellite imagery with detection rates on par or better than traditional threshold-based methods and lower false alarms in complex environments like sunglint, industrial hot spots and volcanic plumes (Zhang et al., 2019). Semantic segmentation models (U-Net, fully convolutional networks) can produce fire perimeter maps from satellite imagery at pixel resolution, enabling critical evacuation planning and operational resource allocation (Huot et al., 2022).

Recurrent models, in particular Long Short-Term Memory (LSTM) and Gated Recurrent Units (GRUs), can capture temporal patterns in weather and fuel condition time series to forecast the fire weather index up to several days in advance (Sharma et al., 2023). Spatiotemporal models such as ConvLSTMs, which combine convolutional and recurrent functions, have shown promise in fire spread modelling (Prapas et al., 2021). Recently, models based on transformers and self-attention mechanisms have outperformed previous models on sub-pixel fire detection in multispectral imagery, and are promising candidates for few-shot adaptation to regional data (Li et al., 2023).

Notwithstanding these gains, there are important operational trade-offs for deep-learning models in wildfire EWS that should be clearly recognised. They are data-intensive, needing extensive labeled imagery (or time-series data) that is not available in many at-risk developing regions. They have high computational demands, limiting real-time operation in resource-constrained or edge settings. More importantly, deep learning models are often black boxes: their decision-making processes are not readily explainable to human users, undermining operational confidence in high-stakes, emergency management applications. The reliability of models under extreme fire weather conditions, which are of most operational interest, is not always well understood due to limited training data. Overcoming these challenges will require not only methodological improvements in explainability and uncertainty characterisation (see Section 6.2) but also a concerted effort to build diverse, globally representative training data.

3.3 Integrating Remote Sensing and Satellite Data

Satellite remote sensing is integral to wildfire EWS. NASA's MODIS and VIIRS sensors provide near-real-time active fire detections at 375 m-1 km resolution with global coverage at least every day (Giglio et al., 2016). The geostationary satellites GOES-16 and GOES-17 provide fire detections every five minutes across the Americas, supporting early detection of new fires at the crucial temporal resolution (Lindsey et al., 2020). ESA's Copernicus Emergency Management Service provides additional capabilities: 10 m Sentinel-2 facilitates fire perimeter mapping and post-fire impact analysis, and Synthetic Aperture Radar (SAR) Sentinel-1 can be used to map fire scars through smoke and cloud. Landsat and Sentinel-2 vegetation index time series support drought stress and dynamic fuel mapping (Chuvieco et al., 2020).

Artificial Intelligence (AI) algorithms are key to unlocking the potential of these sensor networks. Data fusion models combine the high revisit rate of MODIS/VIIRS with the resolution of Sentinel-2 and near real-time updates from GOES to create superior fire detection and fire characterisation products that are not possible with any one sensor. Cross-source data fusion, coordinated by AI, is supporting near-real-time wildfire risk maps at relevant scales. The launch of commercial small satellite constellations, which can capture sub-meter scale images at hourly revisit, adds to the data landscape and enables more detailed AI fire intelligence (albeit at high cost, which constrains access by poorer nations).

3.4 Edge AI and IoT Sensor Networks

IoT sensor networks are important complements to satellite data by offering continuous, high-temporal-resolution observations of fire-relevant environmental conditions (temperature, moisture, wind speed and direction, fuel moisture) at spatial and temporal resolutions not accessible to satellite data. Sensor nodes with onboard cameras and fire detection using algorithms for smoke and flames can perform optical fire detection with sub-second latency, compared to satellite overpass times (Gaur & Kumar, 2020).

Edge AI, or the use of machine learning inference on low-resource sensor nodes, is especially important for wildfire EWS in rural and remote areas with poor connectivity. Models trained using quantization, pruning, and knowledge distillation can run on microcontrollers, allowing for on-device fire detection and risk assessment, eliminating the need to transmit raw data to servers, thereby saving latency, bandwidth, and costs (Farasin et al., 2021). The ALERTWildfire camera system in the western United States is an example of such a system: AI-assisted edge inference of images from hundreds of cameras provides smoke detection within minutes of ignition (He et al., 2019). AI-powered 360-camera networks, such as PanoAI, detect smoke in less than 10 minutes after ignition (Gaur & Kumar, 2020). Despite the promise, edge AI systems struggle to keep models up-to-date as fire conditions shift seasonally and in response to climate change, and are not uniformly deployed across fire-prone landscapes, with the best-endowed areas tending to be the most strategic.

3.5 Natural Language Processing and Social Media Intelligence

Social media intelligence and natural language processing (NLP) allow wildfire EWS to capture actionable fire intelligence from unstructured text sources, such as social media feeds, emergency service operational logs, and news sources. Social media - Twitter/X, Facebook and messaging apps - often include eyewitness reports of fire that can be used to inform fire detection in populated regions, offering an additional real-time intelligence source (Wukich, 2016). AI NLP systems can filter, classify and geolocate fire-related content to supplement sensor-based detection with near real-time fire intelligence.

NLP approaches such as sentiment analysis and topic modeling of social media during wildfires can shed light on fear responses, compliance with evacuation orders, and information needs, guiding emergent communication. Large language models (LLMs) have been used to integrate diverse information sources - sensor warnings, satellite observations, official warnings and social media reports - into natural language situation reports for emergency managers and the public (Li et al., 2023). But there are significant risks associated with NLP-based techniques that need to be considered: social media data is messy, skewed towards densely populated and highly connected regions, and prone to the spread of mis- and disinformation. LLMs are prone to hallucination, which may lead to incorrect information being included in official situational awareness reports. Therefore, rigorous validation and human oversight measures are critical for NLP applications in wildfire EWS.

Table 1. Comparative Analysis of AI Techniques for Wildfire Early Warning Systems

AI Technique	Category	Strengths	Key Limitations	Primary EWS Application
Random Forest	Machine Learning	High interpretability; handles mixed data types; robust to outliers; computationally efficient	Susceptible to overfitting with heavily imbalanced fire/non-fire datasets; limited spatial generalization across climatic regions	Fire danger rating; ignition probability mapping
Gradient Boosting (XGBoost)	Machine Learning	Superior accuracy on imbalanced datasets; built-in feature importance; fast inference	Prone to distributional shift when trained on historical data that does not reflect future climate regimes; requires careful hyperparameter tuning	Post-fire severity prediction; fuel moisture estimation
Convolutional Neural Network (CNN)	Deep Learning	Automated spatial feature extraction from imagery; strong detection performance on satellite/camera data	Data-hungry; computationally expensive to train; limited interpretability; prone to false alarms in visually similar conditions (e.g., sunglint, dust)	Active fire detection from satellite imagery; smoke detection from cameras
LSTM / GRU	Deep Learning (RNN)	Captures long-range temporal dependencies; effective on meteorological time series	Vanishing gradient issues over very long sequences; high memory requirements; difficult to interpret temporal attention	Multi-day fire weather forecasting; fire behavior time-series prediction
ConvLSTM	Spatiotemporal Deep Learning	Jointly models spatial and temporal dynamics; well-suited for fire spread prediction across grids	Very high computational cost; requires large volumes of labeled spatiotemporal data; prone to error propagation in multi-step forecasts	Fire propagation prediction; spatiotemporal risk mapping
Vision Transformer (ViT)	Deep Learning (Transformer)	Superior long-range dependency modeling in imagery; strong performance on sub-pixel fire detection	Extremely data-hungry; high compute requirements; limited applicability in low-resource or real-time edge environments	Sub-pixel fire detection; multi-spectral satellite image analysis

continued on following page

Table 1. Continued

AI Technique	Category	Strengths	Key Limitations	Primary EWS Application
NLP / LLM	Natural Language Processing	Rapid extraction of fire intelligence from unstructured text; multilingual capability; automatic report generation	Hallucination risk; sensitivity to input noise in social media; limited geolocation precision; may amplify misinformation	Social media fire reporting; automated situation reports; multilingual community alerts
Graph Neural Network (GNN)	Geometric Deep Learning	Natively models spatial connectivity; captures topographic fire spread pathways not representable in grid architectures	Requires graph construction from complex topographic data; limited operational deployment to date; scalability challenges over large areas	Fire spread modeling along topographic networks; evacuation route optimization

Note: EWS = Early Warning System; CNN = Convolutional Neural Network; LSTM = Long Short-Term Memory; GRU = Gated Recurrent Unit; ViT = Vision Transformer; LLM = Large Language Model; GNN = Graph Neural Network.

4. DESIGN AND MODULES OF AN AI-BASED WILDFIRE EWS

4.1 End-to-End System Design

A successful AI-based wildfire EWS encompasses several layers of functionality in its operation: data acquisition, preprocessing, wildfire risk modeling, decision-making, and alert. Data acquisition layers acquire inputs from satellite data, IoT sensor networks, weather stations, social media APIs, and land cover data. Preprocessing layers conduct quality assurance, data imputation, spatial resampling and data fusion. Artificial intelligence risk modeling systems transform the preprocessed data into risk maps, fire danger ratings, ignition probability, and fire spread projections. Decision support tools convert models into information for fire managers and emergency planners. Warning systems translate detections and risk ratings into warnings via mass notification systems, emergency broadcast systems, mobile apps and social media platforms.

Cloud computing platforms offer the capacity and computational resources needed for country- and continent-scale AI inference. Systems such as Google Earth Engine, Amazon Web Services and Microsoft Azure Planetary Computer are now the backbone of large-scale wildfire monitoring systems, providing processing of petabyte-scale satellite data archives and analytics in real-time streams (Prapas et al., 2021). Hybrid cloud-edge architectures, using cloud computing for model training and edge deployment for operational use (quick detection and alerting) are increasingly recognised as the best design for operational wildfire EWS, offering the scale of computing resources needed for efficient model training and low-latency alerting through edge deployment.

4.2 Fire Danger Rating and Prediction

AI fire danger rating models integrate downscaled weather forecasts with dynamic fuel moisture content to produce spatially and temporally variable fire danger maps. Numerical weather models (e.g., ECMWF-IFS and NOAA-GFS) are corrected and downscaled with machine learning to enhance fine-scale predictions in rugged landscapes (Abatzoglou & Williams, 2016). Fuel moisture, a key predictor of fire ignition and behaviour, is modeled from satellite vegetation indices and weather data using random forest and neural network models (Chuvieco et al., 2020).

Innovation in operationally-relevant day-to-week fire danger forecasting allows for resource allocation and readiness advice. LSTM and transformer models trained on past fire weather and fire occurrence data have been shown to predict periods of high fire risk, and have been found to outperform traditional fire weather indices in comparative assessments (Sharma et al., 2023). Fire managers are increasingly using probabilistic forecast products, which convey uncertainty ranges instead of single deterministic values, for risk-based decision-making to guide resource allocation, moving towards impact-based warnings as recommended by the World Meteorological Organization.

4.3 Real-time Detection and Mapping

Fire detection relies heavily on sensor fusion. AI fusion algorithms weight the certainty of detections from various sensors (satellite, cameras, IoT sensors) based on sensor precision, environmental factors, and sensor inter-corroboration to reduce false alarms while maximising the likelihood of fire detection (Lindsey et al., 2020). Automated fire perimeter mapping, using time-series analysis of satellite data and fire behavior models, offers near real-time assessment of fire size, spread and direction - key information for evacuation and firefighting resource deployment.

Camera AI systems can detect smoke in the first minutes of fire development, when fires can be put out, before they develop into uncontrollable blazes. Combining LiDAR-based fuel structure data with physics-based and AI fire models allows three-dimensional fire spread modelling, including crown fire and spotting phenomena in complex terrain - an improvement over traditional two-dimensional models.

4.4 Impact forecasting and evacuation decision making

AI-based EWS are increasingly also capable of impact forecasting, predicting the likelihood of exposure of specific communities, infrastructure or ecosystems to fire over time periods. Human exposure models integrate fire spread predictions with population and transportation network models to forecast time of evacuation and road bottlenecks (Cova et al., 2005). AI-driven evacuation routing models can provide dynamic evacuation advice as fire perimeters shift and roads become blocked, decreasing the risk that communities will be "cut off" by rapidly spreading fires.

Smoke models (combining atmospheric transport models with AI-based estimates of fire emissions) can predict air quality impacts for communities downstream of active fires to inform public health warnings and emergency medical services (Johnston et al., 2012). Linking impact forecasting with alerting systems allows for risk-based alerting, supplying specific evacuation advice to communities at greatest risk while guiding preparedness measures in lower risk communities - a major improvement over blanket, area-based warnings.

4.5 Alerting, Communication, and Last-Mile

Alert dissemination is arguably the most critical component of all EWSs: technical fire risk assessment is useless if it is not communicated to communities in a timely, understandable and actionable format, and reaches all vulnerable members through the most effective channels. AI plays multiple roles in this component. Natural language generation algorithms can automatically generate human-readable warning messages, tailored to the specificity and urgency of the model-based risk forecast, to eliminate the delay between computer-generated risk analysis and human written responses (Wukich, 2016). Demographic and behavioural profiling algorithms can target warning message content and channels - such as issuing specific evacuation routes for families with children, and shelter-in-place messages for the mobility challenged.

Omni-channel alert distribution frameworks, incorporating wireless public alert systems, social media, stand-alone mobile apps, email subscribes, and community-based sirens, help ensure maximum population coverage. AI-based tracking of alert uptake via message open rates, social media sharing, and emergency calls provides real-time assessment of communication performance and dynamic adjustment of message strategies during events (Xu et al., 2020). Modern state-of-the-art systems include the Integrated Public Alert and Warning System (IPAWS) in the USA and the EU-Alert system in Europe. The enduring challenge of last-mile communication (reaching digitally marginalised households) requires hybrid systems that retain traditional communication channels such as AM radio broadcasts, community sirens and community messengers, in addition to AI-driven digital channels.

5. OPERATIONAL CASE STUDIES AROUND THE WORLD

5.1 United States: National Wildfire Management using AI

The U.S. has one of the most sophisticated wildfire management programs with the National Interagency Fire Center (NIFC) and state agencies. The GOES-16/17 Fire Detection and Characterization Algorithm delivers real-time fire detection information every five minutes, with AI-based post-processing algorithms to minimise false alarms and support finer-scale characterisation of fire pixels (Lindsey et al., 2020). The National Fuel Moisture Database combined with machine learning models enables operational landscape-scale assessment of live and dead fuel moistures (Chuvieco et al., 2020). California-based utility-based wildfire prevention programs, such as Pacific Gas and Electric (PG&E) and Southern California Edison, have adopted AI-based risk assessment of grid-wildfire risk to inform dynamic de-energisation strategies. San Diego County's AI-powered warning system integrates camera, satellite, and social media data to support near-real-time situational awareness and public warning. However, equity challenges remain in last-mile delivery, while model performance under increasingly unfamiliar fire weather extremes - such as large pyroconvective fires - remains an operational challenge.

5.2 Australia: Black Summer Learnings and Fire Management

Australia's Black Summer fires of 2019-2020 spurred significant investment in AI-driven early warning and fire management. The Bureau of Meteorology's Fire Weather Services offer gridded fire danger ratings (McArthur FFDI), with research to enhance local resolution using ML downscaling (Boer et al., 2020). The Australasian Fire and Emergency Service Authorities Council (AFAC) is trialling AI fire behavior prediction systems using real-time weather, fuel moisture and topography. AI triage algorithms are also being used to improve Australia's Emergency Alert system. A key insight from the Black Summer was the multiple pyroconvective episodes that produced fire-induced thunderstorms, and which highlighted the boundaries of current fire behavior models and inspired an updated AI model architecture capable of expressing extreme fire-atmosphere exchanges (Boer et al., 2020). Operational implementation of AI fire behavior prediction is a work in progress, highlighting the lag between research and operations.

5.3 European Union: Copernicus Emergency Management Service

The Copernicus Emergency Management Service (CEMS) and its European Forest Fire Information System (EFFIS) offer integrated wildfire monitoring across Europe, including fire danger rating from weather forecast, satellite fire detection and fire damage assessment in EU member countries and Mediterranean partner nations (Chuvieco et al., 2020). EFFIS uses AI algorithms for fire danger rating, active fire management and burned area mapping, and provides coordination support in major fire events across borders. The Joint Research Centre's Global Wildfire Information System (GWIS) provides global coverage by AI-driven integration of satellite fire products. Research projects like the Horizon 2020 FIRE-RES project are implementing AI-based prevention and preparedness at landscape and regional scales. An ongoing challenge is the lack of integration between alert systems across EU countries, with variable incorporation of AI-based alert optimization at country level.

5.4 Opportunity in Emerging Economies

Developing economies with high fire risks (Brazil, Indonesia, India, sub-Saharan Africa) have limited wildfire EWS capabilities compared to the risks. INPE's Amazon Fire Monitoring System uses AI algorithms to process near-real-time MODIS and VIIRS fire detections for use by environmental enforcement agencies (van der Werf et al., 2017). AI risk mapping using open data (satellites and meteorological reanalysis) has been proven to be viable in South Asia with low infrastructure needs (Sharma et al., 2023). Reaching the last mile using SMS and community radio to supplement AI-based fire warnings offers opportunities to communicate in regions with low smartphone use (Wukich, 2016). However, these areas significantly suffer from poor infrastructure: limited meteorological networks, institutional technical capacities, digital connectivity, and the poorest not having access to digital warning systems. Addressing these challenges requires international investment in data infrastructure, technical capacity and multi-channel inclusive communication architecture. Table 2 captures the key features and challenges of AI-based wildfire EWS around the world.

Table 2. Synthesis of AI-Based Wildfire EWS: Global Case Studies

Region	System / Programme	AI Components Deployed	Key Achievements	Documented Gaps / Limitations
United States	NIFC; GOES-16/17 FDCA; ALERTWildfire; San Diego AI Warning System	CNN-based fire detection; ML fuel moisture modeling; NLP social media fusion; AI grid-risk assessment (PG&E)	Detection latency reduced to ~5 min; improved false alarm suppression; dynamic de-energization decisions	Equity gaps in last-mile reach; variable model performance under novel climate extremes
Australia	BoM Fire Weather Services; AFAC AI Fire Behavior Prediction; Emergency Alert System	ML-enhanced FFDI; AI fire behavior simulation; satellite-based pyroconvection detection	Post-Black Summer fire behavior understanding improved; AI-assisted evacuation alert triage underway	AI fire behavior models not yet fully operational; 2019–20 pyroconvective events exposed model boundary conditions
European Union	Copernicus CEMS / EFFIS; JRC Global Wildfire Information System (GWIS); FIRE-RES H2020	AI-enhanced fire danger rating; multi-satellite data fusion; burned area mapping algorithms	Pan-European harmonized fire monitoring; cross-border situational awareness; improved burned area products	Alert dissemination fragmentation across member states; limited AI integration in last-mile warning
Brazil	INPE Amazon Fire Monitoring System	MODIS/VIIRS AI fire detection; near-real-time active fire mapping	Comprehensive Amazon fire surveillance; near-real-time alerts to enforcement agencies	Low-bandwidth constraints; limited community-level dissemination; deforestation complicates fuel models
South Asia / Sub-Saharan Africa	Emerging open-access AI EWS pilots	Free-access satellite data (MODIS/VIIRS); meteorological reanalysis-based ML risk mapping; SMS/ radio dissemination	Cost-effective risk mapping demonstrated in data-sparse settings; transfer learning from data-rich regions showing promise	Severe data gaps; limited observational networks; low institutional capacity; last-mile communication highly constrained

Note: NIFC = National Interagency Fire Center; BoM = Bureau of Meteorology; AFAC = Australasian Fire and Emergency Service Authorities Council; CEMS = Copernicus Emergency Management Service; EFFIS = European Forest Fire Information System; INPE = Instituto Nacional de Pesquisas Espaciais; MODIS = Moderate Resolution Imaging Spectroradiometer; VIIRS = Visible Infrared Imaging Radiometer Suite; FFDI = Forest Fire Danger Index.

6. CHALLENGES, LIMITATIONS, AND ETHICAL CONSIDERATIONS

6.1 Data Integrity, Deficiency and Representation

The effectiveness of AI models is limited by the quality and representativeness of available data. Wildfire occurrence data are highly skewed: fire data are more complete in countries with well-funded fire management programs, introducing geographical biases in model transferability (Jain et al., 2020). Distributional shift, in which climate change shifts the statistical distribution of fire-related variables outside of the bounds of historical data upon which models are trained, undermines model performance in precisely those future high-risk scenarios for which EWS is urgently needed (Huot et al., 2022). Low-frequency, high-impact events such as extreme fire weather episodes and pyroconvective fires are under-represented in training data, yet prediction of these events is most important for life-safety.

To overcome these challenges, we need continued investment in data systems: scaling up and increasing the density of ground-based observational networks, digitising historical fire data, robust methods for synthetic data generation, and international data-sharing agreements. Techniques to transfer knowledge from data-rich "source" regions to data-scarce "target" regions (known as "transfer learning") hold significant promise for bringing AI-EWS to emerging markets and remote, high-risk environments (Li et al., 2023).

6.2 Transparency, Trust and Human-AI Partnership

Deep learning models are often described as "black boxes" - their reasoning is not directly accessible to human users, limiting trust and accountability in critical applications. For example, emergency managers who issue evacuation warnings or pre-position resources need to not only know the predicted risk, but also the reasoning behind the prediction in order to evaluate its credibility and apply their human expertise (Molnar, 2020). Explainable AI (XAI) approaches such as SHAP-ley Additive exPlanations (SHAP) values, Local Interpretable Model-agnostic Explanations (LIME), and attention weights in transformer models can explain model predictions after a model has been trained, without compromising model performance. But these explanations are only approximate and can be deceiving, especially for models trained on non-representative data.

Human–AI collaborative systems, where AI is used as a tool to support human decision-making, are considered best practice for wildfire management. Human-in-the-loop systems, where AI predictions are presented with uncertainty metrics and explanations to experienced human agents who maintain final decision-making power, can harness the computational power of AI while retaining the contextual awareness, domain expertise and moral agency of expert fire managers (Gaur & Kumar, 2020). Training fire managers, emergency managers, community leaders and others in the use of AI and involving them in system design is key to unlocking this collaborative potential.

6.3 Equity, Accessibility, and the Digital Divide

AI-powered wildfire EWS will only widen inequities in disaster risk reduction if access to warnings is mediated through the ownership and use of a smart phone and internet connectivity. Rural and indigenous communities in fire-prone areas face compound vulnerability: high risk of wildfire, poor connectivity, lack of transportation for evacuation, and a past history of exclusion from emergency risk management (Whittaker et al., 2020). EWS that only use mobile app alerts or social media to deliver warnings systematically exclude the most vulnerable communities - as required by the Sendai Framework.

Designing inclusive EWS demands careful consideration of multi-channel warning dissemination, avoiding abandonment of traditional channels (sirens, AM radio, messengers) in favour of digital technologies. Co-design processes that engage indigenous and marginalised groups in EWS development can uncover culturally relevant warning thresholds, languages and options for response. Harnessing indigenous fire knowledge and traditional ecological knowledge (TEK) in AI models (such as using Aboriginal burning calendars and country-specific fuel maps in fire management in Australia) can lead to more culturally sensitive and locally relevant early warning (Moosavi, 2017).

6.4 Governance, Accountability and Regulation

The use of AI in life-safety applications raises basic governance and accountability concerns. If an AI-based evacuation alert is wrong - either a false alarm creating unjustified disruption, or a failure to detect, leading to avoidablc livcs lost - questions of liability and responsibility arise. Most jurisdictions' framework of AI governance is not well suited to automated decision support systems in the emergency management domain (Ceccato et al., 2019). The EU AI Act, which categorises AI-based emergency management applications as high risk, and mandates transparency, resilience testing and human oversight, provides one regulatory model, albeit requiring further development and tailoring to regional jurisdictions for application in wildfire EWS operational settings.

A further governance consideration relates to privacy: AI-driven EWS that integrate personal location data, social media content and behavioural data need effective data governance procedures to define principles of data minimization, data retention and community consent. Regulation harmonisation of AI governance norms in disaster risk applications, potentially coordinated by UNDRR, WMO or regional mechanisms, would contribute to the harmonisation of AI governance norms and enable cross-border data sharing required for transnational fire risk monitoring (UNDRR, 2017).

6.5 Validation, Performance and Uncertainty

An area where the wildfire AI research community should do better is in the systematic reporting of model validation and performance. Published studies often report area under the receiver operating characteristic curve (AUC-ROC) or overall accuracy, but fail to report precision-recall tradeoffs, false alarm rates, or disaggregated performance by fire size, weather pattern or geographic location. False alarm rate matters: high false alarm rates reduce trust and adherence to future warnings, and high missed detection rates lead to loss of life. Research that seeks to tune thresholds to reduce false alarms may introduce more missed detections in operational scenarios that are not represented in the data used for model validation.

Quantifying uncertainty is an under-investigated but critical need for AI-EWS. Stochastic forecasts that convey uncertainty ranges (as opposed to single point estimates) facilitate risk-informed fire management. Deep learning with Bayesian inference, Monte Carlo dropout, and ensembles can yield uncertainty estimates, but computational demands and interpretability need to be carefully considered for operational applications. Out-of-sample testing using independent test data (as opposed to random training-test splits) from distinct geographical and climatic regions is essential to evaluate model generalizability, and should be considered a minimum requirement for studies purporting to be relevant for operational fire management.

6.6 Economics, Scalability and Feasibility

The costs and economics of AI-EWS development and operation are under-represented in the literature. Up-front development costs (sensor networks, satellite data subscriptions, cloud computing, and system integration) can be steep, and ongoing operating costs to train models, update software and replace hardware are an ongoing financial burden that many fire protection agencies in developing nations are unable to absorb without external assistance. The cost-benefit argument for AI-EWS investment is compelling: early warning systems typically have benefit-to-cost ratios of 4:1 to 10:1, and major fires cause far greater financial losses than the cost of EWS. However, institutional allocation of EWS funding can be difficult, especially in places where fire risk is perceived as occasional rather than endemic.

Scalability poses technical challenges different from those of installation: systems must continue to perform as volumes of data increase, as operational areas increase in size, and as climatic conditions change. The institutional constraints to deployment, such as the reluctance of experienced fire managers to adopt AI tools, inconsistent data governance across agencies, and the lack of interoperability standards, often pose more significant challenges to deployment than technical challenges. Overcoming these institutional barriers requires change management approaches, co-design processes that enhance trust among operational fire managers, and policy frameworks that mandate data interoperability and system interoperability across fire management agencies.

7. FUTURE DIRECTIONS: BUILDING AI WILDFIRE EWS RESILIENT TO A CHANGING CLIMATE

7.1 Future AI Architectures and Data Fusion

A number of technological advancements will drive the future of AI wildfire EWS. Foundation models, large-scale, pre-trained AI models that are fine-tuned for specific earth observation tasks, are emerging as viable alternatives to fire detection, fuel mapping and fire risk prediction without large, task-specific, labeled data (Li et al., 2023). Multimodal models that can natively process satellite imagery, meteorology time series, text and structured tabular data within a single model framework, provide opportunities for more holistic fire risk assessment. Generative AI and diffusion models could potentially generate plausible fire spread scenarios under new climate regimes, potentially compensating for lack of training data for extreme events.

Graph neural networks (GNNs) are particularly promising for capturing fire spread as a spatial process that explicitly models effects of topography and connectivity of a fuel network on fire behavior, which are difficult to model in grid-based frameworks (Prapas et al., 2021). Reinforcement learning approaches are being applied to dynamic fire management such as resource allocation and pre-positioning for firefighting in rapidly changing risk environments. Physics-informed neural networks, which embed physical equations governing fire behavior as constraints, could improve the model's capacity to predict fire behavior under new conditions, beyond the statistical range of the input training data - an essential quality in the face of climate change-driven shifts in fire regimes.

7.2 Enhanced Satellite Constellations and Observations

Improved satellite observations will greatly improve wildfire EWS over the next decade. Second-generation geostationary satellites, such as the Geostationary Operational Environmental Satellite (GOES)-U, will offer better resolution and new spectral information for fire detection and monitoring (Lindsey et al., 2020). Commercial small-satellite constellations such as Planet Labs, Maxar and Satellogic provide hourly revisit times and sub-meter resolution monitoring, allowing near-continuous perimeter mapping of active fire fronts at operational scales. Synthetic Aperature Radar (SAR) constellations from the European Space Agency (ESA), the Japanese Aerospace Exploration Agency (JAXA) and commercial vendors like Capella Space enable fire scar mapping and perimeter tracking through smoke and cloud cover. Aerial drones with thermal sensors enable on-call, high-resolution fire detection and perimeter mapping for incident management, complementing satellite data. Data fusion using AI techniques integrates these multiple perspectives to produce stratified fire intelligence, surpassing any individual sensor's capabilities.

7.3 Indigenous Knowledge and Community-based Design

Harnessing indigenous fire knowledge systems alongside AI-based EWS is a critical and untapped area. Across the world, indigenous communities have deep historical knowledge of fire ecology, behavior and management that represents a form of environmental intelligence that can be encoded and integrated with sensor data (Moosavi, 2017). Co-design methods that engage indigenous communities as co-producers, rather than consumers, of EWS can lead to systems more closely attuned to local ecological and social contexts. AI techniques that can encode and reason about indigenous ecological calendars, landscape classifications, and fire behavior knowledge hold potential for more culturally relevant and locally relevant wildfire management. Such methods also have inherent ethical value in honouring indigenous intellectual sovereignty and promoting self-determined community empowerment.

7.4 Compatibility with the Sendai Framework and Climate Change Agreements

AI-based wildfire EWS provide a practical technological solution to implement Sendai Framework Priority Area 2 (strengthening disaster risk governance) and Priority Area 3 (investing in disaster risk reduction for resilience), including the target of 100% multi-hazard early warning coverage by 2030. The UN Secretary-General's Early Warnings for All (EW4All) initiative, in particular, calls for the use of AI and digital technologies to address warning coverage gaps, including in low-income countries and vulnerable communities (UNDRR, 2017). Paris Agreement financing for climate adaptation (via the Green Climate Fund or the Adaptation Fund) should be made available to invest in AI-EWS infrastructure in at-risk countries, given the high value and cost-effectiveness of early warning as a climate adaptation measure, with reported benefit-to-cost ratios. Incorporating wildfire EWS into National Adaptation Plans and Nationally Determined Contributions would enhance the policy and finance connections between technology investment and climate risk mitigation.

7.5 Funding, Training and the Path to Operationalisation

To expand the application of AI-driven wildfire EWS from research to national operational systems, investment is needed in three interrelated areas: technology, human capital, and policy. In terms of infrastructure, reliable high-speed telecommunications networks (such as satellite-based broadband networks like Starlink for remote fire-prone regions) are essential to transfer data in real-time from sensors to the central decision-making center to the community. Fire weather observation networks need to be densified and upgraded: the AI model is only as good as the observations.

The human dimension is often overlooked in AI-EWS development. Training of fire agency staff, emergency managers and community leaders is not only needed to use AI-based tools, but interpret their predictions - how uncertainty is reported, what it means when the models "break" and to make professional judgment calls when AI predictions differ from observations (Molnar, 2020). Fire science programs with machine learning and data science training, and training programs for practitioners co-designed with national meteorological and fire agencies, are vital to ensuring the human capacity needed for operational AI-EWS. International capacity-building initiatives supported by WMO, UNDRR, and regional bodies such as ASEAN and the African Union can accelerate the uptake of best practices in nations most vulnerable to wildfires, and with the greatest capacity-building needs.

CONCLUSION

Wildfires are one of the most rapidly growing impacts of climate change and among the most critical issues in global disaster risk reduction. The intersection of machine learning, deep learning, remote sensing, IoT sensor networks, edge computing, and NLP is enabling revolutionary new capabilities for detecting, predicting, forecasting and alerting for wildfire risk. Case studies of operational use cases in the United States, Australia, Europe, and in developing countries show that AI-powered

EWS can substantially accelerate detection-to-alert timelines, improve forecast skill, enable forecasts for impacts, and inform more targeted alerting to communities - all of which save lives and money.

Yet, as this chapter's critical appraisal shows, operationalising this potential across society depends on addressing issues beyond model development. Data quality issues and distributional change under climate change limit model performance in the most critical ways. A lack of algorithmic transparency constrains operational confidence in critical emergency scenarios requiring human oversight. Digital equity challenges risk disproportionately enriching resource- and network-rich communities with AI-EWS, potentially exacerbating protection gaps for vulnerable groups. Policies and governance are not well aligned with the accountability requirements of AI for life safety. Current methods for model validation in the research literature do not adequately describe performance in operational applications, and the economic and institutional constraints on operational deployment are grossly underestimated.

The most valuable AI wildfire EWS will not be the most accurate systems in research validation, but those that accept from the start the constraints of operational deployability, inclusivity, interpretability, and resilience to the new and extreme conditions that climate change is bringing more often. This requires an interdisciplinary venture - bringing together computer scientists, fire behaviour scientists, emergency managers, social scientists, indigenous knowledge systems, and impacted communities - and a governance framework that considers early warning as a public good that requires long-term investment, rather than a by-product of research.

The research and operational communities have an ambitious challenge to fast-track AI capabilities to save lives in the wild. The most urgent steps include: speeding the integration of validated AI models into fire agency operations; building data infrastructure to support AI-EWS deployment in data-scarcity, high-risk fire-prone regions; establishing inclusive multi-channel dissemination systems; developing emergency-management specific AI governance frameworks; and prioritising EWS investments consistent with the Sendai Framework, Paris Agreement and global Early Warnings for All initiatives. The stakes - human lives, livelihoods, ecosystems and climate stability - could not be greater and the time to act is running out with each fire season (Stephens et al., 2013).

REFERENCES

Abatzoglou, J. T., & Williams, A. P. (2016). Impact of anthropogenic climate change on wildfire across western US forests. *Proceedings of the National Academy of Sciences of the United States of America, 113*(42), 11770–11775. DOI: 10.1073/pnas.1607171113 PMID: 27791053

Boer, M. M., Resco de Dios, V., & Bradstock, R. A. (2020). Unprecedented burn area of Australian mega forest fires. *Nature Climate Change, 10*(3), 171–172. DOI: 10.1038/s41558-020-0716-1

Bowman, D. M. J. S., Williamson, G. J., Abatzoglou, J. T., Kolden, C. A., Cochrane, M. A., & Smith, A. M. S. (2017). Human exposure and sensitivity to globally extreme wildfire events. Nature Ecology & Evolution, 1(3), 0058. DOI: 10.1038/s41559-016-0058

Ceccato, P., Flasse, S., & Gregoire, J. M. (2019). Designing a spectral index to estimate vegetation water content from remote sensing data. *Remote Sensing of Environment, 82*(2–3), 188–197.

Chuvieco, E., Mouillot, F., van der Werf, G. R., San Miguel, J., Tanasse, M., Koutsias, N., & Mbow, C. (2020). Historical background and current developments for mapping burned area from satellite Earth observation. *Remote Sensing of Environment, 225*, 45–64. DOI: 10.1016/j.rse.2019.02.013

Cova, T. J., Dennison, P. E., Kim, T. H., & Moritz, M. A. (2005). Setting wildfire evacuation trigger points using fire spread modeling and GIS. *Transactions in GIS, 9*(4), 603–617. DOI: 10.1111/j.1467-9671.2005.00237.x

de Groot, W. J., Flannigan, M. D., & Cantin, A. S. (2015). Climate change impacts on future boreal fire regimes. *Forest Ecology and Management, 294*, 35–44. DOI: 10.1016/j.foreco.2012.09.027

Farasin, A., Colomba, L., & Garza, P. (2021). Double-step U-Net: A deep learning-based approach for the estimation of wildfire damage severity through Sentinel-2 satellite data. *Applied Sciences (Basel, Switzerland), 10*(12), 4332. DOI: 10.3390/app10124332

Flannigan, M., Cantin, A. S., de Groot, W. J., Wotton, M., Newbery, A., & Gowman, L. M. (2013). Global wildland fire season severity in the 21st century. *Forest Ecology and Management, 294*, 54–61. DOI: 10.1016/j.foreco.2012.10.022

Gaur, A., & Kumar, M. (2020). A comprehensive study on machine learning and deep learning techniques for wildfire prediction. *International Journal of Environmental Research and Public Health, 17*(2), 602. DOI: 10.3390/ijerph17020602 PMID: 31963490

Giglio, L., Schroeder, W., & Justice, C. O. (2016). The collection 6 MODIS active fire detection algorithm and fire products. *Remote Sensing of Environment, 178*, 31–41. DOI: 10.1016/j.rse.2016.02.054 PMID: 30158718

Huot, F., Hu, R. L., Goyal, N., Sankar, T., Ihme, M., & Chen, Y. F. (2022). Next day wildfire spread: A machine learning dataset to predict wildfire spreading from remote-sensing data. *IEEE Transactions on Geoscience and Remote Sensing, 60*, 1–13. DOI: 10.1109/TGRS.2022.3192974

Jaafari, A., Zenner, E. K., Panahi, M., & Shahabi, H. (2019). Hybrid artificial intelligence models based on a neuro-fuzzy system and metaheuristic optimization algorithms for spatial prediction of wildfire probability. *Agricultural and Forest Meteorology, 266*, 198–207. DOI: 10.1016/j.agrformet.2018.12.015

Jain, P., Coogan, S. C., Subramanian, S. G., Crowley, M., Taylor, S., & Flannigan, M. D. (2020). A review of machine learning applications in wildfire science and management. *Environmental Reviews, 28*(4), 478–505. DOI: 10.1139/er-2020-0019

Johnston, F. H., Henderson, S. B., Chen, Y., Randerson, J. T., Marlier, M., DeFries, R. S., Kinney, P., Bowman, D. M. J. S., & Brauer, M. (2012). Estimated global mortality attributable to smoke from landscape fires. *Environmental Health Perspectives, 120*(5), 695–701. DOI: 10.1289/ehp.1104422 PMID: 22456494

Jolly, W. M., Cochrane, M. A., Freeborn, P. H., Holden, Z. A., Brown, T. J., Williamson, G. J., & Bowman, D. M. J. S. (2015). Climate-induced variations in global wildfire danger from 1979 to 2013. *Nature Communications, 6*(1), 7537. DOI: 10.1038/ncomms8537 PMID: 26172867

Kasischke, E. S., & Turetsky, M. R. (2006). Recent changes in the fire regime across the North American boreal region. *Geophysical Research Letters, 33*(9), 2006GL025677. Advance online publication. DOI: 10.1029/2006GL025677

Li, X., Zhang, Y., Li, M., & Ouyang, Y. (2023). A transformer-based approach for wildfire smoke detection using remote sensing imagery. *International Journal of Wildland Fire, 32*(4), 501–516. DOI: 10.1071/WF22130

Lindsey, D. T., Schmit, T. J., MacKenzie, W. M., Pavolonis, M. J., Seaman, C. J., & Movva, L. (2020). 10.35 µm GOES-16 ABI channel application for fires. *Journal of Applied Remote Sensing, 12*(4), 046016. DOI: 10.1117/1.JRS.12.046016

Molnar, C. (2020). Interpretable machine learning: A guide for making black box models explainable. Leanpub. https://christophm.github.io/interpretable-ml-book/

Moosavi, V. (2017). Delineating wildfire risk areas using linked weights-of-evidence and GIS—Application in the Zagros forests, western Iran. *International Journal of Wildland Fire*, *26*(6), 556–565.

Moritz, M. A., Batllori, E., Bradstock, R. A., Gill, A. M., Handmer, J., Hessburg, P. F., & Sugihara, N. G. (2012). Learning to coexist with wildfire. *Nature*, *515*(7525), 58–66. DOI: 10.1038/nature13946 PMID: 25373675

Nauslar, N. J., Abatzoglou, J. T., & Marsh, P. T. (2018). Structure, meteorology, and climatology of the Diablo and Santa Ana wind events of northern and southern California. *Natural Hazards*, *91*(3), 1475–1490. DOI: 10.1007/s11069-018-3141-y

Prapas, I., Bountos, N. I., Kondylatos, S., Michail, D., Camps-Valls, G., & Papoutsis, I. (2021). Deep learning for global wildfire forecasting. arXiv preprint arXiv:2211.00534. https://doi.org//arXiv.2211.00534DOI: 10.48550

Sharma, S., Jeet, V., Misha, S., & Pansari, J. K. (2023). Prediction of forest fire using deep learning-based approaches. *Computers & Electrical Engineering*, *108*, 108719. DOI: 10.1016/j.compeleceng.2023.108719

Stephens, S. L., Agee, J. K., Fulé, P. Z., North, M. P., Romme, W. H., Swetnam, T. W., & Turner, M. G. (2013). Managing forests and fire in changing climates. *Science*, *342*(6154), 41–42. DOI: 10.1126/science.1240294 PMID: 24092714

UNDRR. (2017). Terminology on disaster risk reduction. United Nations Office for Disaster Risk Reduction. https://www.undrr.org/terminology

van der Werf, G. R., Randerson, J. T., Giglio, L., van Leeuwen, T. T., Chen, Y., Rogers, B. M., Mu, M., van Marle, M. J. E., Morton, D. C., Collatz, G. J., Yokelson, R. J., & Kasibhatla, P. S. (2017). Global fire emissions estimates during 1997–2016. *Earth System Science Data*, *9*(2), 697–720. DOI: 10.5194/essd-9-697-2017

Vilar, L., Camia, A., & San-Miguel-Ayanz, J. (2016). A comparison of remote sensing products and forest fire statistics for improving fire information in Mediterranean Europe. *European Journal of Remote Sensing*, *49*(1), 185–201.

Whittaker, J., Blanchi, R., Haynes, K., Leonard, J., & Opie, K. (2020). Experiences of sheltering during the Black Saturday bushfires: Implications for policy and research. *International Journal of Disaster Risk Reduction*, *23*, 119–127. DOI: 10.1016/j.ijdrr.2017.05.002

Wukich, C. (2016). Social media use in emergency management. *Journal of Emergency Management (Weston, Mass.)*, *14*(4), 271–285. DOI: 10.5055/jem.2016.0263 PMID: 26312654

Xu, R., Yu, P., Abramson, M. J., Johnston, F. H., Samet, J. M., Bell, M. L., Haines, A., Ebi, K. L., Li, S., & Guo, Y. (2020). Wildfires, global climate change, and human health. *The New England Journal of Medicine*, *383*(22), 2173–2181. DOI: 10.1056/NEJMsr2028985 PMID: 33034960

Yang, J., Tian, H., Tao, B., Ren, W., Kush, J., Liu, Y., & Wang, Y. (2014). Rare and extreme events: The case of Canadian forests. *Environmental Research Letters*, *9*(7), 075005.

Zhang, G., Wang, M., & Liu, K. (2019). Forest fire susceptibility modeling using a convolutional neural network for Yunnan province of China. *International Journal of Disaster Risk Science*, *10*(3), 386–403. DOI: 10.1007/s13753-019-00233-1

Zhang, P., Sutley, E. J., & Hamideh, S. (2023). Wildfire risk and social vulnerability: Mapping compound risks in rural communities of the American West. *International Journal of Disaster Risk Reduction*, *88*, 103614.

Chapter 2
AI-Enabled Climate-Resilient Smart Agriculture for Sustainable Food Systems

Samruddhi Pandit
http://orcid.org/0009-0000-9225-9236
Vishwakarma Institute of Information Technology, India

Anuja Mukherjee
Vishwakarma Institute of Information Technology, India

Rugved Dani
Vishwakarma Institute of Information Technology, India

Abhishek Solanke
Vishwakarma Institute of Information Technology, India

ABSTRACT

Climate change is increasingly threatening agricultural productivity, food security, and sustainability of global food systems. Artificial intelligence (AI) and digital agriculture technologies offer new opportunities to enhance climate resilience and optimize resource use. This chapter proposes a five layer AI enabled smart agriculture framework integrating climate inputs, sensing infrastructure, predictive analytics, decision support, and sustainability outcomes. The framework supports climate adaptation, greenhouse gas mitigation, and long term food system resilience. Key AI applications including precision farming, smart irrigation, soil health monitoring, pest forecasting, and yield prediction are examined. Case studies highlight

DOI: 10.4018/979-8-3373-9978-2.ch002

implementation challenges such as infrastructure limitations, data governance, and digital inequality. Policy and governance strategies are discussed to support inclusive adoption. The chapter contributes a structured framework for AI driven climate resilient agriculture supporting sustainable food systems and climate action.

1. INTRODUCTION

The agricultural sector is one of the most vulnerable to climate change, as it is highly dependent on environmental and climatic conditions. Rising global temperatures, erratic rainfall patterns, extreme weather events, and prolonged droughts are increasingly disrupting agricultural productivity and supply chains. These climate-induced challenges contribute to declining crop yields, soil degradation, water scarcity, biodiversity loss, and increased greenhouse gas (GHG) emissions. At the same time, the growing global population is placing additional pressure on agricultural systems to produce more food using limited natural resources. As a result, traditional farming practices are becoming insufficient to sustain agricultural productivity under rapidly changing climatic conditions. Artificial intelligence (AI) and digital agriculture technologies present significant opportunities for transforming agricultural systems into adaptive, data-driven, and climate-resilient ecosystems.

AI-enabled smart agriculture facilitates predictive modelling, precision resource management, real-time environmental monitoring, and automated decision-making to support climate adaptation strategies. By leveraging machine learning, sensor networks, satellite imagery, and agricultural data analytics, intelligent systems can optimize water use, improve soil health, predict pest outbreaks, and support adaptive crop management strategies. These capabilities enable agricultural systems to transition from reactive approaches to proactive, climate-resilient farming practices. Beyond productivity improvements, AI-driven smart agriculture plays a critical role in enhancing climate mitigation and long-term food-system resilience. Intelligent resource optimization can reduce fertilizer overuse, lower irrigation-related energy consumption, and support carbon monitoring and greenhouse gas reduction strategies. Furthermore, AI-enabled decision-making improves supply-chain stability, adaptive planning, and climate-risk management, thereby strengthening food-system resilience.

The integration of AI into agriculture also supports the water–energy–food nexus, where intelligent management of water resources, energy consumption, and agricultural productivity contributes to sustainable resource utilization and climate adaptation. Climate-resilient food systems refer to agricultural systems that are capable of absorbing environmental shocks, adapting to changing climatic conditions, and maintaining productivity over time. AI-driven smart agriculture enhances resilience by enabling early-warning systems, predictive climate modelling, adaptive planting strategies, and optimized resource allocation. These capabilities support both climate adaptation, by adjusting agricultural practices to climate impacts, and climate mitigation, by reducing environmental impacts and resource inefficiencies. Consequently, AI-driven agriculture emerges as a key enabler of sustainable food-system transformation.

This chapter explores how AI-enabled smart agriculture can support sustainable and climate-resilient food systems through an integrated conceptual framework. The chapter proposes a five-layer AI-driven smart agriculture framework that integrates climate inputs, data acquisition, AI analytics, decision support, and sustainability outcomes. This framework aims to bridge the gap between isolated AI applications and comprehensive climate-resilient agricultural ecosystems. The chapter further examines key AI applications, implementation strategies, case studies, and policy implications necessary to support climate action through intelligent agricultural transformation. By integrating technological innovation, environmental sustainability, and governance considerations, this chapter contributes to the development of adaptive and sustainable agricultural systems aligned with global climate goals.

2. BACKGROUND

Agricultural system transformation requires an interdisciplinary understanding of environmental factors, environmental dynamics, technological developments, and sustainability principles. This background chapter establishes an understanding of the relationships between climate change and agricultural vulnerability; how digital agriculture has evolved as a response to climate change; the theoretical foundations of AI (artificial intelligence) within agricultural systems; and principles of sustainable and climate-resilient food systems. This context will provide a foundation for the understanding of AI-based smart agriculture as it relates to the broader global climate actions.

Figure 1. Climate Change–Agricultural Impact Pathway Model

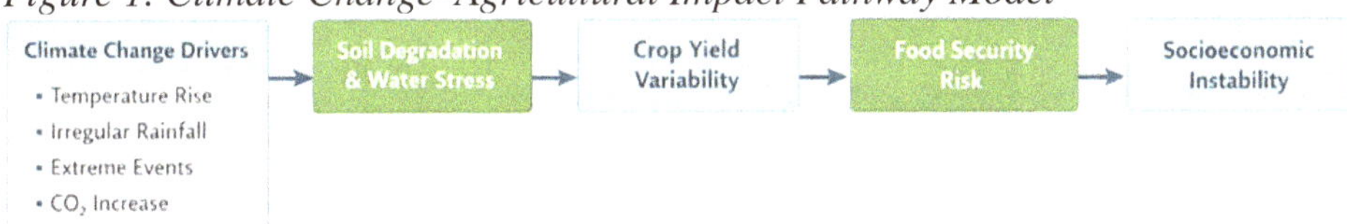

2.1 Climate Change and Agricultural Vulnerability

Given that agriculture is fundamentally reliant on climate, it is particularly sensitive to environmental fluctuations. Crop production is directly impacted by changes in temperature, precipitation patterns, atmospheric carbon concentrations, and the occurrence of extreme weather events. A prolonged drought reduces soil moisture and crop yields, while excessive rainfall can lead to flooding, soil erosion, and nutrient loss from the soil. Increased temperatures also increase the rate of evapotranspiration from plants and change the phenology of crops, which can result in unstable yields. Climate change is also a stressor on biotic pests. The warmer temperatures and change in ecosystems will lead to increased insect infestations and new plant diseases. These biotic stressors will further increase the risk associated with agricultural production and therefore require an adaptive management strategy. Smallholder farmers are especially vulnerable because they have limited financial resources, do not have access to technology, and depend upon rainfall for their agricultural production. In addition to the challenges of producing food, climate change impacts the entire agricultural value chain from the time that food is harvested until it is consumed (storage, transportation, and market stability). The volatility of food prices and the potential for supply disruptions pose a threat to food security, particularly in climate-sensitive regions. Therefore, agriculture's vulnerability should be viewed as both a biophysical issue and a socio-economic and systemic challenge, deserving of integrated technological and policy solutions.

Table 1. Major Climate Stressors and Their Impacts on Agricultural Systems

Climate Stressor	Direct Agricultural Impact	Long-Term Systemic Effect	Climate Resilience Implication
Rising Temperature	Heat stress, reduced crop productivity	Altered crop phenology	Need for adaptive crop varieties
Irregular Rainfall	Water stress, flooding	Yield instability	Smart irrigation systems
Drought	Soil moisture depletion	Land degradation	Water optimization strategies
Extreme Weather Events	Crop destruction	Economic losses	Predictive early-warning systems
Increased CO_2 Levels	Altered plant growth cycles	Nutrient imbalance	AI-driven nutrient modelling

2.2 Evolution of Digital Agriculture

Digital Agriculture (also referred to as "digital farming") was born out of the need to use technology as a way to increase productivity in agriculture and remain competitive. Digital agriculture has evolved over the past decades through three different phases: mechanization, precision agriculture, and now intelligent agriculture . Mechanization introduced machines and automated processes into farming, increasing efficiency in how far you can prepare the land, plant the seed, and harvest the crop. Precision agriculture took advantage of technology such as GPS, GIS, and remote sensing to allow farmers to implement site-specific crop management. This allowed famers to apply necessary inputs (water, fertilizers, pesticides, etc.) to their crops at a decrease in both the amount of input used and the potential for negative impacts on the environment. The most recent phase (Agriculture 4.0 or smart agriculture) is about taking advantage of the Internet of Things (IoT), cloud computing, big data, advanced machine learning models, and real-time data collection (via sensors embedded in the soil and crops) to produce predictive and prescriptive analysis. Sensors embedded in your soil and crop will collect real-time data that can include moisture levels, nutrient levels, temperature (soil and air), and humidity. By collecting this information, farmers will be able to implement the proper inputs at the right time and apply the proper amount of each input based on empirical evidence rather than relying on their past experiences. Farmers also have access to satellite images of their properties to gain insight into their entire growing area rather than just focusing on one small area (field).

Figure 2. Evolution of Agricultural Systems: From Mechanization to AI-Driven Smart Farming

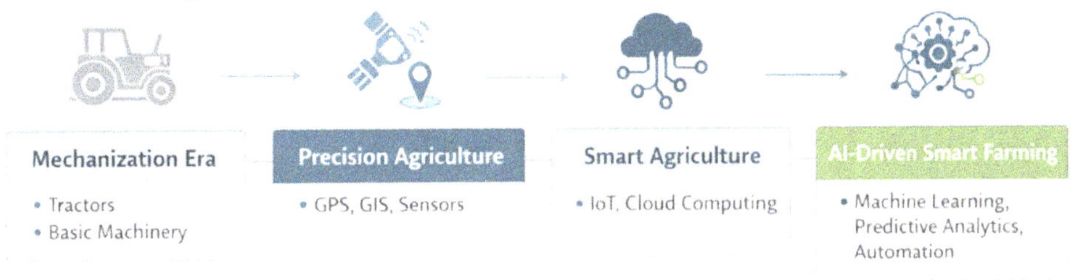

Digital agriculture thus represents a shift from reactive farming practices to data-driven, proactive decision-making. This transition lays the technological foundation for integrating artificial intelligence into climate-resilient agricultural systems.

2.3 Foundations of Artificial Intelligence in Agriculture

Artificial Intelligence (AI) refers to computational systems capable of performing tasks that typically require human intelligence, including learning, pattern recognition, prediction, and decision-making, but not limited to, pattern recognition, learning, prediction, and decision-making. The introduction of AI-based technologies to agriculture has allowed for the analysis of the many and diverse types of large and complicated data associated with agricultural production: data generated from environmental sensors, satellite imagery, weather models, and historical production data. In order to provide data analysis services and ultimately predict future agricultural events, machine learning algorithms look to identify patterns within agricultural data that relate to crop yield predictions, irrigation scheduling, and soil health assessments. Crop, plant, plant product, or animal disease detection typically utilizes supervised machine learning models that allow for classification; regression-type machine learning models allow for crop yield estimation. Additionally, image analysis related to pest and disease identification may incorporate deep learning methodologies – in particular, convolutional neural networks – based upon images taken through either drones or smartphones.AI-based decision support systems incorporate predictive analytics with optimization methodologies to develop efficient resource allocation recommendations. As an example, optimization modelling for irrigation scheduling incorporates weather forecasts, soil moisture levels, and crop growth stages to provide the best possible water application. Similarly, AI-based nutrient management models allow for the proper amount of fertilizer to be applied based upon the nutrient requirements of the crop being considered and the current soil composition.

Figure 3. Artificial Intelligence Technology Stack in Smart Agriculture

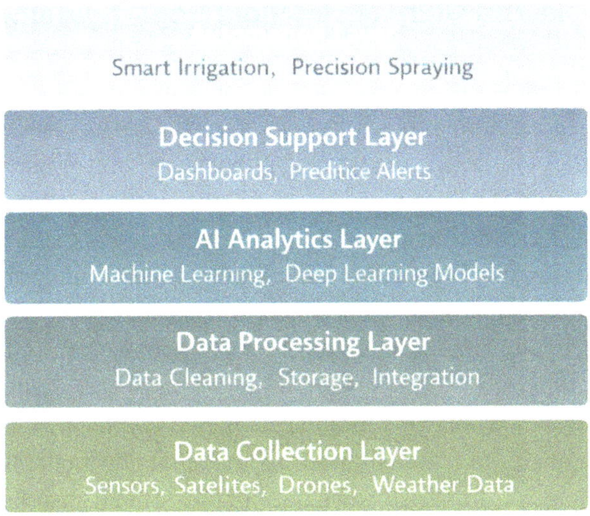

AIL in agriculture has a theoretical foundation based on systems theory and data-driven models, which is the way in which an area of Agriculture is made up of changing interconnected systems that can all be affected by environmental, biological, etc. (Air, Water, Soil); data from such interdependent systems enhance the ability to adapt and the ability to be resilient to climatic changes on the area of Agriculture.

Table 2. Artificial Intelligence Technology Stack in Smart Agriculture

AI Technique	Method Type	Agricultural Application	Sustainability Contribution
Machine Learning (Regression)	Supervised Learning	Crop yield prediction	Reduced uncertainty in production
Classification Algorithms	Supervised Learning	Disease detection	Minimized pesticide overuse
Deep Learning (CNN)	Neural Networks	Image-based pest identification	Early intervention capability
Reinforcement Learning	Adaptive Optimization	Irrigation scheduling	Water conservation
Clustering Algorithms	Unsupervised Learning	Soil health zoning	Efficient fertilizer allocation

2.4 Sustainable and Climate-Resilient Food Systems

A food system that is sustainable addresses present and future food security and nutrition needs while minimizing adverse effects on the environment and conserving natural resources. A climate-resilient food system incorporates adaptation capacity, risk management, and long-term ecological stability into how agricultural systems function when faced with climate variability. Aspects of agricultural sustainability can be examined through three dimensions: Environmental Integrity – this dimension is associated with the maintenance and protection of soil health, biodiversity, and water resources while minimizing greenhouse gas emissions (GHG); Economic Viability – this dimension addresses providing farmers with stable incomes, and ensuring that agricultural markets provide economic resilience; and Social Equity – this dimension relates to access to food, affordable food resources, and equitable access to technology. A climate-resilient agricultural system absorbs shocks, adapts to changing conditions, and recovers from disruptions caused by climate change. Adaptive strategies include crop diversification; climate-based planting schedules; efficient irrigation systems; and monitoring the impacts of environmental stressors in real-time. The use of AI technologies enhances all of these strategies by providing predictive insight and decision-making tools that enable dynamic decision-making. By integrating AI into the sustainable food system, both mitigation and adaptation efforts can be enhanced. Mitigation is achieved through optimized input utilization, and GHG monitoring; adaptability through prediction of climate impacts and intelligent risk assessments. As such, AI-enabled smart agriculture will provide the foundation for climate-resilient food systems to meet global sustainability goals.

3. LITERATURE REVIEW

Recent years have witnessed a significant increase in research exploring the role of artificial intelligence (AI) and digital technologies in agriculture, particularly in response to climate variability and sustainability challenges. Existing studies highlight the potential of smart farming technologies to improve productivity, optimize resource use, and enhance environmental sustainability. However, despite rapid technological advancement, the literature reveals fragmentation in approaches, limited integration of climate-resilience objectives, and insufficient emphasis on system-level sustainability outcomes. This section critically reviews existing research on AI applications in agriculture, climate-smart agriculture frameworks, predictive analytics, and identifies research gaps that motivate the proposed conceptual framework.

3.1 AI Applications in Agriculture: Existing Studies

Artificial intelligence has emerged as a transformative technology in modern agriculture through the development of Smart Farming Technologies (SFTs). Early studies emphasize the role of big data analytics, sensor-based monitoring, and IoT-enabled systems in improving agricultural decision-making and operational efficiency (Wolfert et al., 2017). These technologies enable the collection of large-scale agricultural data from soil sensors, weather stations, drones, and satellite imagery, allowing farmers to make data-driven decisions.

Machine learning techniques have been widely applied to crop yield prediction, disease detection, and irrigation management (Liakos et al., 2018). Deep learning models, particularly convolutional neural networks, have demonstrated high accuracy in image-based crop disease detection and plant health monitoring (Kamilaris & Prenafeta-Boldú, 2018). Similarly, remote sensing technologies and unmanned aerial systems (UAS) have enhanced precision agriculture by providing high-resolution spatial crop data (Zhang & Kovacs, 2012).

Recent studies also highlight AI's role in optimizing agricultural inputs and reducing environmental impacts. Balafoutis et al. (2017) reported that precision agriculture technologies can reduce fertilizer use, optimize water allocation, and contribute to greenhouse gas mitigation. However, despite these technological advances, adoption remains uneven due to infrastructure limitations, economic barriers, and lack of technical capacity, particularly in developing regions (Klerkx et al., 2019).

While these studies demonstrate strong agreement regarding the productivity and efficiency benefits of AI-driven agriculture, there remains limited research addressing long-term climate resilience, socio-economic inclusivity, and sustainability at the food-system level. Most studies focus on isolated AI applications rather than integrated climate-resilient agricultural ecosystems.

3.2 Climate-Smart Agriculture Frameworks

Climate-Smart Agriculture (CSA) has emerged as a key strategy for addressing climate change impacts in agriculture. The Food and Agriculture Organization defines CSA based on three core objectives: improving productivity, enhancing resilience, and reducing greenhouse gas emissions (FAO, 2017). These objectives emphasize sustainable intensification and climate adaptation strategies.

Reports from the Intergovernmental Panel on Climate Change indicate that climate change is already reducing crop productivity in climate-sensitive regions (IPCC, 2022). Additionally, Ray et al. (2013) demonstrated that current yield growth rates are insufficient to meet projected global food demand by 2050. These findings highlight the urgency of adopting climate-resilient agricultural strategies.

Although CSA frameworks provide valuable policy guidance, many existing approaches rely primarily on agronomic practices such as crop diversification, improved irrigation, and climate-informed planting schedules. These frameworks often lack integration with advanced AI-driven predictive technologies and real-time decision-support systems. As a result, CSA frameworks remain largely static and reactive rather than adaptive and predictive.

Furthermore, current CSA models rarely address system-level issues such as data infrastructure, digital governance, and interoperability. This gap suggests the need for integrated AI-enabled CSA frameworks that combine predictive intelligence, automation, and sustainability evaluation.

3.3 Predictive Analytics in Crop and Resource Management

Predictive analytics has become a core component of AI-driven agriculture. Machine learning models utilize historical weather data, soil information, and crop growth patterns to predict yields and resource requirements (Liakos et al., 2018). AI-based early-warning systems have also been developed to forecast pest outbreaks and climate-induced crop stress (Sharma et al., 2021). Predictive irrigation models have demonstrated improved water-use efficiency without reducing yield stability (Sharma et al., 2021).

Additionally, AI-based nutrient management systems optimize fertilizer application and reduce environmental contamination. Big data platforms integrating satellite imagery, IoT sensors, and weather forecasts enable real-time decision support for farmers (Wolfert et al., 2017).

However, despite these advancements, several limitations remain. Predictive models often rely on region-specific datasets, limiting transferability across different agro-climatic zones. Data quality, interoperability, and model uncertainty remain significant challenges. Furthermore, many predictive systems focus on technical accuracy without considering practical adoption constraints, farmer decision-making behavior, and institutional barriers. These limitations highlight the need for integrated frameworks that combine predictive analytics with governance, infrastructure, and sustainability considerations.

Table 3. Summary of Existing Studies on Artificial Intelligence Applications in Agriculture

Author(s)	Year	Focus Area	Methodology	Key Findings
Wolfert et al.	2017	Smart farming & big data	Literature review	Data-driven farming improves efficiency
Kamilaris & Prenafeta-Boldú	2018	Deep learning in agriculture	Survey study	High accuracy in disease detection
Liakos et al.	2018	ML in agriculture	Review analysis	Improved yield forecasting
Sharma et al.	2021	Precision irrigation	Predictive modelling	Water savings achieved

3.4 Critical Research Gaps in AI-Driven Climate-Resilient Agriculture

Although existing research demonstrates the potential of AI in agriculture, several critical gaps remain. First, most studies focus on isolated AI applications such as crop prediction, irrigation optimization, or disease detection. There is limited research on integrated AI-driven ecosystems that combine sensing, prediction, optimization, and sustainability evaluation within a unified framework.

Second, current literature largely emphasizes productivity improvements rather than climate mitigation and long-term resilience. There is insufficient discussion on how AI can support greenhouse gas reduction, carbon monitoring, and sustainable resource management.

Third, governance and ethical considerations remain underexplored. Issues such as data ownership, algorithmic bias, digital inequality, and farmer dependency on proprietary systems require further investigation.

Fourth, smallholder farming contexts are inadequately addressed in existing research. Many AI solutions assume access to advanced infrastructure, which may not be available in developing regions.

Finally, interoperability challenges across sensors, platforms, and data systems limit scalability of AI-enabled agriculture. Without standardized frameworks, integration of climate-resilient technologies remains difficult.

3.5 Contribution of This Chapter

To address these limitations, this chapter proposes a five-layer AI-enabled smart agriculture framework that integrates climate and environmental inputs, data acquisition and sensing infrastructure, AI analytics and predictive intelligence, decision-support and optimization mechanisms, and sustainability and resilience outcomes.

This framework moves beyond isolated applications by providing a comprehensive ecosystem for climate-resilient agriculture. The proposed model integrates technological, environmental, and governance considerations, contributing to sustainable and adaptive food systems under climate change. This integrated framework addresses fragmentation in existing literature by combining predictive intelligence, governance considerations, and sustainability outcomes within a unified climate-resilient agricultural ecosystem.

4. CONCEPTUAL FRAMEWORK

This section presents the conceptual foundation and structured framework for AI-driven smart agriculture designed to support climate-resilient and sustainable food systems. The proposed framework integrates sustainability theory, systems theory, and data-driven intelligence to create a comprehensive model for adaptive agricultural decision-making under climate variability. Unlike traditional precision agriculture approaches that primarily focus on productivity and efficiency, the proposed framework emphasizes climate resilience, sustainability, and long-term food-system stability through intelligent automation, predictive analytics, and adaptive decision-support mechanisms. The conceptual model bridges technological innovation and agricultural sustainability by integrating climate inputs, digital sensing infrastructure, predictive analytics, and sustainability outcomes within a continuous feedback-based system.

4.1 Theoretical Foundations

The proposed framework is grounded in three major theoretical perspectives: sustainability theory, systems theory, and data-driven decision theory. These theoretical foundations provide a comprehensive basis for integrating artificial intelligence into climate-resilient agricultural systems. Sustainability theory emphasizes balancing environmental protection, economic viability, and social equity within agricultural systems. This "triple bottom line" approach requires maintaining soil health, conserving water resources, reducing greenhouse gas emissions, and ensuring stable farmer livelihoods. AI-driven smart agriculture supports sustainability by optimizing resource usage, minimizing waste, and improving productivity within ecological limits. Through predictive modelling and intelligent optimization, AI technologies enable precision fertilizer application, efficient water management, climate-risk prediction, and carbon monitoring. These capabilities contribute to both climate mitigation and adaptation, strengthening long-term sustainability of agricultural systems. Systems theory recognizes agriculture as a complex socio-ecological system involving interconnected environmental, technological, economic, and biological components. Climate change affects multiple agricultural subsystems simultaneously, including crop growth, soil health, water availability, and pest dynamics. Artificial intelligence enhances system coordination by integrating diverse data sources such as weather data, soil information, crop health indicators, market trends, and environmental conditions. This integrated approach enables holistic decision-making and improves agricultural resilience. By adopting systems theory, the proposed framework treats agriculture as a dynamic and adaptive ecosystem rather than isolated farming operations. Data-driven decision theory emphasizes predictive analytics and evidence-based decision-making under uncertainty. Climate variability introduces uncertainty in agricultural planning, but AI models can identify patterns from historical and real-time data. Machine learning algorithms support yield forecasting, irrigation optimization, pest prediction, and resource allocation. These predictive capabilities reduce uncertainty and enable adaptive agricultural planning. Together, these theoretical foundations justify the integration of artificial intelligence as a systemic solution for climate-resilient agriculture.

4.2 Proposed AI-Driven Smart Agriculture Framework

The proposed conceptual framework consists of five interconnected layers that collectively support climate-resilient agricultural decision-making. These layers include climate and environmental inputs, data acquisition and sensing infrastructure, AI analytics and intelligence, decision-support optimization, and sustainability-resilience outcomes. The climate and environmental inputs layer represents external drivers such as temperature, rainfall variability, humidity, carbon dioxide levels, and extreme weather events. These factors directly influence agricultural productivity, resource requirements, and risk exposure. The data acquisition and sensing layer collects real-time agricultural information through IoT sensors, satellite imagery, drones, weather stations, and remote sensing technologies. These digital infrastructures enable continuous monitoring of environmental and crop conditions. The AI analytics and intelligence layer processes structured and unstructured data using machine learning, deep learning, predictive analytics, and optimization algorithms. These models generate forecasts related to crop yield, irrigation scheduling, soil nutrient status, and pest outbreaks. The decision-support and optimization layer converts predictive insights into actionable decisions, including precision irrigation, fertilizer optimization, crop planning, and automated farm management. Finally, the sustainability and resilience outcomes layer evaluates system performance in terms of productivity improvement, greenhouse gas reduction, resource efficiency, climate adaptation, and long-term sustainability. Figure 4 illustrates the conceptual five-layer framework for AI-enabled climate-resilient smart agriculture. The proposed model demonstrates a continuous adaptive cycle where climate inputs influence predictive intelligence and decision-making, enabling dynamic agricultural management. Continuous feedback loops allow performance outcomes to improve future predictions, creating an adaptive and resilient agricultural ecosystem.

Figure 4. Conceptual Five-Layer Framework for AI-Enabled Climate-Resilient Smart Agriculture

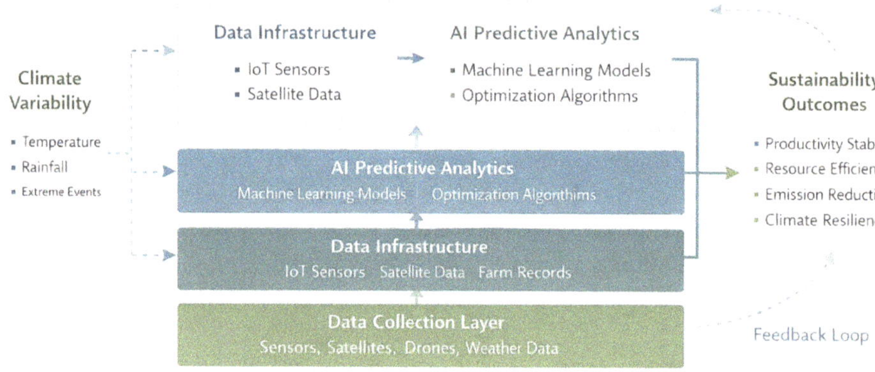

The proposed five-layer framework demonstrates a continuous adaptive cycle where climate inputs influence predictive intelligence, enabling dynamic and resilient agricultural decision-making. The framework emphasizes continuous feedback loops, where performance outcomes inform subsequent data inputs, enabling dynamic learning and system adaptation.

4.3 Framework Components and Interrelationships

The proposed conceptual framework consists of several interrelated components that collectively support climate-resilient and sustainable agricultural systems. These key constructs include climate variability, data infrastructure, AI predictive capability, optimization efficiency, and sustainability performance. Climate variability acts as an external driver that influences agricultural productivity, resource availability, and environmental risks. Data infrastructure, including sensor networks, satellite systems, and digital platforms, enables continuous collection of real-time environmental and agricultural data. This data supports AI predictive capability, which forms the analytical core of the framework by generating forecasts related to crop performance, irrigation requirements, soil health, and climate risks. Improved predictive accuracy enhances optimization efficiency by enabling intelligent resource allocation, adaptive crop planning, and automated decision-making. Optimization efficiency subsequently contributes to sustainability performance by reducing input waste, minimizing greenhouse gas emissions, improving water-use efficiency, and strengthening agricultural resilience.

Continuous feedback loops further strengthen system adaptability by allowing outcomes from agricultural decisions to inform future predictions and system adjustments. These interrelationships demonstrate that improved data quality enhances predictive accuracy, predictive intelligence improves optimization efficiency, and optimized decision-making supports sustainability outcomes. This interconnected and iterative structure reflects the adaptive nature of AI-enabled climate-resilient agriculture and supports long-term sustainable food-system transformation.

Table 4. Constructs and Definitions of the Proposed Conceptual Framework

Construct	Definition	Functional Role in Framework	Expected Outcome
Climate Variability	Environmental fluctuations affecting farming	External driver	Risk exposure
Data Infrastructure	Sensor & digital data systems	Enabling mechanism	Accurate data collection
AI Predictive Capability	Analytical processing of data	Core intelligence	Forecasting accuracy
Optimization Efficiency	Resource allocation performance	Operational output	Reduced input waste
Sustainability Performance	Environmental & productivity balance	Final outcome	Climate resilience

This interconnected structure reflects the systemic and iterative nature of climate-resilient smart agriculture.

4.4 Model Assumptions

The proposed framework assumes the availability of digital infrastructure, access to reliable datasets, farmer adoption capability, predictable climate patterns, and a supportive policy environment. Additionally, the framework assumes responsible and ethical AI implementation, equitable access to technology, and institutional support for long-term deployment. These assumptions highlight the importance of governance structures, infrastructure readiness, and inclusive deployment strategies for successful implementation.

4.5 Novelty of the Proposed Framework

Unlike existing climate-smart agriculture frameworks, the proposed five-layer model integrates predictive intelligence, adaptive decision-making, and sustainability outcomes within a continuous feedback loop. Traditional precision agriculture models primarily focus on improving productivity and input efficiency, whereas the proposed framework incorporates climate mitigation, resilience, and sustainability as core objectives. The proposed framework integrates real-time sensing infrastructure, AI predictive analytics, adaptive decision-support systems, and sustainability performance evaluation into a unified ecosystem. This system-level integration enables dynamic adaptation to climate variability and supports long-term food-system sustainability. Furthermore, the model extends beyond farm-level optimization by incorporating greenhouse gas monitoring, water-energy optimization, and sustainable resource management strategies. The framework also emphasizes smallholder inclusivity, data governance, and institutional deployment considerations, which are often overlooked in existing models. Therefore, the proposed five-layer framework contributes a comprehensive and adaptive model for AI-enabled climate-resilient agriculture, bridging the gap between technological innovation and sustainable food-system transformation.

5. APPLICATIONS OF AI IN CLIMATE-RESILIENT AGRICULTURE

Modern agriculture has been greatly altered by artificial intelligence (AI) through the combination of forecast modelling, optimization algorithms, and real-time data analysis to overcome the uncertainties generated due to climate change. The use of AI will improve the ability of farmers to adapt to changing environmental conditions, reduce the negative environmental impact of agriculture, and support sustainable intensification of food systems. In this section, we identify the main application areas where AI provides substantial benefits regarding climate resilience and resource efficiency as well as ensuring the long-term sustainability of food systems.

5.1 Precision Farming Systems

Precision agricultural practices are different from traditional farming systems because they are designed to be more effective. Agricultural variability can affect many aspects of agriculture on the same property, including soil moisture content, the availability of nutrients, and crop growth over time. Precision agriculture uses advanced technology such as drone imagery, satellite imagery, and soil sensor data to examine how agricultural systems are affected by change over space and time. Machine learning algorithms process multispectral images of farm fields to identify crop stress, nutrient deficiencies, and crop growth problems. Based on these analyses, precision agriculture provides farmers with variable rate application maps that guide them in better utilizing fertilizers, pesticides, and water. This allows farmers to reduce greenhouse gas emissions from excess fertilizer application, as well as reduce the risk of nutrient runoff that causes degradation to soil and water. Using artificial intelligence to analyse spatial data will also allow farmers to adaptively plant by combining historical yield results with historical climate trends. The result of this data-driven approach is improved productivity while maintaining ecological balance. Finally, by using precision agriculture, farmers can proactively manage their farms in anticipation of variable climatic conditions, therefore increasing the resilience of their production systems to the impacts of climate change.

5.2 Smart Irrigation and Water Optimization

One of the biggest issues facing agriculture today is the water shortage caused by climate change. Climate change has led to an increase in temperatures, which increases the rate of evapotranspiration. Erratic precipitation patterns make irrigation difficult to manage by farmers. Smart irrigation systems powered by artificial intelligence (AI) collect data from soil moisture sensors, weather forecasts, and crop growth stages to create a highly optimized irrigation schedule. In addition, predictive modelling uses real-time data on current climatic conditions as well as past trends of evapotranspiration to estimate the amount of water required to grow crops. Reinforcement learning allows these systems to adjust irrigation times dynamically to account for both under-irrigation and over-irrigation. Therefore, irrigation systems can be made significantly more efficient with regard to water consumption and energy consumption related to pumping. Optimization algorithms also distribute water supplies evenly on multiple fields using a systematic approach that gives priority to areas of greater stress. This approach allows farmers to conserve water while maintaining yield stability even during droughts.

Figure 5. AI-Based Smart Irrigation Decision and Optimization Model

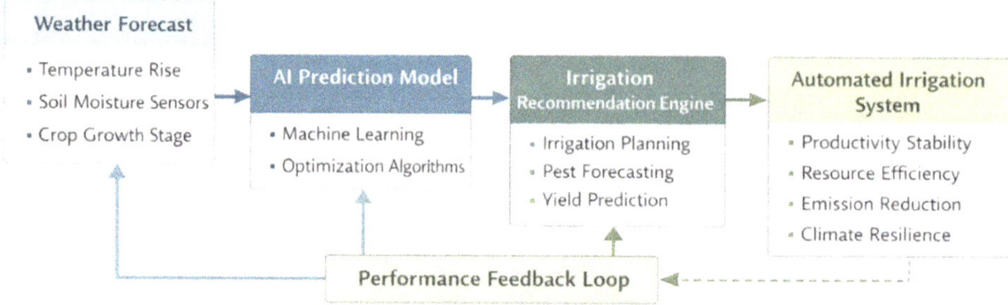

By integrating predictive analytics with automated irrigation control, AI systems enhance adaptive capacity and mitigate water-related climate risks.

5.3 Soil Health and Nutrient Management

Soil degradation and erosion, in addition to nutrient depletion, will continue to be aggravated by the impact of climate variability and unsustainable agricultural practices. AI-based soil health monitoring systems are capable of analysing chemical, biological and physical parameters for different soils and providing precise nutrient recommendations. Machine learning models assess soil composition, organic carbon, pH balance and historical fertilization records to determine the optimal rates of nutrient application. Predictive nutrient modelling enables the reduction of excessive fertilizer application rates, thereby lowering nitrogen oxide emissions and limiting environmental contamination. In addition to this information, AI-driven decision support systems will be able to establish long-term trends in soil fertility through the analysis of climatic and cropping pattern data and assist with sustainable land management practices such as crop rotation planning and regenerative farming interventions. The overall effect of improving soil management is the development of resilient soils that maintain their productivity even in environmentally stressed climates.

5.4 Pest and Disease Forecasting

The changes in climate affect pest life cycles, locations, and patterns of disease emergence. Warmer temperatures and more humid environments are causing many of these new infestations, which make crops more vulnerable than ever before. Pest and disease forecasting systems that use AI and its associated deep learning algorithms for the generation of predictive models are based upon photographs of leaves taken by drones or smartphones. The leaf photographs have been analysed through convolutional neural networks (CNNs), which allow for the rapid identification of diseases on the leaves of crop plants with a high degree of accuracy. Epidemiological predictive models use climatic data, phenological data on crops and historical data on previous pest infestations to help predict when and where a pest will invade an area prior to its invasion. The use of intelligent forecasting systems enables growers to take proactive measures to protect their crops from damage due to pests before they have spread widely. Intelligent forecasting systems assist in reducing the financial losses suffered by farmers due to pests, as well as reducing the amount of pesticides applied to their crops which helps to promote environmentally sound agricultural practices.

5.5 Crop Yield Prediction and Climate Modelling

Crop yield variability resulting from changes in climates is a major effect. To predict future yields, AI-based yield prediction models are developed to analyse weather data, soil measurements, satellite images, and crop production history. Regression-based machine models are used to estimate yields with varying weather conditions and deep learning models are used to consider non-linear relationships between environmental conditions. Ensemble modelling enhances the robustness of making predictions using several different algorithms. Climate simulation models are also employed to estimate long-term trends for yield due to future anticipated changes in temperature and rainfall. These simulations can help policymakers and farmers choose climate-resistant crop varieties and adapt planting schedules based on weather conditions. By accurately forecasting crop yield, supply chain integrity, market stability and resilience to food insecurity can be improved in the face of climate change.

6. IMPLEMENTATION STRATEGIES

To successfully implement AI-based smart agricultural systems, technology needs to be innovative; however, other factors are necessary as well, such as sufficient infrastructure scalable deployment methods, and institutional preparedness. The implementation of AI in agriculture will require strategies to assist with issues of data quality; interoperability with other systems; access for producers and farmers who will use these systems; and developing the skills to use AI tools to ensure the long-term viability and climate resiliency. This part of the report outlines the essential infrastructure, technology and human resources for successful deployment of AI-based smart agriculture systems.

6.1 Data Collection and Infrastructure Requirements

High quality, real-time multi-dimensional datasets are the foundation of AI driven agricultural systems. A layered infrastructure which collects, transmits, stores and processes agricultural data across different climatic environments is essential for successful implementation.

The main sources of data are:

1) Soil moisture & nutrients
2) Weather monitoring stations
3) Remote sensing/satellite imagery
4) Crop monitoring via drones
5) Historical yield and farm management data

A digital infrastructure that integrates sensor networks, reliable internet access and cloud technology is vital for establishing a robust agricultural technology system. A dedicated, reliable and high speed internet connection will be important across all geographic locations but may be especially vital for those in rural or semi arid regions who are at high risk for climate change impacts. Data standardization will also be required in order to allow for interoperability between platforms and devices. Infrastructure that is able to scale must have the capacity for both real-time analysis and forecasting of future climate conditions, with the processing power to achieve both as well as adequate secure data architectures to manage those data. Cybersecurity systems will also be crucial for safeguarding sensitive data about agriculture and farmers. The infrastructure design is focused on continuous data-flow from the sensor to the user and back through an adaptive feedback loop, which effectively creates a dynamic decision-making process based on current and likely future climate conditions.

6.2 Integration of IoT, Cloud, and Edge Technologies

Intelligent agricultural ecosystems are supported by the combination of Internet of Things (IoT), cloud and edge computing technologies. The three components accomplish different but interrelated tasks that help to increase overall efficiency and response time for agricultural enterprises. IoT devices gather real-time environmental data as well as crop-specific data derived from sensor networks installed in multiple locations. This type of sensor network is capable of collecting measurements for temperature, humidity, soil moisture, nutrient levels and plant health. All of these types of data require a method to transfer them to a centralised cloud storage facility. The challenge with centralised storage is latency and bandwidth issues associated with large volumes of data. Edge computing provides a way to overcome latency and bandwidth issues by process data locally or with a small distance from the source of the data. Data will be filtered to remove irrelevant data, anomalies will be detected, and decisions will be made quickly at the edge of the network for critical issues such as controlling irrigation or warning of a pest outbreak. In addition, the cloud provides long-term storage of large volumes of data, creating large, multi-dimensional datasets that can be used for model development and predictive analytics over time. Machine learning algorithms are trained on the cloud using large, historical datasets to properly train the algorithm. Cloud storage also allows agricultural enterprises to integrate external databases, climate projections, and agricultural compliance monitor systems. The combination of IoT, cloud computing, and edge computing technologies enables agricultural enterprises to create scalable solutions that reduce latency and improve reliability, particularly in uncertain climates.

6.3 Deployment Models for Smallholder vs Large Farms

Effective AI adoption strategies must accommodate specific features of the farm (size, resources, and level of technology readiness). Generally, smallholder farmers have limited access to financial and infrastructure resources as well as digital literacy. In comparison, larger commercial agricultural operations generally have access to a greater amount of capital and advanced technologies. Larger commercial farms use comprehensive centralised AI platforms that bring together precision machinery, unmanned equipment, and enterprise-level analytics dashboards. In contrast, smallholders' deployment strategies must rely on low-cost, mobile-friendly, and modular solutions. These deployment models will only be successful if they are coordinated through public-private partnerships, state and local government subsidies, or through cooperative data sharings. These methods are necessary to provide equal access for all to AI-enabled tools and services

Table 5. Comparative Deployment Strategies for Smallholder and Large-Scale Farms

Dimension	Smallholder Farms	Large Commercial Farms
Investment Capacity	Low	High
Technology Access	Mobile-based advisory tools	Integrated AI platforms
Data Infrastructure	Shared or cooperative systems	Private IoT networks
Scalability	Community-supported expansion	Enterprise-driven scaling
Climate Adaptation Strategy	Risk mitigation focused	Predictive automation driven

6.4 Capacity Building and Farmer Training

Technology by itself isn't enough to provide the foundation for AI. The successful integration of AI depends on having both human resources as well as the necessary levels of digital competency and institutional support systems available.

Capacity building strategies must include:

- A training programme for reading/interpreting results from AI
- Campaigns to increase awareness about climate-smart agriculture
- Digital literacy workshops for rural farmers
- AI extension services (agriculture) through advisory systems
- Technical support networks for help troubleshooting and maintaining AI systems

The partnerships between educational institutions, agri-tech start-ups and government agencies are critical for developing and disseminating knowledge about building agriculture through AI, especially because developing skills aligns with sustainable development goals and climate change adaptation policies.

Table 6. Skill Requirements for AI Adoption in Agriculture

Skill Category	Description
Digital Literacy	Basic use of mobile apps and data dashboards
Data Interpretation	Understanding predictive recommendations
Climate Awareness	Knowledge of climate risks and adaptation strategies
Technical Maintenance	Sensor handling and device management
Decision-Making Skills	Applying AI insights to farm operations

6.5 Institutional and Governance Requirements for AI Deployment

Successful implementation of AI-driven smart agriculture depends not only on technological infrastructure but also on institutional readiness, governance structures, and long-term sustainability mechanisms. While digital technologies can improve agricultural productivity and climate resilience, their effectiveness is shaped by how these technologies are deployed, governed, and maintained over time. One important factor influencing AI adoption is the role of agricultural extension systems. Traditional extension services have historically supported farmers through knowledge transfer and advisory support. Integrating AI-driven decision-support systems into agricultural extension networks can improve farmer access to predictive analytics and climate-risk information. Government-supported digital advisory platforms, mobile-based AI tools, and cooperative technology-sharing models can enhance accessibility, particularly for smallholder farmers. Without extension system integration, advanced AI technologies may remain underutilized or inaccessible to vulnerable farming communities. Another critical consideration is data governance and ownership. AI-driven agriculture relies heavily on farm-level data, including soil information, crop yield history, and environmental conditions. When this data is managed by private agritech companies, concerns arise regarding farmer autonomy, data privacy, and vendor dependence. Farmers may become reliant on proprietary platforms for decision-making, potentially limiting their control over agricultural operations. Establishing transparent data governance frameworks and open-data standards can help ensure equitable access and reduce dependency on private vendors. Financing models also play a significant role in enabling AI adoption. High upfront costs associated with sensors, connectivity infrastructure, and AI platforms can limit accessibility for smallholder farmers. Innovative financing approaches such as cooperative ownership models, subscription-based services, government subsidies, and public-private partnerships can reduce financial barriers. Shared digital infrastructure and community-based AI services can further enhance affordability and scalability. Long-term sustainability of AI-enabled agricultural systems also requires continuous model updating, technical support, and infrastructure maintenance. Climate conditions evolve over time, and AI models must be retrained using updated datasets to maintain predictive accuracy. Without institutional support and long-term maintenance strategies, AI systems risk becoming outdated or ineffective. Therefore, sustainable deployment requires ongoing collaboration among governments, research institutions, technology providers, and farming communities. Overall, successful implementation of AI-driven agriculture depends on a combination of technological innovation, institutional support, governance mechanisms, and inclusive deployment strategies. Addressing these factors strengthens climate resilience and ensures that AI-enabled smart agriculture contributes to sustainable and equitable food systems.

6.6 Policy and Governance Integration for Climate-Resilient AI Agriculture

The successful deployment of AI-driven smart agriculture depends not only on technological innovation but also on supportive policy frameworks and governance mechanisms. Climate-resilient agriculture requires coordinated policies that integrate agricultural transformation, climate adaptation, and digital innovation.

Without appropriate governance structures, AI-enabled systems may face challenges related to data ownership, accessibility, and long-term sustainability. Governments play a crucial role in enabling AI-driven agriculture through investments in digital infrastructure, rural connectivity, and data-sharing platforms. Expansion of broadband networks, development of open agricultural datasets, and support for climate monitoring systems can enhance the accessibility of AI-based decision-support tools. These policies are particularly important in developing regions where limited infrastructure restricts the adoption of advanced technologies. Policy frameworks must also address data governance and farmer rights. AI-driven agriculture relies on farm-level data, including soil information, crop yields, and environmental conditions. Clear policies defining data ownership, privacy protection, and interoperability standards are necessary to prevent vendor lock-in and ensure equitable access to digital agricultural platforms. Open-data standards and public digital infrastructure can help reduce dependency on proprietary systems. Financial policy mechanisms are also critical for promoting inclusive AI adoption. Government subsidies, cooperative financing models, and public-private partnerships can reduce the financial burden associated with adopting AI technologies. These mechanisms enable smallholder farmers to access predictive analytics, smart irrigation systems, and climate advisory tools that would otherwise remain inaccessible. In addition, climate-focused agricultural policies should integrate AI-driven decision-support systems into national climate adaptation strategies. AI-enabled forecasting systems can support drought preparedness, water resource management, and climate-resilient crop planning. These technologies also contribute to greenhouse gas mitigation through optimized fertilizer application, reduced energy consumption, and improved soil carbon management. Cross-sector collaboration between agricultural, environmental, and technology ministries is essential for developing integrated climate-smart agriculture strategies. Policy alignment across sectors ensures that AI-driven agriculture contributes to broader sustainability goals, including food security, climate adaptation, and environmental protection. Integrating governance frameworks into AI-driven agriculture strengthens institutional capacity, promotes equitable access, and enhances long-term sustainability. Policy-supported digital agriculture ecosystems are more resilient to climate variability and better positioned to support global climate action goals.

7. CASE STUDIES AND PRACTICAL INSIGHTS

The application of artificial intelligence in agriculture has expanded rapidly across both developed and developing regions. However, the success of AI-driven agricultural systems depends not only on technological advancement but also on infrastructure availability, governance mechanisms, affordability, and farmer adoption. This section evaluates selected case studies to examine how AI-enabled solutions contribute to climate resilience, sustainability, and agricultural productivity. The discussion focuses on how these initiatives address climate adaptation, implementation challenges, scalability, and long-term sustainability within agricultural systems.

7.1 Technology-Driven Agriculture Initiatives

Several national and international initiatives demonstrate the growing role of AI in supporting climate-resilient agriculture. One prominent example is the IBM Watson Decision Platform for Agriculture, which integrates satellite imagery, weather forecasting, soil data, and machine learning algorithms to support precision farm management. The platform provides predictive insights related to irrigation scheduling, crop health monitoring, and climate-risk management. By enabling farmers to anticipate weather variability and optimize resource allocation, the platform improves yield stability and reduces climate-related risks. However, despite its technological sophistication, adoption remains limited primarily to large commercial farms due to high implementation costs, data infrastructure requirements, and reliance on stable internet connectivity. These limitations highlight the challenges associated with scaling advanced AI systems across resource-constrained agricultural environments. Another major initiative is Microsoft's AI for Earth program, which uses cloud-based analytics, geospatial data, and machine learning models to support sustainable agricultural practices and climate adaptation. This initiative enables farmers and policymakers to monitor water resources, assess soil conditions, and evaluate environmental risks. The platform contributes to climate resilience by supporting long-term planning, resource optimization, and environmental monitoring. However, the effectiveness of such cloud-based systems depends heavily on data availability, technical expertise, and digital infrastructure. These factors can limit accessibility for smallholder farmers, particularly in developing regions where digital connectivity remains limited. In contrast, the International Crops Research Institute for the Semi-Arid Tropics (ICRISAT) has developed a Climate-Smart Digital Advisory System designed specifically for smallholder farmers. This system integrates localized weather data, crop recommendations, and climate-risk information to support farmers in semi-arid and drought-prone regions. The advisory platform enhances climate adaptation by providing timely guidance on planting schedules, irrigation planning, and crop selection. Unlike enterprise-level AI platforms, this approach emphasizes accessibility and localized decision-making. However, challenges such as limited digital literacy, infrastructure constraints, and funding dependencies continue to affect scalability and long-term sustainability. These case studies illustrate that AI-enabled agricultural solutions vary significantly in terms of accessibility, scalability, and climate-resilience outcomes. Enterprise-level platforms offer advanced predictive capabilities but face adoption barriers due to cost and infrastructure requirements. Public-sector and research-based initiatives improve inclusivity and local adaptation but often face scalability challenges. These differences highlight the importance of developing integrated frameworks that balance technological sophistication with socio-economic accessibility.

7.2 AI-Based Climate Adaptation Models

AI-based climate adaptation models focus on predictive simulation and optimization to reduce climate-related agricultural risks. Yield forecasting models integrate temperature patterns, rainfall variability, and soil characteristics to estimate agricultural productivity under changing climate scenarios. These predictive systems allow farmers to make adaptive decisions regarding crop selection, planting schedules, and resource allocation. Reinforcement learning-based irrigation systems dynamically adjust water distribution based on evapotranspiration forecasts, improving water-use efficiency under drought conditions. Similarly, ensemble machine learning techniques are used to predict drought probability and pest outbreaks, enabling early-warning systems that reduce crop losses and minimize excessive pesticide use. Another emerging development involves climate-risk dashboards that combine long-term climate projections with real-time farm data. These decision-support systems assist both farmers and policymakers in making short-term operational decisions and long-term planning strategies. However, the effectiveness of these models depends on data quality, infrastructure availability, and model interpretability. Farmers are more likely to adopt AI-based recommendations when systems provide transparent explanations and reliable performance across different agro-climatic regions.

7.3 Comparative Analysis of Implementation Outcomes

Comparative analysis of AI-driven agricultural initiatives reveals both opportunities and limitations across different deployment contexts. Large-scale commercial farms benefit from predictive analytics, automation, and integrated decision-support systems, resulting in improved productivity and climate resilience. However, smallholder farmers often face barriers related to affordability, infrastructure availability, and technical capacity. Public-sector initiatives and research-based programs tend to improve accessibility and inclusivity but may lack the technological sophistication and scalability of enterprise-level platforms. Table 7 presents a comparative analysis of IBM Watson, Microsoft AI for Earth, and ICRISAT digital advisory initiatives, highlighting differences in target users, technology focus, climate resilience contributions, scalability, and limitations. The comparative assessment indicates that enterprise-level AI platforms offer high predictive capability and automation, while research-based initiatives improve accessibility and climate awareness. Smallholder-focused systems enhance localized adaptation but require institutional support and infrastructure investment to scale effectively. These findings demonstrate that successful AI deployment in agriculture depends on digital infrastructure maturity, cross-sector collaboration, and farmer training initiatives. Integrated models that combine technological innovation with socio-economic inclusivity are more likely to achieve long-term climate-resilient and sustainable agricultural outcomes.

Table 7. Comparative Case Study Analysis

Dimension	IBM Platform	Microsoft AI for Earth	ICRISAT Digital Advisory
Target Users	Commercial farms	Global research & policy stakeholders	Smallholder farmers
Core Technology	AI analytics & satellite integration	Cloud-based geospatial AI	Weather-based advisory models
Climate Focus	Yield & irrigation optimization	Environmental monitoring	Drought adaptation
Key Benefit	Precision resource use	Large-scale climate insights	Localized adaptive guidance
Scalability	High (enterprise-driven)	High (cloud-supported)	Moderate (region-specific)
Limitations	High implementation cost	Data dependency	Infrastructure constraints

The comparative assessment indicates that:

1) Enterprise-level AI platforms produce a greater degree of automated processes and predictive capability.
2) Initiatives or projects funded by public entities or through research have greater inclusivity and increase the awareness of climate change.
3) Smallholder focused systems are able to improve local adaptation but require infrastructure support in order to scale.

Successful deployment of AI is strongly related to the maturity of digital infrastructure, collaboration among institutions, and training initiatives for farmers. Integrated models that provide both technological sophistication and socio-economic inclusivity generate the longest-term sustainable results.

8. CHALLENGES AND BARRIERS

Artificial Intelligence-enabled smart farming can potentially change the future of agriculture but has a number of structural, technological, economic and regulatory barriers to widespread implementation that prevent scalability, inclusivity or long-term sustainability. In order to create responsible and equitable climate-resilient systems for agricultural production, we must first understand these barriers. In this section we critically evaluate key challenges to implementation related to data governance, affordability, technical feasibility and policy alignment.

Figure 6. Multidimensional Artificial Intelligence Adoption Barrier Framework

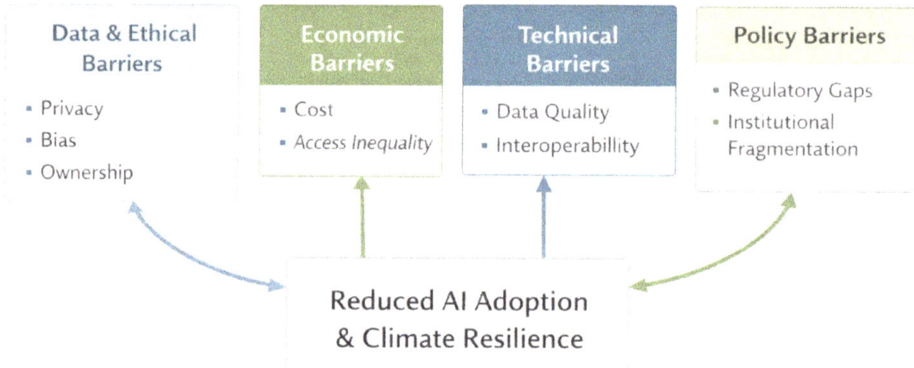

8.1 Data Privacy and Ethical Concerns

Significantly relying on farm-level data, such as soil type, crop yield, land ownership information and financial statements, creates some concerns regarding privacy and ownership of this data. When farm-level data is centralised and aggregated, it raises a number of privacy issues and questions of farmer's refusal to provide operational data to companies unless there are clear governance frameworks governing the use of the data. Technology companies have more information than farmers, which can place farmers in a position of disadvantage. When farmers provide their data to technology companies to access a proprietary artificial intelligence (AI) application that analyses and provides the farmers with predictive recommendations derived from their data, the farmers no longer have control over what type of recommendations the technology firms provide or how the recommendations will be made. Additionally, the complexity of the algorithms used to generate the recommendations makes it difficult for farmers to understand the logic behind the recommendations and do not provide farmers with the information necessary to make informed decisions based on the recommendations. Ethical concerns can arise from algorithmic biases generated by a lack of input data from farmer's locale. For instance, the AI algorithms developed using predominantly (or exclusively) datasets from developed regions may generate inaccurate or inequitable recommendations when applied to other geographical locations due to the predominant use of developed region datasets. Predictive systems that do not use oversized data sources risk perpetuating systemic inequality. Thus, establishing opportunities for the use of AI technologies, AI-driven agriculture must be emphasised by the creation of responsible AI frameworks with principles that prioritise transparency, consent to collect data, data security, and established rights for farmers to own their data. Principles of ethical governance are also essential to facilitate the alignment of agricultural technology innovation with the principles of sustainability and social equity.

8.2 Algorithmic Bias and Digital Inequality in AI-Driven Agriculture

Artificial intelligence systems in agriculture rely heavily on data-driven models, and the quality and diversity of training data significantly influence their effectiveness. One major ethical concern is algorithmic bias, which occurs when AI models are trained on datasets that do not adequately represent diverse agricultural conditions. For example, AI systems developed using data from large commercial farms or developed regions may generate inaccurate recommendations when applied to smallholder farms or different agro-climatic environments. This can lead to unequal benefits, where technologically advanced regions gain more advantages while resource-constrained regions remain underserved. Digital inequality also presents a significant challenge in the adoption of AI-driven agriculture. Farmers in rural and low-income regions often lack access to reliable internet connectivity, digital infrastructure, and technical training. As a result, AI-enabled decision-support systems may primarily benefit large-scale farms and technologically advanced agricultural systems. This imbalance risks widening the digital divide between large commercial farms and smallholder farmers, potentially reinforcing existing inequalities in agricultural productivity and income distribution. Another important concern is farmer dependency on proprietary AI platforms. Many AI-driven agricultural systems are developed and maintained by private agritech companies. When farmers rely heavily on these platforms for decision-making, they may lose autonomy over farm management decisions. Additionally, limited transparency in AI algorithms can make it difficult for farmers to understand how recommendations are generated. This "black-box" nature of AI systems may reduce trust and create long-term dependency on technology providers. The political economy of digital agriculture also plays a critical role in shaping technology adoption. Large technology companies often control digital infrastructure, data platforms, and AI models. This concentration of technological power may influence agricultural markets, data ownership, and decision-making processes. Without appropriate regulatory frameworks, digital agriculture may prioritize commercially valuable crops and large-scale farming operations, potentially marginalizing smallholder farmers and agroecological diversity. Addressing these ethical challenges requires the development of responsible AI governance frameworks. Transparent algorithms, inclusive datasets, open-data standards, and farmer-centered design approaches can reduce bias and promote equitable access. Public-sector involvement, cooperative data-sharing models, and policy interventions can further support inclusive AI adoption. Ensuring equitable deployment of AI-driven agriculture is essential for building sustainable, climate-resilient food systems that benefit diverse agricultural communities.

8.3 Technical Limitations

There are two key challenges facing agricultural AI: data quality and model interpretability. Poor quality data (e.g., incomplete, incorrect, or noisy) can produce inaccurate and unreliable machine learning models. Additionally, due to the heterogeneous nature of agricultural environments, as well as regional differences, there are significant difficulties with generalizing from a single model across different geographies. Similarly, there is a great deal of climate uncertainty that results in random variations in weather patterns (such as floods or heat waves) which may fall outside of the range of weather that has ever been recorded. Therefore, it will be difficult to accurately forecast agricultural activities using existing machine learning models, because of the inherent limitations of those models. Interoperability issues hinder the integration of data from multiple agricultural technologies. These technologies typically operate separately within incompatible platforms, leading to fragmentation of data. As a result, there is significant difficulty in achieving seamless integration across IoT devices, cloud services, and analytics tools. Additionally, due to their "black box" nature, many deep learning models are not easily interpretable by farmers and policymakers; therefore, improving the explainability of machine learning algorithms will be very important if farmers and policymakers are going to be able to trust the models and use them for decision-making.

8.4 Policy and Regulatory Issues

The regulatory landscape for agriculture technology in the context of AI is still at an early stage and is developing rapidly. Countries around the world generally lack comprehensive and clear regulatory guidelines for the use of agricultural data, the transfer of data across national borders, as well as standards for AI accountability. Without common guidelines it is difficult to develop interoperable and scalable solutions in digital agriculture. In addition, there is a lack of coordination between the climate, agricultural, and technology sectors – resulting in gaps in institutional support Additionally, the structure of government subsidies provide little incentive to invest in innovative digital agricultural technology by primarily supporting traditional agricultural inputs. Furthermore, current intellectual property regulations pose barriers to sharing technology and collaboratively innovating.

To build smart, climate-resilient agriculture the regulators must create a coordinated and comprehensive regulatory framework for:

- Promoting ethical use of AI,
- Protecting the data rights of farmers,
- Enabling innovation via incentives,
- Supporting the creation of inclusive digital infrastructure.

Table 8. Barriers to Artificial Intelligence Adoption and Corresponding Mitigation Strategies

Barrier Category	Core Issue	Impact on Adoption	Mitigation Strategy
Data Privacy	Lack of ownership clarity	Low farmer trust	Clear data governance policies
Economic	High implementation cost	Limited adoption	Subsidies & cooperative models
Technical	Poor data quality	Reduced model accuracy	Standardized data protocols
Infrastructure	Weak rural connectivity	System inefficiency	Broadband expansion programs
Policy	Regulatory fragmentation	Slow innovation	Integrated AI-agriculture frameworks

9. FUTURE RESEARCH DIRECTIONS

There is great potential for the integration of AI into agricultural systems that are climate-resilient and sustainable due to rapid advancements in both AI and digital technologies. There are many positive outcomes associated with current implementation; however, further development is needed in terms of new emerging technologies, ethics-related issues, and long-term effects on the environment. The next part of this document discusses research pathways essential for enhancing the use of AI in adaptive and sustainable food systems.

Figure 7. Future Smart Farm Ecosystem

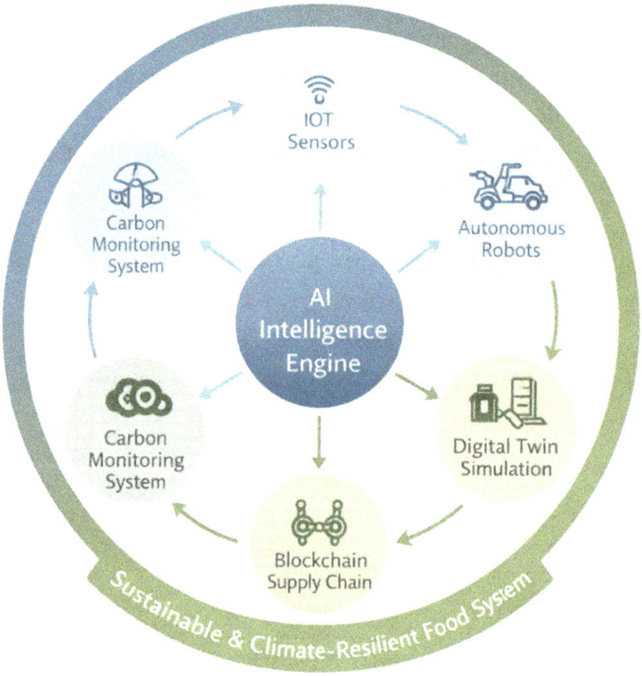

9.1 Explainable AI in Agriculture

AI-oriented agricultural research priorities centre largely on making predictive models more transparent and interpretable. Advanced ML/deep learning models are "black boxes" because they create predictions without explaining how they came to be. Given that farmers' lives (or food security) often depend on the outcomes of agricultural decisions, it is important to establish trust and accountability in AI models used by farmers. Future research will develop Explainable AI (XAI) model types that will produce an interpretable understanding of predicted plant yields, forecasted pest outbreaks, and irrigation advice. Other methods, including using visualization-based tools for producing explanations, feature importance analysis, and/or hybrid rules-based systems to explain model predictions, will also help address the transparency gap. XAI models are particularly important for climate adaptation, as farmers will need to know how various environmental variables impact the outputs of the models they are using. Future research should also focus on addressing how the interpretability of XAI models impacts their adoption rate, as well as farmers' confidence in making decisions using them.

9.2 Autonomous Farming Systems

The automation of agricultural processes is an exciting possibility for the future of climate-resilient agriculture. Autonomous farming technology that integrates robotics, computer vision, and AI to operate autonomously can perform tasks such as planting, weeding, harvesting, and precision spraying with reduced human input. Future research can investigate integrating autonomous drones and robotic platforms into crop monitoring systems that are capable of responding adaptively in real time to changing weather conditions. Reinforcement learning (RL) algorithms will allow the development of machines that can optimize field operations based on real-time feedback about environmental conditions. If successfully implemented, autonomous systems could eliminate much dependency on manual labour, reduce waste of inputs, and improve efficiency through better operational management. More research must still be conducted to determine how much energy these systems may consume, the costs associated with scaling them up for large-scale operations, and potential trade-offs to the environment that might occur with widespread adoption of automated technologies in the agricultural sector.

9.3 Digital Twins and Simulation Models

Digital twins technologically transform how agricultural ecosystems are modelled. Essentially, a digital twin is a virtual representation (or 'mirror') of an actual farm that receives up-to-date measurements from environmental inputs and sensors. In the future, research must focus on integrated digital twin models that represent the different components of agriculture, such as crop growth, soil variations, water cycles, and climate changes, both as they happen now and how they would happen in; creating predictive simulations of the climate and using those simulations to develop adaptive management approaches. Researchers will also need to look into scalability issues when integrating these AI-based predictive analytics with simulation models into digital twin models for farms and agricultural policy planning and to address the technical issues associated with deploying digital twins in various agro-climatic environments, such as the challenges related to data synchronization and computing resources.

9.4 AI-Enabled Carbon Monitoring

Agriculture has a complex relationship with climate change; it is both a contributor to greenhouse gas emissions and holds promise for sequestering carbon using regenerative agricultural methods. Future research needs to focus on developing artificial intelligence (AI) based carbon accounting systems that measure soil carbon, emissions from fertilizer use, and energy consumption in real-time. Using remotely sensed data and predictive modelling together will allow for the accurate measuring of a farm's carbon footprint and of carbon at the regional level. AI-enhanced carbon measurement systems will enable carbon credit markets, sustainability certification systems, and compliance with climate change policy. Standardization of methodology and validation protocols will help to ensure the reliability and transparency of these measurement systems. This ecosystem model integrates predictive intelligence, automation, simulation, and sustainability assessment within a unified digital agricultural framework. It represents the convergence of emerging technologies toward holistic climate action in food systems.

Table 9.Emerging Technologies and Research Opportunities

Emerging Technology	Research Focus Area	Climate Resilience Impact	Key Challenges
Explainable AI	Model transparency & interpretability	Improved trust & adoption	Balancing accuracy and simplicity
Autonomous Robotics	AI-driven farm automation	Reduced resource waste	High capital cost
Digital Twins	Real-time farm simulation	Scenario-based adaptation planning	Data synchronization
Carbon Monitoring AI	Emission tracking & sequestration modeling	Climate mitigation support	Standardization & validation
AI + Blockchain	Transparent supply chain tracking	Reduced food loss & fraud	Regulatory integration

REFERENCES

Balafoutis, A. T.. (2017). Precision agriculture technologies positively contributing to greenhouse gas emissions mitigation. *Sustainability*, *9*(8), 1339. DOI: 10.3390/su9081339

Kamilaris, A., & Prenafeta-Boldú, F. X. (2018). Deep learning in agriculture: A survey. *Computers and Electronics in Agriculture*, *147*, 70–90. DOI: 10.1016/j.compag.2018.02.016

Klerkx, L., Jakku, E., & Labarthe, P. (2019). A review of social science on digital agriculture. *NJAS Wageningen Journal of Life Sciences*, *90–91*(1), 100315. DOI: 10.1016/j.njas.2019.100315

Liakos, K. G., Busato, P., Moshou, D., Pearson, S., & Bochtis, D. (2018). Machine learning in agriculture: A review. *Sensors (Basel)*, *18*(8), 2674. DOI: 10.3390/s18082674 PMID: 30110960

Ray, D. K., Mueller, N. D., West, P. C., & Foley, J. A. (2013). Yield trends are insufficient to double global crop production by 2050. *PLoS One*, *8*(6), e66428. DOI: 10.1371/journal.pone.0066428 PMID: 23840465

Sharma, A., Jain, A., Gupta, P., & Chowdary, V. (2021). Machine learning applications for precision agriculture. *Agricultural Water Management*, *241*, 106–123. DOI: 10.1016/j.agwat.2020.106547

Wolfert, S., Ge, L., Verdouw, C., & Bogaardt, M. J. (2017). Big data in smart farming: A review. *Agricultural Systems*, *153*, 69–80. DOI: 10.1016/j.agsy.2017.01.023

Zhang, C., & Kovacs, J. M. (2012). The application of small unmanned aerial systems for precision agriculture. *Precision Agriculture*, *13*(6), 693–712. DOI: 10.1007/s11119-012-9274-5

Zhang, Y., Wang, G., & Li, Y. (2022). Artificial intelligence in smart agriculture: A systematic review. *Computers and Electronics in Agriculture*, *195*, 106800. DOI: 10.1016/j.compag.2022.106800

Chapter 3

AI-Based Biodigesters and University Engagement for Biodiversity Conservation in East Africa

Shashi Kant
http://orcid.org/0000-0003-4722-5736
St. Joseph University in Tanzania, Tanzania

Thompson Xavier Ananth
St. Joseph University in Tanzania, Tanzania

Shashank Mittal
http://orcid.org/0009-0007-8643-4350
O.P. Jindal Global University, India

Duncan Mwakipesile
St. Joseph University in Tanzania, Tanzania

ABSTRACT

The use of AI in environmental conservation, especially biodiversity conservation and sustainable resource management, makes a desirable contribution to the optimisation of the allocation of environmental resources and enhances predictive analytics. This is done with the help of SEM-based mediation analysis, a sample size of 248, and studies in the context of Tanzania that can process meaningful datasets which can be used to track environmental trends, optimise the use of resources, and enable

DOI: 10.4018/979-8-3373-9978-2.ch003

more informed decision-making. The synthesis of big data and AI further optimises these capabilities, allowing for the creation of predictive models for the purposes of monitoring habitats and managing species. Such technological developments play a required function in meeting the complex challenges of biodiversity conservation, especially in areas that are subject to major ecological pressures. Developing cost-effective models for resource sharing and forming strategic partnerships with local technology companies and universities.

INTRODUCTION

The rising global biodiversity crisis requires integrating technological solutions with high-end technology to help reinforce conservation efforts (Ullah et al., 2024). AI offers a transformative potential in this area in terms of enhanced ecological data gathering, analysis, prediction and decision-making processes (Ziemba et al., 2024). Specifically, AI applications can enhance species detection and monitoring greatly, facilitate managing invasive species and optimize conservation planning using predictive analytics and optimized resource distribution. These AI driven methodologies have shown an accuracy rate over 90% for Urban wildlife tracking and Invasive Species management, as well as enhancing its prediction rates by 18.5% compared to conventional conservation methodologies. As well as, the AI-driven platforms that rely on machine learning and computer vision enable real-time detection and classification of fauna, human activity and poaching-related objects from various visual and thermal infrared data sources. This technological integration supports the comprehensive biodiversity assessment and habitat monitoring to establish a synergy tool between innovative AI tools and established conservation practice. Such capabilities Wuhan pandas extend to mitigate human-wildlife conflict by enhancing monitoring and predictive efficiency and community engagement. It is required in order to address a critical challenge in biodiversity hotspots (Ojija et al. 2025). The use of AI-based biodigesters leads to further expansion of this paradigm by presenting sustainable approaches derived from waste management, which in turn provide sustainable energy sources and a reduced environmental impact through indirect protection of biodiversity through enhanced ecological health and reduced human encroachment (Ojija et al., 2025; Chisom et al., 2024).

Given the rising threat of extinction to 1 million species, majorly caused by anthropogenic factors such as destruction of habitats and pollution by human activities, the tactical use of AI-based systems is essential for effective biodiversity monitoring and assessment. This urgency only goes to show the urgent need for new methodologies, ones that go beyond traditional methods, especially with the use of cutting-edge AI-based technologies for enhanced data analysis and predictive modelling in conservation (Rathoure & Ram, 2024). Traditional ecological monitoring techniques have an inherent limitation in terms of scalability and data currency, so AI-powered tools like the one developed using machine learning and deep learning are crucial to perform a rather complex assessment for biodiversity on a massive amount of spatial and temporal data (Ullah et al., 2024; Ziemba et al., 2024).

Although more work has been done on artificial intelligence in conserving biodiversity and numerous biodigester technologies are being applied, there is a gaps in research on how AI-based biodgesters combined with university engagement can synergistically increase the effectiveness of conservation of biodiversity in the East African setting.

This paper posits that the introduction of AI-powered biodigesters alongside active university interactions is the key to the biodiversity protection through the optimization of the sustainable waste management approach, creation of renewable energy resources, and sustainability-driven by digital information environment decisions regarding East Africa.

STUDY BACKGROUND

Modern approaches harness computational approaches which involve methods such as machine learning, deep learning and spatiotemporal data fusion and cloud-based geo-processing which enables applications ranging from automated species distribution modelling to predictive early-warning systems for biodiversity loss (Reynolds et al., 2024). This allows you to gain real-time ecological data and apply and prevent conservation approaches using cutting-edge data analysis and remote sensing technologies (Kyathanahally et al., 2022). Specifically, computer vision and remote sensing technologies with AI can be integrated with advanced sensor technology to provide powerful capabilities for tracking animal migration patterns, monitoring biodiversity metrics and delineating required ecosystem. These leading-edge analytical designs using methods like random forests as well as support vector machines therefore are essential to categorizing degeneration risks and also in supporting specific actions (Ayadi et al., 2025).

Along with, predictive modelling, applying advanced machine learning algorithms, can be used to predict species distribution shifts in response to different environmental changes and help to optimise conservation interventions in order to prevent biodiversity loss proactively. This combination of AI and remote sensing, also LiDAR (Light Detection and Ranging) for the fine analysis of 3D vegetation structure and high-resolution imagery from drones, allows to monitor ecosystem and their inhabitants at a fine scale in a flexible way. As well as, the invention of intelligent biodigesters, with a mix of AI for optimized operational parameters provides a localised solution for waste valorisation, avoiding polluting loads on ecosystem simultaneously (and will produce renewable energy for remote conservation facilities).

Such innovations, with the input and expertise of multidisciplinary scientists and cutting-edge computational tools, are needed to be able to generate and process desirable dates of data at the scale required to implement accurate and timely biodiversity monitoring and conservation interventions (Adanma & Ogunbiyi, 2024; Reynolds et al., 2024). Developing cost-effective models for resource sharing and strategic partnerships with local technology firms and extra academic institutions can considerably slow the financial burden of AI tool subscriptions and infrastructure development (Singhal et al., 2025; Achinas & Quintero, 2025). Within the EC context involving Tanzania and the East African region, the peculiarities of biodiversity conservation, such as the degradation of the habitat, the inadequate technological infrastructure, and institutional barriers, require solutions based on AI and supported by the university at the local level.

ALIGN WITH SDG

This type of alignment is especially relevant in light of the need to meet the of the United Nations, where AI can play an required function in the race to reach environmental goals at a faster pace. Specifically, AI-enabled biodigesters aid in achieving SDG 7 (the process of using waste to produce biogas) and SDG 15 (the process of subjecting waste to sustainable management techniques, reducing the need of ecologically-damaging synthetic fertilizers) (Sandbrook, 2024; Varghese, 2022). Robots based biodigesters can play a direct role in climate action (SDG 13) by helping to reduce the emission of methane into the atmosphere due to organic waste, maximizing the production of biogas with intelligent control systems, and facilitating energy transitions that are associated with low carbon production in areas with limited resources like Tanzania and East Africa (Schwaeke et al., 2025).

Along with, these systems support the SDGs 6 by water pollution mitigation through better wastewater treatment and 13 by reducing greenhouse gas emissions from organic waste decomposition, which will offer climate action (Han et al., 2024). AI's ability to optimize resource management and to enhance ecological monitoring is directly related to SDG 12, encouraging responsible consumption and production patterns, and its use in biodiversity conservation is related to SDG 15, to protect terrestrial ecosystems (Afoma et al., 2025). This all-inclusive effect describes the various contributions of AI-aided biodigester technologies to the status of achieving sustainable development. Along with, the effective implementation of such AI solutions, particularly in developing contexts such as Tanzania, requires a balance and consideration of the unique aspects of these cultures, that is, the technology is not simply imported from a more technologically advanced economy (Greif et al., 2024). This involves not only the imperative of dealing with the monetary costs of adoption, but also the cognitive and institutional costs, in order to promote trust and interpretability among the end-users (Wu & Zhong, 2025).

Along with, inclusivity is critical to ensure in AI systems as both the production and training phases of AI models can have major environmental impacts due to their energy and materials-intensive profiles. Therefore, promoting local research and development programmes as well as strategic partnerships between private industry, science research and governmental organizations is key in order to customize the application of AI solutions to specific agro-ecological areas and socio-cultural frameworks that are predominant throughout Africa. As well as, ethical data procedures, communities' participation and reduction of algorithmic biases have been deemed as the essential factors about the successful and sustained integration of AI-based SDGs projects in Sub-Saharan Africa (Chaudhary, 2023; Mienye et al., 2024). Additionally, there is a need to develop strong partnerships between industry, technology companies, financial institutions, and research institutions which are crucial to the crafting of all-rounded solutions to the multifaceted problems facing African agriculture. Such partnerships are essential to build local AI capacity and nurture AI sovereignty to overcome foreign DI AI models dependency and ensure local appropriate AI technology development of African nations themselves (Ndiaye, 2025).

THEORETICAL FOUNDATION

The Ecological Modernization Theory

This theoretical lens can be summarized summarily, in terms of the political arguments, namely that environmental problems can be solved through technological innovation and the restructuring of society rather than through de-industrialization. Applying this theory to biodigesters that are driven by AI implies that technological developments could help promote environmental sustainability by making the most efficient use of resources and reducing pollution. This view focuses on the possibilities of AI in driving problems with efficiencies on waste conversion and in energy production in such a way that economic development is on par with ecological protection (Begho et al., 2025). Along with, there exists a critical perspective on the concept of ecological modernization, in which a problem-driven approach with the environmental and socio-economic contexts of regions such as Africa informing the creation and implementation of AI solutions is implied, instead of adopting a universal, 'copy-and-paste' technological framework.

Sustainable Development Theory

This pattern highpoints the incorporation of eco-friendly fortification, community equity besides economic feasibility. Such engagement is very required for mitigating possible harms including the entrenchment of digital divides and the perpetuation of systemic biases often embedded in AI technologies (Lehuede, 2024). Further, the Technological Determinism Theory offers a frame of reference for understanding the function of AI as a major force of change that can affect the sustainable development of a nation on economic, environmental, and social levels, thus requiring a multidisciplinary perspective of its application and implications.

Diffusion of Innovations Theory

This theory originated by Everett Rogers and describes the way in which new technologies and ideas spread through social systems in such terms as the factors influencing the speeds of adoption, including perceived attributes of the innovation, communication channels, time, and the nature of the social system itself. This perspective is required in analyzing the adoption of biodigesters with AI in cases such as assumed in rural Tanzania because many circumstances that affect its adoption such as the perceived relative advantage, compatibility with indigenous practices and how complicated the technology is are relevant in helping to understand its adoption among the target communities. Specifically, this theory identifies the importance of the perceived usefulness and ease of use of the AI-driven biodigesters on their continued adoption, and calls for inclusive and participatory implementation models to overcome socio-economic and institutional factors. The Technology Acceptance Model goes further by adding that the choice for an individual to adopt a technology is determined mainly by their perceived usefulness and perceived ease of use.

Stakeholder Theory

This theory highlights the importance of including all communities that will be affected by a project, such as local communities, farmers, governmental bodies, and technology providers to ensure that the AI-based biodigesters are implemented ethically and equitably. Effective communication with stakeholders and establishing clear priorities are essential for creating successful collaborative groups in such types of initiatives (Razak et al., 2024). Along with, the resource-based view views AI abilities as strategic assets that can assist organizations reach environmental, social, and governance objectives, as well as give them a competitive advantage. Besides, the Knowledge-Based View is a re-conceptualization of AI as a knowledge infrastructure that enriches the absorptive, integrative, and transformative capacities in the firms to engage challenges in sustainability and this creates a system through the conversion of uncodified environmental insights into codified knowledge as well as strengthening feedback loops from data analytics into sustainability decision-making. This raises the attention that the potential of AI to think in terms of ecology value is reliant on the institutionalization of sound governance institutions that effectively interconnects digital knowledge with credible organizational practices (Bawa & Asongu, 2026). The Technology Acceptance Model presents a complementary model by explaining the idea of perceived ease of use and usefulness and shows how these factors directly influence the adoption and continued use of such AI-driven systems.

The theoretical framework has been closely matched with the SEM constructs and hypotheses, with AI-based biodigesters serving as the technological driver, university engagement being a mediating institutional mechanism, with biodiversity conservation being the outcome variable that is tested in the structural model.

The theoretical framework is simplified, as only the most concerned theories have been selected and clearly correlated to the model of the study, in which, collectively, they provide the explanations of interdependence between AI-based biodigesters, university activity, and biodiversity protection.

LITERATURE REVIEW

AI-Based Biodigesters & Biodiversity Conservation

A systematic review of available literature shows a rising interest in using AI to maximize the performance (Slimani et al., 2025) and increase the applicability of bio digestion technologies as this relates particularly to addressing problems related to waste management, energy production along with and environmental conservation (Raihan et al., 2024). Specifically, AI is increasingly being used in real-time monitoring, adaptive management, and predictive analytics to enhance the efficiency of biodigesters and thus contribute to the Sustainable Urban Development in and of Ecosystem Resilience (Schwaeke et al., 2025). This means both application of AI-powered predictive analytics and optimization methods that tune the renewable energy systems into optimal efficiency levels, driving the push to a low carbon energy future (Ukoba et al, 2025). In the context of biodiversity, AI-driven solutions are especially effective at processing extensive amounts of data gathered by remote sensors, camera traps, and satellite images with respect to wildlife population and environment changes or threats, and enabling time-bound interventions. These AI-driven ecological modelling systems can evaluate meaningful amounts of data about how different species are distributed, which habitats are suitable to sustain them, and various nature-based indicators to identify areas of high biodiversity value and prioritise conservation efforts (Varghese, 2022).

AI Based Biodigesters and University Interaction

The use of AI in biodigester technology, especially in a context of university engagement, is a new form of resolving problems of sustainability as well as biodiversity promotion. This method takes advantage of the fact that AI can be used to optimize anaerobic digestion processes for enhanced biogas production and simultaneously process waste and reduce greenhouse gas emissions (Ukoba et al, 2025). University engagement, particularly, can increase drastically the development and deployment of AI-based biodigester technologies through interdisciplinary research, development, capacity building and community reach out. This collaborative model encourages the creation of good AI algorithms to optimise the performance of biodigesters, predict feeing quality or maximise the yield of biogas, thus waste becomes a useful energy opportunity causing less negative environmental impact. The use of AI in bio-technological processes (continuous monitoring and real-time adaptation of the environment of bioreactors, which further increases the efficiency of waste decomposition and resource recovery). In addition, AI algorithms can be used to optimize the use of renewable energy systems, enhance energy efficiency, and accelerate the shift towards sustainable energy practices, which can have a direct impact on reducing efforts to combat climate change (Schwaeke et al., 2025).

As well as, the processing and interpretation of the vast datasets produced in the field of biodiversity studies are hotly in demand for AI, and its capabilities can be harnessed for species identification, behaviour tracking, and threat detection using image recognition. This analytical prowess also extends to that of evaluating ecosystem services as well as biodiversity benefits, which allows towards making policy decisions and land use planning by simulating different scenarios in order to understand long term environmental impacts. By allowing for the exact control over the environmental variables, AI-powered systems can vastly enhance microbial activity inside of the biodigesters and cause the waste to be decomposed much faster and in a more efficient manner.

This control of the processes allows and not only the maximum production of biogas, but also supports the circular economy through the efficient use of organic waste and the efficient production of renewable energy and nutrient-rich digestate. Along with, it can also play its function in cutting greenhouse gas emissions through the identification and quantification of emission sources using satellite data and machine learning, as well as in the optimization of renewable energy deployment and management. Such advancements not only promise to revolutionize the conversion of waste into energy, but they also account for a desirable piece of the climate change mitigation measure puzzle by reducing the emissions of Methane from organic waste and displacing the usage of fossil fuels (Al-Raeei, 2024; Balakumar et al., 2025).

University Engagement & Conservation of Biodiversity

Engaging universities and the development of biodigesters using AI opens a symbiotic relationship with academia supporting technological innovation and the world serving as a useful data collection space (Hamdan et al., 2024). These partnerships can result in the development of specialised AI models that are able to analyse meaningful amounts of environmental data to assess the ecosystem services as well as biodiversity benefits for the environment, informing policy decisions and land use planning processes. Along with, the cross-platform AI-driven decision support systems that emerge from these collaborations can even simulate different scenarios for the policy makers that can help with understanding the long-term effects of their decision on ecosystems as well as biodiversity (Ziemba et al., 2024).

University engagement is conceptualized as a multidimensional construct encompassing research collaboration, knowledge transfer, community outreach, and capacity-building activities, which collectively facilitate the adoption and effective implementation of AI-based biodigesters for biodiversity conservation (Al-Raeei, 2024). This analytical capability is further extended to predict success of conservation and restoration efforts, providing environmental policymakers with choices that can be based empirically. Such predictive capabilities provide for proactive resource management, as conservation funds may be allocated for those types or combinations of interventions where there is the most promising probability that those interventions will have a positive ecological outcome (Stellinga et al., 2025). This integrated approach, using the knowledge of the University and the AI technology, also allows for the formulation of efficient frameworks for waste and resources and respectively an optimized strategy for recycling and reducing waste, further promoting the circular economy, reducing anthropogenic pressures on the natural ecosystems. The implementation of AI based tools through university initiatives also goes a long way in strengthening conservation efforts through analysing real-time data from various sources such as remote sensing and sensors network, which is not only required for identifying threatened ecosystems and wildlife populations (Schwaeke et al., 2025).

Biodigesters, AI-Based Technology & University Engagement

The capacity of AI to search meaningful-scale data sets provides required information's concerning the comprehension and conservation of biological diversity. By using AI, it will become possible to monitor environmental changes in real-time using satellite imagery and sensor networks to spot patterns and anomalies that are desirable to biodiversity protection (Hamdan et al., 2024). This enables the targeted interventions towards threat mitigation and conservation of the vulnerable species and habitats to be implemented in time. The integration of AI in environmental conservation also optimises resource management and facilitates advanced monitoring and analysis to reduce the impacts in the environment. The use of AI-powered biodigesters is essential in reducing emissions and controlling climate change as it will utilize organic waste to harvest methane, utilize intelligent algorithms to optimize the anaerobic digestion process, and eliminate the use of fossil fuel to obtain energy as biogas (Abduljabbar et al., 2025). Specifically, AI models, including genetic algorithms, artificial neural networks, and Bayesian networks are central in solving biodiversity problems at both spatial-temporal scales.

These AI-based methods allow us to predict ecosystem services, an required factor of the most effective biodiversity conservation efforts. Such sophisticated predictive modelling, using AI, plays an required function in predicting vulnerable species, habitat suitability and shifting of biodiversity in the face of climate change (Ukoba et al, 2025). Besides, explainable AI techniques such as Shapley values provide a powerful tool to interpret complex AI models so to increase transparency and trust in such a predictive computation for environmental decision-making. This transparency is required to enable adoption of AI systems in environmental monitoring, particularly in situations where decisions have major social and ecological consequences (Schwaeke et al., 2025). This allows the stakeholders to have knowledge on the underlying rationale behind the AI-generated insights, leading to better collaboration and more effective deployment of conservation policies (Chisom et al., 2024). In the context of biodiversity conservation in Tanzania, university engagement proves crucial for the development and deployment of AI immediately-based bio-digesters comparatively, to make sure that such technological interventions are culturally appropriate and are ecologically sound (Reynolds et al., 2024).

METHODOLOGY

This study adopted a quantitative research method in investigating the role of AI-based based biodigesters in biodiversity preservation by involving universities in East Africa with specific mention to the St. Joseph University in Tanzania. St. Joseph University in Tanzania was selected as the study site due to its active involvement in sustainability and biodigester initiatives, while the findings are cautiously interpreted to reflect broader East African relevance within similar institutional and environmental contexts. A cross sectional design was adopted considering the fact that data were gathered at one point in time from practitioners involved in biodigester projects in the university. A cross-sectional design refers to data collected from respondents at a single point in time, which allows analysis of relationships among variables but limits the ability to establish causal inferenccs. The target population included people directly involved in planning, implementation and management of biodigesters initiatives such as project coordinators, technical staff, sustainability officers and academic researchers. Purposive sampling was used to identify relevant people with relevant knowledge and a practical experience concerning AI - integrated biodigesters systems as key informants. Based on these criteria a total sample of 248 respondents was seen as sufficient for statistical analysis and structural modelling. Primary data were gathered from a structured questionnaire that included previously validated scales taken from the literature and then modified to the local context of AI based biodigesters and biodiversity conservation. The instrument was closed ended items that were measured using a five point Likert scale and comprised of key constructs such as adoption of AI technology in biodigester use by farmers, biodiversity conservation outcomes and university engagement. Prior to the main data collection, a pilot study was carried out on a small group of respondents to ensure the clarity, reliability and validity of the instrument. Necessary changes were implemented according to the feedback received. Study operationalized the constructs by employing a validated multi-item scale, scaled to their context, which is the use of AI-based biodigesters, and measured the constructs on a five-point Likert scale, assessing the results of technological acceptance, involvement of the university, and conservation of biodiversity.

Data collection was conducted using direct mode of distribution and online survey techniques to achieve a higher response rate. Ethical considerations were strictly adhered to such as informed consent, confidentiality, and voluntary participation of respondents. Upon collecting data; the responses were checked for missing values, outliers and normal using statistical software. The data analysis was performed by using the Structural Equation Modeling (SEM). The analysis was conducted in two steps involving the measurement model assessment and structural model evaluation. The reliability and validity of the measurement model were determined in terms of Cronbach's alpha, composite reliability, variance average extracted (AVE), and validity discriminant using Fornell-Larcker criterion and HTMT ratio. The structural model was assessed in terms of the following model statistics: path coefficients, t values, p values, and confidence intervals derived from the use of bootstrapping procedures. Additionally, the coefficient of determination analysis (R2), effect size (f2) and predictive relevance (Q2) were conducted to evaluate the explanatory and predictive power of the model.

The results of the SEM analysis gave information on the relationship between implementation of biodigesters using AI, engagement from the universities and the biodiversity conservation achievement. The methodology approach ensured strength, reliability and validity of the results, which make this study appropriate for academic and policy implications for sustainable environmental management in East Africa.

RESULTS

Table 1. Stratified Distribution

S. No.	Strata (Practitioner Category)	Estimated Population (N)	Proportion (%)	Sample Size (n)
1	Project Coordinators	60	24.2%	60
2	Technical Staff (Engineers/ Operators)	80	32.3%	80
3	Academic Researchers	50	20.2%	50
4	Sustainability Officers	30	12.1%	30
5	Field Assistants / Support Staff	28	11.2%	28
Total		248	100%	248

Source: authors

The stratified distribution table 1 represents the distribution of the entire sample of 248 respondents into various groups of practitioners involved in biodigester projects. Each group was considered as a distinct stratum to make sure there is a proper representation of all pertinent categories in the study. The number of samples for all the different groups were allocated according to their percentage in the entire population; this helps in improving the accuracy and fairness of the results. From the table, the meaningful part, with 32.3% of the total sample, that is 80 people respondents, are the technical staff (engineers/operators). This means that they play major function in biodigester projects and their input is of high importance in the study. Project coordinators were the second meaningful group with 24.2% (60 respondents) and revealed their considerable engagement in the planning and managing of such projects. Academic researchers accounted for 20.2% (50 respondents) representative of their contribution to research, innovation and knowledge development in relation to AI-based biodigesters. Sustainability officers' amount to 12.1% (30 respondents): This represents their function in ensuring environmental and sustainable practices. Lastly, field assistants and support staff then made up 11.2% (28 respondents), emphasising their function in operational and 'on the ground' activities. In the real sense, the table shows that the sample was proportionally distributed among all the practitioner categories. This helps to ensure that every group's perspective is considered, which helps to reduce bias and makes the study findings more reliable.

Table 2. Reliability and Validity Analysis

Construct	KMO	Cronbach's Alpha	Composite Reliability (CR)	AVE	Factor Loadings Range
AI-Based Biodigesters	0.79	0.88	0.91	0.63	0.72 – 0.85
University Engagement	0.76	0.85	0.89	0.61	0.70 – 0.83
Biodiversity Conservation	0.81	0.90	0.92	0.66	0.74 – 0.87

Source: authors

The results of reliability and validity of the constructs are presented in the table 2, which shows that all the constructs utilized in this study have a strong and acceptable result for analysis. First, the KMO's for the variables AI Based Biodigesters (0.79), University Engagement (0.76) as well as biodiversity Conservation (0.81) are all greater than the points outed value of 0.70. This implies that the sample data is suitable for a factor analysis and variables are correlated well. Looking at Cronbach's Alpha, all the values are above 0.80 and this is good internal consistency. This means that the things we use to measure each construct are reliable; i.e. they give us consistent results. Similarly, the Composite Reliability (CR) values are all greater than 0.70, which confirms that the constructs are reliable and well-measured.

The Average Variance Extracted (AVE) values of all the constructs are above 0.50 and hence indicating good convergent validity. In practical words, this implies that the items under each construct are highly related to each extra and measure the same idea well. Additionally, the factor loadings for all items are within accepted ranges (above 0.70) that indicate all items provide a good contribution for its construct. On the whole, the outcomes confirm the reliability and validity of the measurement model, which is appropriate for further analysis.

Table 3. Discriminant Validity

Construct	AI-Based Biodigesters	University Engagement	Biodiversity Conservation
AI-Based Biodigesters	0.79	—	—
University Engagement	0.49	0.78	—
Biodiversity Conservation	0.38	0.53	0.81

Source: authors

The results in the table 3 indicates that all the variables that are used in the study are reliable and valid for the analysis. The KMO values for AI Based Biodigesters

(0.79), University Engagement (0.76), as well as biodiversity Conservation (0.81) are all higher than 0.70 which means the data is appropriate for factor analysis and the sample is sufficient. The Cronbach's Alpha values are all above 0.80 indicating that the items are consistent and reliable. This means that respondents responded to the questions in a stable and dependable manner. Similarly, the Composite Reliability (CR) values are also above 0.70 which further supports the measurement reliability of the constructs. The AVE values are all above 0.50, indicating that each construct explains a good amount of its items, = good validity. The factor loadings are also found to be above 0.70 which implies that all the items are representing their respective variables to a strong degree. In somewhat practicalr terms, the results prove good, reliable, valid measurement model that can be confidently used for further SEM analysis.

Table 4. HTMT Ratio

Construct	AI-Based Biodigesters	University Engagement	Biodiversity Conservation
AI-Based Biodigesters	—	0.61	0.48
University Engagement	0.61	—	0.69
Biodiversity Conservation	0.48	0.69	—

Source: authors

The HTMT ratio table 4 demonstrates the level of DV between the constructs in the study. Discriminant validity refers to the distinction between different constructs, that is, each construct should be different and measure something distinct. In this table, all the HTMT values are less than the recommended limit of 0.85, which means that the issue of overlap between the variables does not exist. Specifically, the HTMT value between AI-Based Biodigesters and University Engagement is equal to 0.61, between AI-Based Biodigesters as well as biodiversity Conservation is 0.48 and between University Engagement to Biodiversity Conservation is equal to 0.69. Since for all these values are well below 0.85, it confirms the fact clearly distinguishing each construct from the extra. In practical terms, this refers to the fact that the variables in the study are not measuring the same thing, and each one is a different concept. Therefore, the model has a good discriminant validity, and is applicable for a further SEM analysis.

Figure 1. Structure equation model

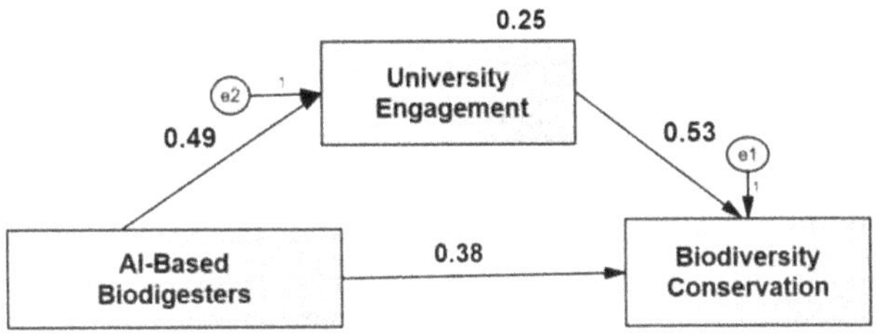

Table 5. Structural Model

H	Relationship	Path Coefficient (β)	t-value	p-value	95% CI (LL–UL)
H1	AI-Based Biodigesters → Biodiversity Conservation	0.38	6.12	0.000	0.25 – 0.50
H2	AI-Based Biodigesters → University Engagement	0.49	8.45	0.000	0.36 – 0.60
H3	University Engagement → Biodiversity Conservation	0.53	9.21	0.000	0.41 – 0.63

Source: authors

The results of the structural model in figure 1 are presented in the form of table 5 on how the relationships between key variables in the study and whether the hypotheses relationships proposed in this study. For H1, the relationship between AI-Based Biodigesters as well as biodiversity Conservation has a path coefficient (b) value of 0.38, which is a positive effect. The t-value and p-value i.e., 6.12 and 0.000 respectively indicates that, this relationship is statistically desirable. The confidence interval (0.25-0.50) does not include zero, along with, confirming that the effect does mean something. This means that the use of biodigesters based on AI directly contributes to the enhancement of biodiversity conservation.

For H2, AI-Based Biodigesters got more positive impact on the University Engagement (path co-efficient 0.49). The high t-value 8.45 and p-value 0.000 shows it is a very desirable relationship. The confidence interval (0.36-0.60), along with, also supports this finding. This further implies that the implementation of AI-biogas digesters promotes higher engagement and participation of universities in sustainability activities.

For H3, the strongest is the relationship between University Engagement as well as biodiversity Conservation with a path coefficient of 0.53. The t-value (9.21) and p-value (0.000) show that this relationship is very desirable. The confidence interval (0.41-0.63) again does not contain the value zero. This point outs that the engagement with university has a desirable part to play in improving biodiversity conservation outcomes. Looking at the Coefficient of determination (R squared), the value of the University Engagement is 0.26 which is considered moderate. This means that University Engagement can be explained to about 26% by AI-Based Biodigesters. While this is a reasonable level, it also implies that there are extra factors that may not be included in the model that might have an influence on university engagement.

For Biodiversity Conservation, the R2-value is 0.53 in table 6, which is considered to be substantial. This shows the combined explanatory power of the variables AI-Based Biodigesters and University Engagement to be 53% in biodiversity conservation. In the practical terms, the model explains more than half of variable change in biodiversity conservation, which is a strong explanatory power. In the real sense, outcomes from the structural models validate all of the supported hypothesized relationships are positive and statistically desirable. The model has good explanatory strength particularly for biodiversity conservation, and hence is reliable to understand the impact of AI-based biodigesters and universities engagement.

Table 6. Determination (R^2)

Endogenous Construct	R^2 Value	Interpretation
University Engagement	0.26	Moderate
Biodiversity Conservation	0.53	Substantial

Source: authors

The effect size (f2) results indicated in table 6. It helps to understand the practical importance of each relationship, and not just statistical significance. The relationship between AI-Based Biodigesters and University Engagement is also found to have f2 value of 0.26 which is seen at medium effect. This means that the impact of AI-based biodigesters to enhance university engagement is noticeable and desirable. In practical terms, the use of these technologies plays a required function in making universities more involved in sustainability along with and environmental efforts. For the relationship between AI-Based Biodigesters as well as biodiversity Conservation are 0.14 or a small effect. This point outs that although AI-based biodigesters make a contribution towards biodiversity conservation, it is relatively small. Along with, this effect is nevertheless required and should not be ignored as it adds value to extra influencing factors.

Table 7. Effect Size

Relationship	f^2 Value	Effect Size
AI-Based Biodigesters → University Engagement	0.26	Medium
AI-Based Biodigesters → Biodiversity Conservation	0.14	Small
University Engagement → Biodiversity Conservation	0.31	Meaningful

Source: authors

In table 7 the relationship between University Engagement as well as biodiversity Conservation has the highest effect size, as f2 value is 0.31, which is meaningful effect size. This implies there is a powerful and desirable influence of university engagement on the outcomes of biodiversity conservation. In practical words, the higher the engagement of universities, the greater the impact on conservation of biodiversity. In the real sense, the outcome points outs that the university engagement contributes to the key function in driving biodiversity conservation and AI-based biodigesters contributes both directly and indirectly (through the university engagement).

Table 8. Predictive Relevance

Endogenous Construct	Q^2 Value	Interpretation
University Engagement	0.18	Medium Predictive Relevance
Biodiversity Conservation	0.29	Strong Predictive Relevance

Source: authors

The table 8 of predictive relevance (Q2) results indicate how good the model is at predicting the results of the endogenous variables. In the case of SEM analysis, Q2 values greater than zero mean that the model has predictive relevance, which means that the model can be used to effectively predict the data. For University Engagement, Q2 value is 0.18 which is medium predictive relevance. This means that the model has a moderate ability in predicting changes in university engagement and based on AI bio-digesters. In layman's lingo, the model can adequately explain and predict how universities will be involved in sustainability activities. For Biodiversity Conservation, the Q2 value is 0.29, which indicates that there is a high level of predictive relevance. This implies that the model has a high capability to make predictions on the outcome of biodiversity conservation. In extra words, the combination of AI-based biodigesters and universities offers a solid foundation for predicting the enhancement in biodiversity conservation. In the real sense, the results confirm that the model has good predicting ability, especially for the conservation of biodiversity.

DISCUSSION

The findings suggest that the model not only has statistical sufficiency but also has significant predictive value, specifically when it comes to explaining the effect of AI-based biodigesters and university involvement when working together on the outcomes of biodiversity conservation. The positive Q2 values affirm that the model has an out-of-sample predictive relevance and therefore it has a potential implication on its practical relevance in forecasting outcome of sustainability in such a case. Instead of simply validating statistical hypotheses, these results make it clear that AI integration, and institutional partnership in creating conservation effectiveness has a functional role. In addition, the values of R2 represent a moderate to a high degree of explanatory power, which means that although the model may cover the main factors driving the phenomenon of biodiversity conservation, other context-dependent and environmental drivers may also be helpful causes. It implies that the use of AI-based interventions, backed by the institutional involvement is a significant yet not the sole route to sustainability. All in all, the findings indicate a background on the interaction between technological and organizational mechanisms to create specific measurable benefits on the environment that is a little more insightful than the mere statistical validation.

The results of this paper support the main claim that AI-based biodigesters, with the assistance of active university activity, can become a valuable contributor to biodiversity protection by overseeing the sustainability of waste management, improving the utilization of renewable energy, and promoting the use of data-driven decisions concerning the environment in the context of East Africa.

The research makes a clear distinction among sustainability results as it specifies renewable energy benefits (generation of biogas), waste management benefits (reduced organic waste), biodiversity benefits (reduced ecological pressure), and climate action benefits (reduced greenhouse gas emission) by noting that they depend on each other but they cannot be analyzed analytically.

The results of this work are not to be taken at face value, since they rely on the data of a perceptual survey and not on actual ecological measurements, which could reduce the possibility of forming causal relationships and objectively measuring biodiversity outcomes.

CONCLUSION

This paper has explored the importance of AI-based biodigesters and university participation in improving on biodiversity conservation in East African context based on empirical evidence based on Tanzania. According to the findings of the structural equation model, the effect of AI-based biodigesters on the conservation of biodiversity is significant and has both a direct and an indirect effect directly and indirectly through university involvement. Particularly, engagement in the university was referred to as a powerful contributing factor, which showed that institutional support, knowledge transfer and collaborative innovation play an essential role in producing sustainable environmental results. The findings also show that the model has a significant explanatory and predictive strength which implies that incorporating technological solutions and incorporation of the role of academic participation is a possible way forward in enhancing conservation intentions. Nevertheless, the results are supposed to be understood in the context of the research design since the cross-sectional and perceptual data restricts the causal generalization of the research. In general, this research paper adds to the already increasing literature by showing that AI-powered biodigester systems, with the assistance of an active university involvement, can contribute significantly to the biodiversity conservation and sustainability. The study has viable implications on policy makers, academic institutions and environmental practitioners who want to exploit the use of technology-enabled solutions to manage the ecological situation in developing areas.

FUTURE SOLUTIONS

Aided by these insights, future research should consider the nuances of how AI integration is integrated based on factors such as green technology innovation, human- AI technology integration, and other determinants that have been shown to have a significant impact on sustainability outcomes as well as ESG performance. Further investigation of the mediating role of innovation in driving green enterprise practices through innovation adoption on AI could yield better understanding of how to achieve maximum environmental benefits. Specifically, quantifying the success of the AI capabilities in effectiveness to the efficiency of green innovations, especially in the governance effective framework, could help interpret pathways for environmental stewardship enhancement. And additionally, examining the relationship between AI and ESG performance through examining modulations to strategies such as industry type, regulatory environment, and ownership characteristics will also be crucial for developing tailored sustainability strategies. As well as, exploring the effect of AI adoption in sustainable finance and green financial instrument can reveal positive feedback loops, thereby serving to augment the collective impact of AI discovery and green finance management. Continued academic efforts could also be to optimize environmental performance of biodigester operation such that recommendations arising from data are creative and specific to ecological maximization and operational efficiency.

Based on the above insights, later investigation should also focus on wider implications of AI-digital transformation on the impact of ESG performance, keeping in mind that although digital initiatives help improve sustainability, more variables are pertaining as desirable to the substantive results. Finally, it should be noted that future studies should examine the use of artificial intelligence systems for standardization of ESG assessments and to support predictive decision-making decisions, particularly with the capabilities of data processing that AI has, enabling the processing of large volumes of sustainability-relevant data.

REFERENCES

Abduljabbar, H. M., Abdullah, W. M., Bazool, S. D. S., Muhsin, A. I., Alfalahi, S. T. Y., & Temnikov, V. (2025). AI-Driven Environmental Impact Assessments: Ethical and Legal Considerations. *Environment and Social Psychology, 10*(12). Advance online publication. DOI: 10.59429/esp.v10i12.3994

Achinas, S., & Quintero, M. (2025). Embodiment of machine learning in biogas power industry: An opinion. *Frontiers in Energy Research, 13*, 1589782. Advance online publication. DOI: 10.3389/fenrg.2025.1589782

Adanma, U. M., & Ogunbiyi, E. O. (2024). Artificial intelligence in environmental conservation: Evaluating cyber risks and opportunities for sustainable practices. *Computer Science & IT Research Journal, 5*(5), 1178–1209. DOI: 10.51594/csitrj.v5i5.1156

Afoma, U. M., Singh, S., Mishra, A. K., Sharma, C. K., Gupta, K., Mishra, M. K., Roy, B., Verma, V. V., & Sharma, V. K. (2025). Integrating Artificial Intelligence in Environmental Monitoring: A Paradigm Shift in Data-Driven Sustainability [Review of *Integrating Artificial Intelligence in Environmental Monitoring: A Paradigm Shift in Data-Driven Sustainability*]. *EcoHealth*. Springer Science+Business Media. DOI: 10.1007/s10393-025-01752-8

Al-Raeei, M. (2024). Artificial intelligence for climate resilience: Advancing sustainable goals in SDGs 11 and 13 and its relationship to pandemics. *Discover Sustainability, 5*(1), 513. Advance online publication. DOI: 10.1007/s43621-024-00775-5

Ayadi, R., Forouheshfar, Y., & Moghadas, O. (2025). Enhancing system resilience to climate change through artificial intelligence: A systematic literature review. *Frontiers in Climate, 7*, 1585331. Advance online publication. DOI: 10.3389/fclim.2025.1585331

Balakumar, S., Narayanan, M., Kamaraj, M., & Aravind, J. (2025). Harnessing artificial intelligence for sustainable environmental remediation a review [Review of *Harnessing artificial intelligence for sustainable environmental remediation a review*]. *International Journal of Environmental Science and Technology, 22*(13), 13189. Springer Science+Business Media. DOI: 10.1007/s13762-025-06528-9

Bawa, S., & Asongu, S. A. (2026). Investigating How Firms in Emerging Economies Integrate AI-Driven Systems and ESG Metrics Into Biodiversity Conservation and Circular Economy Strategies. *Business Strategy and the Environment*, bse.70547. Advance online publication. DOI: 10.1002/bse.70547

Begho, T., Daubry, T. P., & Irabor, A. E. (2025). A Scoping Review of Artificial Intelligence and the Future of African Agriculture: Opportunities, Challenges and the Path Forward. *Advances in Agriculture, 2025*(1). DOI: 10.1155/aia/2862074

Chaudhary, G. (2023). Environmental Sustainability: Can Artificial Intelligence be an Enabler for SDGs? *Nature Environment and Pollution Technology*, *22*(3), 1411–1420. DOI: 10.46488/NEPT.2023.v22i03.027

Chisom, O. N., Biu, P. W., Umoh, A. A., Obaedo, B. O., Adegbite, A. O., & Abatan, A. (2024). Reviewing the role of AI in environmental monitoring and conservation: A data-driven revolution for our planet. *World Journal of Advanced Research and Reviews*, *21*(1), 161–171. DOI: 10.30574/wjarr.2024.21.1.2720

Greif, L., Kimmig, A., Bobbou, S. E., Jurisch, P., & Ovtcharova, J. (2024). Strategic view on the current role of AI in advancing environmental sustainability: A SWOT analysis. *Discover Artificial Intelligence*, *4*(1), 45. Advance online publication. DOI: 10.1007/s44163-024-00146-z

Hamdan, A., Ibekwe, K. I., Etukudoh, E. A., Umoh, A. A., & Ilojianya, V. I. (2024). AI and machine learning in climate change research: A review of predictive models and environmental impact [Review of *AI and machine learning in climate change research: A review of predictive models and environmental impact*]. *World Journal of Advanced Research and Reviews, 21*(1), 1999. GSC Online Press. DOI: 10.30574/wjarr.2024.21.1.0257

Han, H., Liu, Z., Li, J., & Zeng, Z. (2024). Challenges in remote sensing based climate and crop monitoring: Navigating the complexities using AI. *Journal of Cloud Computing (Heidelberg, Germany)*, *13*(1), 34. Advance online publication. DOI: 10.1186/s13677-023-00583-8

Kyathanahally, S. P., Hardeman, T., Reyes, M., Merz, E., Bulas, T., Brun, P., Pomati, F., & Baity-Jesi, M. (2022). Ensembles of data-efficient vision transformers as a new paradigm for automated classification in ecology. *Scientific Reports*, *12*(1), 18590. Advance online publication. DOI: 10.1038/s41598-022-21910-0 PMID: 36329061

Lehuedé, S. (2024). An elemental ethics for artificial intelligence: Water as resistance within AI's value chain. *AI & Society*. Advance online publication. DOI: 10.1007/s00146-024-01922-2

Mienye, I. D., Sun, Y., & Ileberi, E. (2024). Artificial intelligence and sustainable development in Africa: A comprehensive review [Review of *Artificial intelligence and sustainable development in Africa: A comprehensive review*]. *Machine Learning with Applications, 18*, 100591. Elsevier BV. DOI: 10.1016/j.mlwa.2024.100591

Ndiaye, S. M. (2025). *Building Trustworthiness as a Requirement for AI in Africa: Challenges*. Stakeholders and Perspectives., DOI: 10.1007/978-3-031-75674-0_3

Ojija, F., Ogwu, M. C., Ally, J. S., John, J. P., Stephano, A., Felix, N., & Tekka, R. S. (2025). Artificial intelligence-driven solutions for mitigating human–wildlife conflict in biodiversity hotspots. *Science Progress*, *108*(4), 00368504251394584. Advance online publication. DOI: 10.1177/00368504251394584 PMID: 41259221

Raihan, A., Paul, A., & Rahman, M. (2024). Artificial Intelligence (AI) for Environmental Sustainability: A Concise Review of Technology Innovations in Energy, Transportation, Biodiversity, and Water Management. *Journal of Technology Innovations and Energy*, *3*(2), 64–73. DOI: 10.56556/jtie.v3i2.953

Rathoure, A. K., & Ram, B. L. G. (2024). Unveiling the marvels of biodiversity: Recent advancements in conservation efforts. *Biodiversity International Journal*, *7*(2), 51–61. DOI: 10.15406/bij.2024.07.00211

Razak, A., Yogarayan, S., Sayeed, M. S., & Derafi, M. I. F. M. (2024). Agriculture 5.0 and Explainable AI for Smart Agriculture: A Scoping Review [Review of Agriculture 5.0 and Explainable AI for Smart Agriculture: A Scoping Review]. *Emerging Science Journal*, *8*(2), 744–760. DOI: 10.28991/ESJ-2024-08-02-024

Reynolds, S. A., Beery, S., Burgess, N. D., Burgman, M. A., Butchart, S. H. M., Cooke, S. J., Coomes, D. A., Danielsen, F., Minin, E. D., Durán, A. P., Gassert, F., Hinsley, A., Jaffer, S., Jones, J. P. G., Li, B. V., Aodha, O. M., Madhavapeddy, A., O'Donnell, S., Oxbury, W. M., … Sutherland, W. J. (2024). The potential for AI to revolutionize conservation: a horizon scan [Review of *The potential for AI to revolutionize conservation: a horizon scan*]. *Trends in Ecology & Evolution, 40*(2), 191. Elsevier BV. DOI: 10.1016/j.tree.2024.11.013

Sandbrook, C. (2024). Beyond the Hype: Navigating the Conservation Implications of Artificial Intelligence. *Conservation Letters*, *18*(1), e13076. Advance online publication. DOI: 10.1111/conl.13076

Schwaeke, J., Gerlich, C., Nguyen, H. L., Kanbach, D. K., & Gast, J. (2025). Artificial intelligence (AI) for good? Enabling organizational change towards sustainability. *Review of Managerial Science*, *19*(10), 3013–3038. DOI: 10.1007/s11846-025-00840-x

Singhal, N., Vardhan, H., Jain, R., Vashistha, P., Pandey, A., Wagri, N. K., & Gaur, A. (2025). Algorithms for nature: Integrating technology, ecology, and society for sustainable conservation. *Environmental Systems Research*, *14*(1), 30. Advance online publication. DOI: 10.1186/s40068-025-00431-5

Slimani, S., Omri, A., & Jabeur, S. B. (2025). When and how does artificial intelligence impact environmental performance? *Energy Economics*, *148*, 108643. DOI: 10.1016/j.eneco.2025.108643

Stellinga, L., Korenhof, P., & Blok, V. (2025). Bio-Centered Artificial Intelligence: Towards a Progressive AI-biosphere Relation Through the Concepts of Poiesis and Mimesis. *Topoi*. Advance online publication. DOI: 10.1007/s11245-025-10270-9

Ukoba, K., Onisuru, O. R., Jen, T., Madyira, D. M., & Olatunji, K. O. (2025). Predictive modeling of climate change impacts using Artificial Intelligence: A review for equitable governance and sustainable outcome. *Environmental Science and Pollution Research International*, *32*(17), 10705–10724. DOI: 10.1007/s11356-025-36356-w PMID: 40183866

Ullah, F., Saqib, S., & Xiong, Y. (2024). Integrating artificial intelligence in biodiversity conservation: Bridging classical and modern approaches. *Biodiversity and Conservation*, *34*(1), 45–65. DOI: 10.1007/s10531-024-02977-9

Varghese, A. (2022). AI-Driven Solutions for Energy Optimization and Environmental Conservation in Digital Business Environments. *Asia Pacific Journal of Energy and Environment*, *9*(1), 49–60. DOI: 10.18034/apjee.v9i1.736

Wu, P., & Zhong, Y. (2025). Artificial intelligence in sustainable agriculture: Towards a socio-technical roadmap. *Smart Agricultural Technology*, *12*, 101578. DOI: 10.1016/j.atech.2025.101578

Ziemba, E., Duong, C. D., Ejdys, J., González-Pérez, M. A., Kazlauskaitė, R., Korzyński, P., Mazurek, G., Paliszkiewicz, J., Stankevičienė, J., & Wach, K. (2024). Leveraging artificial intelligence to meet the sustainable development goals. *Journal of Economics and Management*, *46*, 508–583. DOI: 10.22367/jem.2024.46.19

Chapter 4
AI-Driven Regenerative Intelligence for Ecological Restoration and Climate Action

K. Aravinthan
http://orcid.org/0000-0002-7923-1842
Anjalai Ammal Mahalingam Engineering College, Thiruvarur, India

Anorgul Ashirova
Mamun University, Khiva, Uzbekistan

S. Aarthi
http://orcid.org/0009-0006-9064-2091
Marwadi University, Rajkot, India

R. N. Ravikumar
http://orcid.org/0009-0009-3705-1681
Marwadi University, Rajkot, India

Babamuratov Bekzod
Termez University of Economics and Service, Termez, Uzbekistan

Ulugbek Vosiqov
Kimyo International University in Tashkent, Tashkent, Uzbekistan

ABSTRACT

The chapter explores the way the Artificial Intelligence can go beyond carbon mitigation to ecological regeneration and restoration-oriented governance of climate action. It states that AI can be the most useful in combination with the ecological objectives, adaptive planning, governance safeguards, and long-term resilience. To reinforce this stand, the chapter presents the AERSI framework, which relates environmental sensing, analytical intelligence, adaptive decision support, integration to governance, and sustainability responsibility. The targeted literature review and analytical evidence reveal that the use of AI can reinforce the monitoring, predictive planning, and restoration decisions; yet, it is conditional to the quality of data,

DOI: 10.4018/979-8-3373-9978-2.ch004

institutional capacities, local involvement, and ecological legitimacy. The chapter concludes that AI can only be used as restoration infrastructure through a combination of technological ability, governance responsibility, and regenerative outcomes.

1. INTRODUCTION

Historically, the climate action has been concerned with mitigation measures which include reduction of carbon, control of emission and energy transition. Although these strategies are still considered a necessity, they are being seen to be inadequate to deal with the entire extent of ecological disturbance. Climate change has become highly interconnected with the loss of biodiversity, land degradation, water stress and the loss of resiliency in the ecosystem. Consequently, environmental action needs to shift to less carbon-focused responses to more ecological repair and systemic recovery models (Nasruddin et al., 2025). Here, artificial intelligence is coming out as a possibly revolutionary facilitator of restoration planning, adaptive monitoring, and evidence-based environmental governance. Nevertheless, its place should be positioned in such a way that the technological ability enhances ecological health and social righteousness. This chapter thus makes AI not just a means of efficiency, but rather as an element of a broader regenerative systems approach that incorporates restoration, governance, resilience, and long-term sustainability into climate action.

1.1 Climate Action Limits

The traditional approach to climate action has been focused mostly on mitigation, in particular, decarbonization, carbon accounting, and emissions reduction targets. These are the initiatives, which are needed, yet not sufficient to cover the larger ecological implications of environmental degradation. The fragmentation of forests, soil erosion, biodiversity collapse, coastal vulnerability and water insecurity are indicators that climate disruption is just one element of a bigger systems crisis. Policies that focus solely on carbon neutrality can thus ignore the functionality of the ecosystem and community resilience as well as regenerative capacity in the long term (Pasupuleti, 2024). In most situations, mitigation projects are also not attuned to local ecological conditions and governance requirements. This poses a disconnect between climate ambition and restoration results on the ground. The drawbacks of the conventional climate action indicate the possibility to have a more comprehensive structure that coordinates mitigation with adaptation, ecological recovery, and institutional coordination. These changes create room to have AI-enabled systems that facilitate more integrated, dynamic, and restorative-based environmental decisions.

1.2 Need for Regeneration

The regeneration is coming into the limelight due to the fact that environmental systems need not only protection or reduction of damage but active restoration, renewal and building resilience in the long run. In contrast to methods that enforce compliance, or even sustainability, regenerative thinking focuses on restoring ecosystems in such a way that it becomes functional, diverse, and able to adapt. These involve the restoration of soils, maintenance of healthy watersheds, restoration of habitats and enhancement of social-ecological interactions within communities and institutions. Regeneration is important in climate context since it bridges the gap between mitigation and adaptation and biodiversity recovery as opposed to having them as independent objectives (Bains et al., 2025). It also promotes a more futuristic system of governance where ecological systems are continuously checked, fixed and administered. Regeneration is particularly imperative in landscapes that experience cumulative stress due to urban growth, loss of resources, pollution and extreme weather conditions. A regenerative approach thus offers a better platform through which climate action can be supported by integrating ecological restoration with resiliency, justice, and recovery of planets that are long-term.

1.3 Chapter Contribution

This chapter is a part of the growing discussion on artificial intelligence and climate action by re-conceptualizing artificial intelligence as a facilitator of ecological restoration and not simply as a means to achieve efficiency in mitigation. Its main thesis is that AI could be more helpful in assisting climate responses when it is integrated into the regenerative governance systems linking restoration objectives, adaptive intelligence, and institutional responsibility. The chapter presents the conceptual framework of AERSI as a tool to comprehend the potential of AI-enabled systems to aid ecological monitoring, predictive planning, restoration coordination, and resilience-oriented decision-making. Instead of positioning AI as a solution to all problems, the chapter notes that the utility of AI is context-specific and requires governance protections and valuable involvement of local stakeholders. The input is mainly conceptual and backed by selective evidence of analysis and discussion of governance. In this way, the chapter explains how AI can bring climate action beyond carbon neutrality to a more comprehensive ecological restoration and more responsible approaches to planetary recovery.

2. CONCEPTUAL FOUNDATIONS

The intellectual underpinning of this chapter is based on the overlap of ecological regeneration, intelligent systems, and governance-oriented climate action. These three are not independent themes, but rather parts of a larger regenerative transition which depends on each other. The substantive goal is ecological regeneration with its focus on repair, renewal, and restoration of ecosystem functions. Regenerative intelligence explains how AI and data-driven systems can help facilitate adaptive, context-driven, and long-term restoration. It is in the governance that the technological and ecological processes are made legitimate, accountable and socially-grounded. Combined, these foundations change the discourse of the limited view on technological innovation to a systems-based concept of restoration infrastructure (Wong et al., 2022). This point of view is required since AI in environmental fields cannot be assessed in the context of computational accuracy or predictive performance. It is equally applicable to its relevance with regard to improving ecological recovery, improving decision quality and reinforcing fair and participatory management in various restoration contexts as well as scales.

2.1 Ecological Regeneration

Ecological regeneration is a set of processes that brings back the degraded ecosystems to allow them to regain their structure, function, resilience and long-term productivity. It goes past conservation or lessening impact by addressing renewal of ecological processes and relationships. The regeneration can be in the form of habitat restoration, reforestation, soil reconstruction, wetland recovery, reintroduction of the species, and water system restoration depending on the situation. Notably, regeneration is not merely biophysical, but also social and institutional, since the results of restoration require quality governance, involvement by communities, and stewardship over the long term. Ecological regeneration is an important concept in climate discourse as it connects the issue of emissions with the biodiversity, adaptation, and resilience (Frazier & Song, 2024). It assists in re-setting environmental action beyond its single technical goals and to system repair. The concept thus offers a better foundation to comprehending the possible role of AI in climate governance. In combination with regenerative objectives, AI can aid the ecological change monitoring, priorities of restoration, and adaptive interventions in the context of complex environmental systems.

2.2 Regenerative Intelligence

Regenerative intelligence is the application of AI and data systems in promoting ecological restoration, resilience and adaptive environmental management. In contrast to traditional AI applications that are primarily concerned with efficiency, optimization, or prediction in isolation, regenerative intelligence is geared towards ecosystem restoration and sustainable governance, which are two elements of the long term. It is a set of sensing technologies, data analytics, predictive models, and feedback mechanisms combined, to enhance the understanding and restoration of degraded systems over time. This type of intelligence is regenerative in that it is meant to help in the renewal process as opposed to merely quantifying the loss or mechanizing control (Kuppusamy, 2025). Its usefulness is the possibility to make context-sensitive decisions which capture ecological variability, uncertainty and change. Regenerative intelligence should however, still be restricted by governance safeguards, data ethics and ecological legitimacy. Devoid of these, AI can be technocratic and fail to take local preferences into consideration. Regenerative intelligence, therefore, can be seen as a socio-technical ability that aids regeneration only when ecological ambitions and the accountability of institutions are strategically combined.

2.3 Governance Lens

An ecological restoration and climate action governance lens is needed to comprehend the role of AI in working responsibly in ecology and climate endeavors. Technological systems do not exist in vacuums but they are entrenched in institutions, rules, stakeholder relation and power structure which influence the manner in which decisions are made and priorities given to the interests of stakeholders. Having a governance system that defines how AI promotes transparency, accountability, participation, and ecological legitimacy or reinforces centralized and opaque decision-making in the context of regenerative systems depends on AI. A governance lens thus extends the analysis past the technical capability to encompass stewardship, data ownership, local engagement and ethical oversight. It is particularly relevant in restoration contexts where decisions have impacts on land, livelihoods, biodiversity and resilience of communities in the long term (Rane et al., 2024). Through this prism, the chapter focuses on the fact that AI-assisted restoration should be created as a social and environmental governance issue, rather than a computing one. This outlook assists in aligning digital tools with restoration justice, institutional fidelity and restorative socio-ecological healing.

3. LITERATURE REVIEW

The field of AI-enabled ecological regeneration includes multiple overlapping fields, such as restoration ecology, environmental monitoring, climate governance, and sustainability transitions. Nevertheless, this literature is still in pieces. There are studies that concentrate on ecological restoration and biodiversity rehabilitation and others that study AI as a technical means of sensing, categorizing, and forecasting in the ecological system. Another stream investigates governance, ethics and institutional responsibility of digital environmental management (Ravikumar et al., 2025). However, despite the value of these strands separately, they are hardly combined in one framework that would explain the ways in which AI can help restore as a long-term socio-ecological process but not just as a monitoring tool. This poses a critical paucity in existing research. The literature indicates a great potential of AI in the context of restoration, yet it also indicates conceptual and practical constraints of scale, data quality, governance, and legitimacy (Hamdan et al., 2024). The review needs to be more selective, however, in order to determine where the current knowledge is robust, incomplete, and the contribution of the proposed framework.

3.1 Restoration Studies

Traditionally, restoration studies have focused on ecological restoration via processes like reforestation, wetland rehabilitation, watershed restoration, soil restoration and habitat restoration. This literature is valuable in terms of insights of species recovery, ecosystem services, biodiversity conservation, and resilience building. More recent literature has to conceptualize restoration as a technical environmental practice and also as a socio-ecological process that is institutionally mediated, participatory, and stewardly over the long-term (R. Wang et al., 2024). Such research points out that the success of restoration hinges on the ecological conditions of the area, quality of governance, the sustainability of funding and ability to track the change over time. Nevertheless, a lot of restoration literature continues to revolve around field-based approaches, policy-making or community-based action without a comprehensive involvement of digital intelligence systems as an organizing element of restoration plan. When we speak about technology we tend to think of it as a supportive means of an integrated restorative infrastructure (Kaymaz Mühling, 2023). Here lies one of the opportunities to develop the concept. There is a greater connection that needs to be established between restoration scholarship and intelligent systems research in such a way that ecological recovery can be perceived as a biophysical as well as an adaptive decision-making problem.

3.2 AI and Monitoring

The body of literature on AI and environmental monitoring has been expanded in general, particularly in the area of remote sensing, species detection, land-use classification, disaster prediction, vegetation mapping, and climate-risk assessment (Ravikumar & Aarthi, 2025). Big data analytics through machine learning and other analysis techniques are now more commonly applied to satellite, drone, sensor network, and geospatial data to manage large volumes of environmental data. These papers show that AI is able to enhance the speed of detection, recognising patterns, predicting results, and observing the environment in real-time. These capabilities are most applicable in the context of restoration since they might be useful to detect degraded areas, monitor changes in the ecosystem, measure intervention success, and assist in adaptive reactions. A lot of this literature though is operationally narrow (Kotwal, 2025). The application of AI has been perceived as a technical way of tracking performance, whereas its broader use in the restoration planning, integration of governance, and long-term resilience is not as developed. Moreover, most studies are based on the assumption of fixed conditions of data and do not take into account disproportionate data quality, ecological uncertainty, and institutional constraints (Ravikumar & Aarthi, 2026). This makes the existing literature favorable to the applicability of AI in the area of monitoring, yet it lacks in detailing how AI can be used to support regenerative systems that integrate restoration intelligence and governance accountability.

3.3 Governance Gaps

One of the weaknesses of the current literature is that the literature does not address the governance gaps of AI-supported environmental systems properly. Though most of the studies recognize ethics, accountability or transparency, these issues are usually discussed in an abstract way as opposed to operational. How data governance, stakeholder involvement, local stewardship, model validation, and accountability of decisions would be incorporated into AI-dependent restoration systems is not fully described (Vemana, 2025). This is particularly concerning when dealing with ecology where online choices can affect land stewardship, conservation agendas, resource distribution and the livelihoods of the people. There is also comparatively little discussion of the application of Indigenous and local knowledge in terms of data collection, model development, and evaluation procedures in the literature without being limited to the symbolic incorporation(Aarthi & Ravikumar, 2026). Moreover, the institutional capacity to implement AI responsibly (under various scales of restoration) is seldom discussed in the governance context. Such gaps are important since not only is restoration a technical endeavor, it is also a political, social, and institutional process. A more governance-oriented approach is thus required to make sure that AI is being used to increase ecological legitimacy, fairness and long-term restoration effects. In Table 1 the main themes in the literature are compared and the gap that is filled by the research in the chapter is identified.

Table 1. Review of Regeneration, AI, and Governance Literature

Theme	Main Focus	Common Strength	Main Gap
Ecological Regeneration	Ecosystem repair, biodiversity recovery, resilience, restoration practice	Strong ecological grounding and restoration relevance	Limited integration of AI as a structured restoration system
AI and Environmental Monitoring	Remote sensing, detection, classification, forecasting, monitoring automation	Strong data processing and predictive capability	Often focused only on technical monitoring, not regenerative governance
AI Governance	Transparency, accountability, ethics, participation, data management	Highlights institutional and ethical concerns	Often broad and not specifically linked to restoration systems
Sustainability Studies	Resource efficiency, lifecycle impacts, responsible innovation	Brings attention to environmental cost of digital systems	Rarely connected directly to ecological restoration use cases
Integrated Research Gap	Linking restoration, AI, governance, and sustainability	Shows need for interdisciplinary thinking	Lack of one clear framework connecting all dimensions

4. AERSI FRAMEWORK

It is proposed to use AERSI framework to understand how artificial intelligence can assist in restoring ecologies by utilizing a combination of restoration intelligence and governance. The paradigm does not conceptualize AI as a specific technical tool per se but rather, it is a subset of a larger socio-ecological system where data, prediction, decision-making, and stewardship are constantly interacting. It is aimed at linking the activities of environmental monitoring, restoration planning, adaptive intervention and governance safeguards in a single structured model (Afuwape, 2025). In this way, AERSI can overcome the limitations of the fragmented solutions that consider climate action, restoration and AI deployment as separate. The framework highlights that the regenerative systems should be based on both the analytical capacity and institutional legitimacy, environmental significance, and resiliency in the long term. It therefore emerges as the primary scholarly input to the chapter as it shows how AI can transcend the narrow capabilities of mitigation to a more inclusive role in climate action that is more deterministic towards the restoration of the environment and the governance of the environment in a more regenerative way.

4.1 Framework Logic

The AERSI framework is logically grounded on the concept that ecological restoration should be a continuous process between environmental sensing, analytical intelligence, decision support and governance oversight. Restoration is a dynamic process that does not involve a single intervention but opens up to the altering ecological conditions, stakeholder priorities, and long-term sustainability objectives. The framework consequently structures AI as an element of a cyclical system whereby environmental data are gathered, digested, converted to restoration insights and subsequently applied towards adaptive action. Such actions are then re-appraised with the help of a continuous monitoring and feedback. The associated logic enables restoration initiatives to be more receptive to uncertainty, ecological variability, and constraints of implementation. Meanwhile, the framework acknowledges that intelligence cannot be used, without legitimacy and accountability (Afuwape, 2025). It is then necessary that governance mechanisms are provided to every step of analysis and action. This logic is valuable in its ability to establish a connection between technical capability and ecological purpose and institutional accountability to establish a more consistent model of AI-assisted restoration systems.

4.2 Core Dimensions

AERSI framework is created upon a number of interconnected dimensions that contribute to being a responsible regenerative intelligence. The initial dimension is environmental sensing comprising of gathering of data by using remote sensors, sensors, drones, and ecological observation systems. The second is analytical intelligence where AI techniques are used on environmental data to determine patterns of degradation, areas of restoration, and probable ecological reactions. Adaptive decision support is the third dimension that transforms analytical results into planning and intervention design as well as dynamic restoration management (Yue et al., 2024). The fourth one is the integration of governance, which makes sure that the transparency, accountability, participation, and local ecological legitimacy influence the utilization of AI-assisted systems. Sustainability responsibility is a fifth cross-cutting dimension, and it takes into account the lifecycle, the use of computational resources, and the institutional viability in the long term. These dimensions are not layers that can stand alone but complementing parts of a larger restoration system. Their combined efforts will result in AI making a contribution to the environmental observation and also to ecologically justified, socially acceptable, and practically viable restoration practice. Table 2 displays the key dimensions of the AERSI framework and how it contributes to the restoration of the environment.

Table 2. Core Dimensions of the AERSI Framework

Dimension	Purpose	Role in Restoration
Environmental Sensing	Collect ecological data from landscapes and ecosystems	Supports observation of degradation, recovery, and environmental change
Analytical Intelligence	Process and interpret environmental data using AI	Identifies patterns, risks, and restoration priorities
Adaptive Decision Support	Convert insights into planning and intervention guidance	Helps improve restoration actions over time
Governance Integration	Ensure transparency, accountability, and participation	Aligns AI use with ecological legitimacy and social trust
Sustainability Responsibility	Evaluate lifecycle and institutional feasibility	Ensures restoration intelligence remains environmentally and operationally responsible

4.3 Novelty and Value

The AERSI framework is novel because it incorporates the concepts of ecological regeneration, AI, and governance responsibility into one concept framework of restoration. Current methods tend to focus on either technical surveillance, environmental forecasting, or overall sustainability governance, but are seldom incorporated to offer a framework that describes how AI can be used as restoration infrastructure. AERSI is valuable because it demonstrates that AI should be perceived not only as a driver of the efficiency of analytical tasks, but also as a component of a socio-ecological decision-making framework with objectives of accountability, local involvement, and resilience (Sharma et al., 2025). It makes a conceptual, as well as strategic contribution. It offers a more explicit in-house rationale of how to connect sensing, analysis, intervention, and oversight as well as a differentiation between the general technological potential and the responsible use of the environment. In this way, the framework provides a more stringent framework on which the discussion of AI in climate action could be held, particularly where restoration, adaptation, biodiversity recovery, and governance need to be designed as a set of components rather than as an independent set. Figure 1 Shows environmental sensing, analytical, adaptive decision support, integration of governance and sustainability responsibility collaborates in the ecological restoration.

Figure 1. AERSI Framework for AI-Enabled Ecological Regeneration

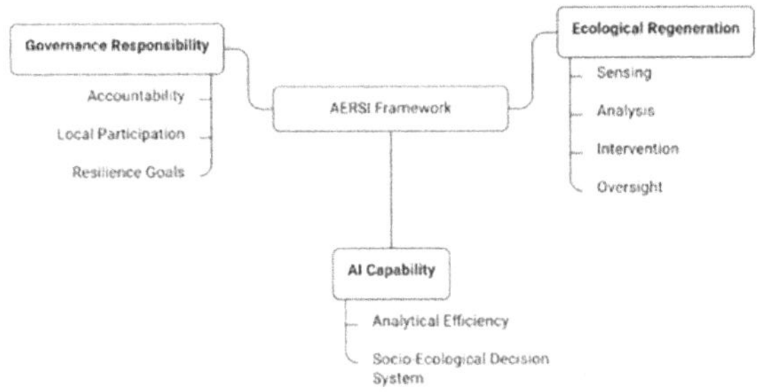

5. AI IN RESTORATION

Its uses in ecological restoration can be to enhance the processes of observation, interpretation, and management of environmental systems over time. Making decisions in restoration contexts can be subject to large and complex data, unknown ecological behavior, and local dynamics. The value of AI in this regard is that, unlike traditional methods, it can identify trends, predict the result, and help make adaptive decisions more quickly due to its capabilities to predict and forecast (Shreha et al., 2025). Nevertheless, it should not be restricted to technical automation. In restoration systems, AI can be most useful when it is a component of a larger decision infrastructure that facilitates monitoring, planning and continual intervention. This implies that its outputs need to be ecologically readable, institutionally pertinent and locally responsive (Elliott et al., 2022). Three key AI roles in restoration are outlined in the discussion below and they include: monitoring of ecological conditions, predictive planning and adaptive decision-making. These functions, collectively, demonstrate how smart systems can enhance restoration activities and stay in line with governance and sustainability issues.

5.1 Monitoring Systems

One of the most direct and feasible methods through which AI can aid in the ecological restoration is monitoring systems. The environmental recovery is based on the opportunity to monitor the state of land, vegetation change, soil health, biodiversity indicators, hydrological changes, and other ecological indicators with time. AI empowers this process by processing large datasets created via satellites, drones, geographic information systems, camera traps and sensor networks. These devices can enhance the rapidity and uniformity of the degradation detection, restoration progress, and signs of ecological stress that can be signalized early. Most restoration settings also require the use of AI-based monitoring to facilitate frequent and scalable monitoring compared to field-based methods (Abubakar et al., 2024). Monitoring however must be done with keen care with regard to the quality of data, ecological interpretation and local context. Raw pattern recognition cannot be enough when the outputs have no relation with the restoration objectives or field realities. As such, monitoring systems are to be construed as the observational basis of regenerative intelligence as opposed to individual technical solutions.

5.2 Predictive Planning

Predictive planning is the application of AI to predict an ecological change and contribute to improved restoration design prior to interventions applied. Uncertainty in restoration planning may be where to intervene, what to give priority, how the ecosystem may react and the risks in the environment that may erode recovery are common. The answers to these questions can be provided with the assistance of AI, which will detect the patterns in the past and present data on the environment, simulate potential scenarios, and predict the probable results under the conditions of various restoration directions (Kumar & Leonardi, 2025). This is capable of enhancing prioritization of degraded areas, allocation of resources, specification of species and risk sensitive planning. Predictive systems are particularly effective when the restoration decisions should take into consideration the variability of climate, land-use pressures, or the resilience of the ecosystem over time. But predictive planning is not to be deterministic. Ecological systems are dynamic and socially engrained in the sense that predictions are subject to change on the quality of data, institutional capacity and change in context. This is why AI-driven predictive planning is the most useful in cases when it enlightens human decision-making, promotes adaptive governance, and is subject to revision by continuous feedback of ecological information (Rathinavel et al., 2025).

5.3 Adaptive Decisions

Restoration is based on adaptive decisions since recovery of an ecology does not take a linear or predetermined direction. Conditions evolve with time as a result of climate change, ecological feedbacks, implementation gaps and local socio-economic pressures. AI can provide support in adaptive decision-making with the constant updating of the information, the detection of changes in the conditions in the system, and indication of possible adjustment of the restoration strategies (Bongarts Lebbe et al., 2021). This enables the restoration managers and institutions to be in a position to react faster to the emerging risks, poor performance, or even to the unexpected ecological reactions. As an illustration, AI can aid in deciding whether there should be intervention priorities, whether the recovery indicators are on track, or whether novel stressors are influencing the restoration success. Nevertheless, automated systems should not be totally in charge of making adaptive decisions (Selvakumar & Natarajan, 2026). Local knowledge and ecological judgment as well as governance oversight is still fundamental in the interpretation of outputs and acting accordingly. In this regard, AI facilitates adaptation, not through agency replacement, but by increasing the ability to make informed, timely, and context-sensitive decisions in restoration in complex environmental systems. Figure 2 Shows the flow between monitoring and predictive planning and adaptive decisions in restoration systems.

Figure 2. AI in Restoration Workflow

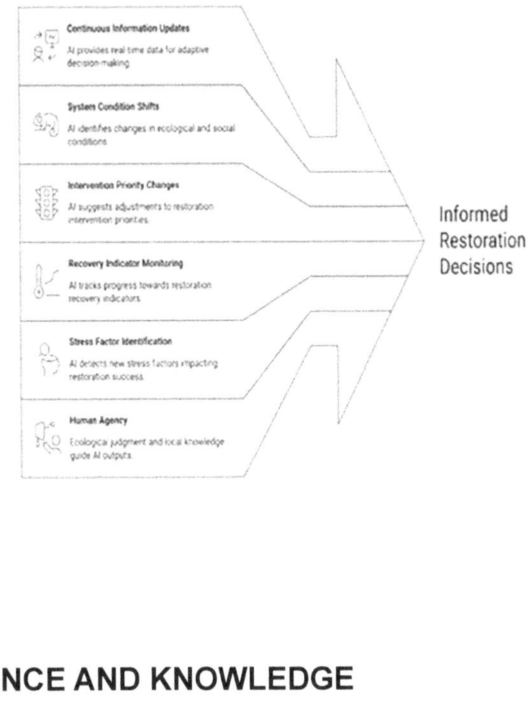

6. GOVERNANCE AND KNOWLEDGE

The integration of governance and knowledge is critical to make sure that AI-assisted systems of restorations are ecologically sound, socially responsible, and institutionally convincing. The ecological regeneration is also a challenge of governance (of rights, responsibilities, participation and stewardship), as well as a technical challenge of data analysis and prediction. In the absence of effective governance, AI systems can become centralized when making decisions, lack accountability, or prioritize technical efficiency over ecological and community values. Similarly, without local and experiential knowledge in data collection, interpretation and decision-making processes, restoration intelligence is incomplete (Sang et al., 2025). This section thus reiterates that regenerative systems that are responsible need to have a capacity to integrate digital capacity with transparent governance frameworks and open knowledge practices. The concept of data governance stipulates the collection, verification, sharing, and security of information. Local involvement will guarantee that local communities define priorities and control (Siirila-Woodburn et al., 2021). The indigenous knowledge also enhances restoration systems by providing long term ecological knowledge, contextual knowledge and culturally based stewardship strategies that cannot be offered by traditional data systems.

6.1 Data Governance

Data governance is one of the most critical requirements of a system that uses AI to contribute to the restoration process since the quality of decisions directly depends on the quality of data, its ownership, accessibility, and legitimacy. Ecological data are usually availed by different sources, which include satellites, sensors, drones, official documents and fieldwork all of which have different levels of reliability, resolution and institutional authority. This ought to be followed by good data governance that will give clear guidelines on data collection, validation, interoperability, storage, sharing and accountability. It should also determine who should have access to data, correction of any errors, protection of sensitive ecological or community related data. Such malpractice of data governance can cause faulty models, inequity in access, and mistrust to digital systems within a restoration environment (Maddipatla, 2026). Through good governance, in turn, transparency is enhanced and there is improved coordination among institutions and stakeholders. It clarifies that AI outputs do not reflect as an actuality, but as a decision depending on the data quality, governmental resolutions, and environmental purpose.

6.2 Local Participation

Local involvement is critical since ecological restoration occurs in inhabited settings in which communities have direct and frequent environmental change experiences and frequently have long-term stewardship liabilities. Restoration planning and monitoring can be enhanced through AI-supported systems, yet the local actors cannot be replaced by the practical knowledge, priorities, and legitimacy that they bring to the decision-making process. It should thus be engaged during the entire restoration processes, including the specification of ecological objectives and the choice of indicators and the interpretation of AI outputs and assessment of intervention outcomes. This assists in making sure that restoration policies are not dictated downwards or in a purely technocratic fashion. It also enhances trust, contextual relevance and ability to implement in the long run (Greiner, 2025). Engagement will go beyond consultation to meaningful involvemcnt, which means having an influence on how data is used, how the planning is planned, and the arrangements to control it. By doing so, local involvement enhances the quality of governance and effectiveness in restoration. It renders AI systems more receptive to socio-ecological facts and minimises the chances of the digital tools promoting abstract efficacy over ecological justice and community strength.

6.3 Indigenous Knowledge

Indigenous knowledge is an important element of restoration governance as it provides ecological knowledge on a long-term and place-based ecological basis, which may be missing in formal datasets alone. This kind of knowledge encompasses the observations of species behavior, seasonal fluctuation, land associations, water patterns and the generation-long sustainable stewardship practices. Indigenous knowledge cannot be regarded as the symbolic addition and the secondary source of data in AI-supported restoration system. Rather it needs to be incorporated in a tangible form of data design, model validation, ecological interpretation, and participatory decision-making (Upadhyay et al., 2024). This implies that knowledge holders are engaged in the definition of what is considered as relevant data, what is a meaningful indicator and whether the AI-generated recommendations are in line with the local ecological realities. Other governance protections that ensure respectful integration entail consent, data sovereignty, cultural rights, and acknowledgment of the ownership of knowledge. Incorporated in a responsible way, Indigenous knowledge enhances the ecological legitimacy and contextual accuracy. It assists in making sure that regenerative intelligence is based on relational stewardship as opposed to depending on abstract computational models. Figure 3 demonstrates the correlation between the data governance, local involvement, and Indigenous knowledge in the responsible restoration systems.

Figure 3. Governance and Knowledge Integration Model

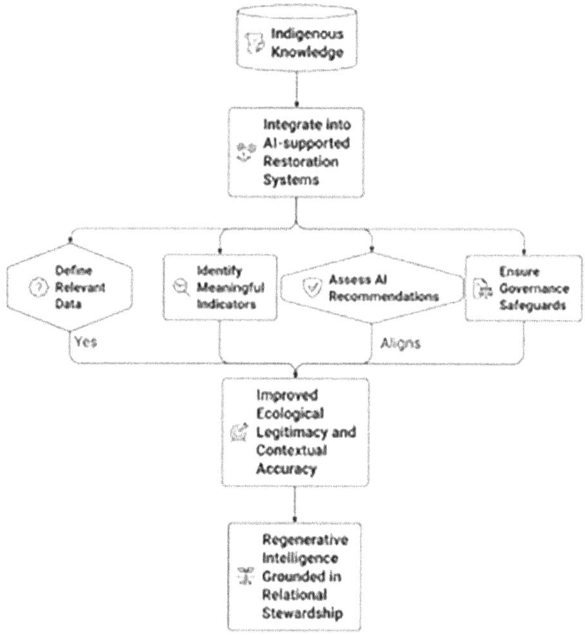

7. ANALYTICAL EVIDENCE

The conceptual discussion of the chapter needs to be based on analytical evidence to put it into practical environmental contexts. Instead of case studies being used as a promotional example, this section addresses them as an analytical resource to aid in understanding what AI can feasibly bring to the restoration systems. This is not to assert the overall success but to explore the functionality of AI in varied ecological, technical, and institutional contexts. This involves taking into consideration the type of data which was utilized, what type of restoration work was being aided, what measurements were taken and what were the limitations which influenced implementation (Asorose, 2025). A comparative approach is particularly essential since restoration projects are very diverse in scope, locality, administration, and ecological intricacy. As such, evidence should be seen as something that is to be interpreted and not generalized as it is. By considering the context of the cases, comparative insight, and limitations of evidence collectively, the section helps to see AI-enabled regeneration in a more balanced light. It aids in separating the proven value, conditional promise, and unsolved problems in restoration-based digital systems.

7.1 Case Context

Contextualization of cases is crucial since the applicability of AI in restoration varies significantly based on the ecological environment, the capacity of institutions, access to data, and objectives of implementation. The process of restoration of forests, wetlands, agricultural landscapes, and coastal ecosystems varies with regard to the monitoring requirements, timelines, and quantifiable results. Equally, the application of AI can be useful in different aspects related to supporting species identification, vegetation mapping, predictive risk forecasting, restoration planning, or adaptive monitoring. Context of the case should then be outlined cautiously to prevent the consideration of all the applications of the environment as similar manifestations of regenerative intelligence. Such contextual factors as the extent of the intervention, ecosystem type, data infrastructure quality, governance structure, and stakeholder involvement level are significant (Y. Wang et al., 2023). In the absence of this information, the statements of success or effectiveness may be exaggerated. The contextual analysis would explain how AI makes a valuable contribution and in which areas its use is still minor or experimental. It also makes sure that analytical evidence is construed with respect to real conditions of restoration as opposed to abstract technological expectations.

7.2 Comparative Insights

Comparative analysis aids in determining the similarities and differences of AI-supported cases of restorations that will be relevant to the interpretation. In general, AI can contribute value to the enhanced environmental detection, faster data processing, and more focused planning or monitoring across cases. The effectiveness of these contributions however, relies on aspects like complexity of the ecosystem, reliability of the data, institutional coordination and availability of human expertise. Certain cases might demonstrate good results related to classification or predicting, but provide limited information concerning ecological recovery over time. It is possible that others will show better operational efficiency without explicitly focusing on governance, legitimacy or community participation (Oloruntoba, 2024). Comparative insight is thus not handy to establish one of the success models but to comprehend the circumstances within which AI will be more or less efficient as a restoration support. It also shows that the technological performance and ecological value are not the same. AI can be used to support restoration operations, but it is important to influence whether the intervention is likely to increase actual ecological performance in plausible and context-relevant governance systems. Table 3 compares the chosen restoration scenarios in the context, the role of AI, the outcome, and the limitations.

Table 3. Comparative Analysis of AI-Supported Restoration Cases

Case Type	Restoration Context	AI Role	Key Outcome	Main Limitation
Forest Restoration	Reforestation and vegetation recovery	Image analysis, land mapping, health monitoring	Better tracking of vegetation change and degraded zones	Long-term ecological impact may remain uncertain
Wetland Restoration	Water system and habitat recovery	Pattern detection, hydrological monitoring, change analysis	Improved monitoring of ecosystem conditions	Data availability and seasonal variability may affect reliability
Coastal Restoration	Mangrove, shoreline, and ecosystem protection	Risk prediction, change detection, adaptive planning	Supports planning for erosion and habitat recovery	High context dependence and model uncertainty
Agricultural Landscape Recovery	Soil health and regenerative land use	Predictive assessment, classification, resource targeting	Helps identify restoration priorities and intervention areas	May overlook social and local governance factors
Cross-Case Insight	Multiple restoration settings	Monitoring, planning, and adaptive support	AI is useful when combined with governance and contextual knowledge	Technical performance does not always equal restoration success

7.3 Evidence Limits

The research on AI in ecological restoration is still positive, yet it is still scarce, and this fact should be brought up. Numerous of the existing cases focus on technical success (accuracy of classification, speed of detection, or support of decisions) but do not offer as strong support on the long-term restoration (or ecological recovery) and social legitimacy. In other cases, the claims of cases are hard to check since they are based on limited periods of observation, limited indicators or even contexts which cannot be compared across ecosystems. The digital interventions are also given a predisposition to be presented as scalable without a thorough investigation of the data costs, institutional preparedness, and complexity of governance. All these are limitations, not suggestions that AI has no value, but need to be interpreted carefully (Lestari et al., 2025). Evidence should however be seen as conditional and contextualised as opposed to being universally applicable. The identification of these limits enhances the analysis accuracy of the chapter. It helps to avoid excessive conclusions and defend a more responsible thesis that AI can contribute to the process of restoration only when the conditions of ecological, governing, and implementing are in harmony.

8. SUSTAINABILITY AND LIMITS

Sustainability and limits should be considered as key to any ecological restoration of the AI-enabled one. Even though AI has the potential to enhance monitoring, prediction, and adaptive decision support, the value of AI to the environment cannot be predetermined. Ecological and operational costs are necessitated in digital systems through the need to use energy, hardware, and data infrastructure, technical maintenance, and institutional coordination. It is because of this that responsible regenerative intelligence should not solely look at what AI is capable of doing, but what AI is consuming, where AI is fitting in, and where its efficacy may be on the decline. In this section, three workable boundaries are pointed out. To begin with, lifecycle sustainability demands the consideration of both the computational and material footprint in the system design and deployment (Feng et al., 2023). Second, there are actual analytic benefits and obvious shortcomings of digital twins in their ability to reflect ecological complexity. Third, institutional capacity identifies how AI tools can be implemented, governed and maintained in significant ways. A combination of these issues makes restoration intelligence be realistic, accountable, and geared towards long-term ecological responsibility.

8.1 Lifecycle Sustainability

Sustainable lifecycle: the nature of AI systems is that they have environmental effects, which need to be taken into account, and the possible environmental restoration benefits of AI systems have to be considered. There is a need to use data processing, cloud storage, model training, hardware production, sensor deployment and continuous system maintenance, which involves the use of energy and material resources. When these resource requirements are not fulfilled, an AI system that is created to aid the ecological recovery can create unseen sustainability trade-offs (Aragani & Thirunagalingam, 2025). Regenerative intelligence needs to be responsible thus lifecycle thinking at the inception of the design up to operation and eventual replacement or upgrade. This involves consideration to energy efficiency, model complexity, hardware durability, data minimization and the environmental cost of digital infrastructure. It also involves assessing the ecological utility of a system in the context of its computation footprint in particular restoration scenarios. Lifecycle sustainability is not just a concept of the future, but also current design value. Restoration systems can be more transparent about these trade-offs to make technological decisions more ecologically responsible, so that AI helps create better regeneration without reinforcing unsustainable trends of digital growth.

8.2 Digital Twin Scope

Digital twins have the potential to aid ecological restoration by offering dynamic models of the environmental systems to assist planners to test scenarios, track change, and predict potential outcomes. They can be applicable in terms of simulating watershed behaviour, vegetation recovery, land-use interactions, habitat change, or climate-related risk patterns, in the context of restoration. Nevertheless, their areas should be narrowed down. Ecological systems are complex, adaptive, and are frequently driven by things that are hard to model in their entirety such as species interactions, social practices, institutional change and uncertain climate impacts. Consequently, the closest equivalent of ecological reality is that digital twins should not be considered fully simulated but instead as decision-support tools (Nasirahmadi & Hensel, 2022). They are valuable based on the quality of the data used in them, assumptions in the model, the resolution of time, and interpretation governance. Excessive use of digital twins can lead to unfounded confidence when the outcomes of simulations are considered final, as opposed to conditional. Thus, digital twins can reinforce the process of restorative planning, although their limits of analysis must be recognized, and their outputs subject to ecological evaluation and validation by the field.

8.3 Institutional Capacity

The issue of institutional capacity can be a determinant of the ability of AI-supported restoration systems to be useful in the long term. Even a well-developed digital tools need organizations, agencies, and communities that are able to handle data, interpretation of the results, infrastructural maintenance, and accountable choices. Institutional constraints like poor coordination, insufficient technical skills, insufficient funding, divided governance, or unreliable policy assistance in most restoration contexts can diminish the usefulness of AI (Tian et al., 2024). Capacity also involves the capacity to make digital systems to meet ecological objectives, community needs, and regulatory demands. In the absence of this alignment, AI could be experimental, not utilized, or unrelated to the actual restoration practice. Institutional readiness is thus not just a technical adoption, but a governance design, staffing, training, budget continuity and sector based collaborative arrangements (Ravikumar et al., 2026). The acknowledgement of capacity as an essential state of being pays off in not overstating the scalability. It also reinforces the argument of the chapter that regenerative intelligence does not rely on computational power, but on the institutions that can manage and maintain digital systems that are oriented towards restoration.

9. POLICY AND PRACTICE

Policy and practice are essential to transform AI-enabled regenerative intelligence conceptual potential into a substantial restoration action. Implementation of ecological goals and technical systems is subject to governance decisions, operational priorities and context-specific decision-making, despite the well-design of ecological goals and technical systems. This implies that the innovation logic cannot solely inform restoration-oriented AI; rather, it has to be influenced by the public accountability, the ecological legitimacy, and viability. The policy frameworks are required to establish guidelines on the data governance, participation, transparency, and sustainability whereas the decisions at the practice level are made on how the systems are implemented, monitored, and adapted in reality (Esiri et al., 2024). The most important issue is to make sure that AI enhances the ability to restore without replacing local knowledge, ignore institutional limits, or propose overly technocratic options. This section thus dwells on how governance concerns, implementation options and responsible use can entrench AI in the systems of restoration. These aspects, combined, contribute to the linking of the conceptual argument in the chapter to the real-world decision-making in the fields of environmental governance and ecological recovery.

9.1 Governance Priorities

The AI-supported restoration systems must be designed with governance priorities in mind and cannot be implemented after the technical implementation. The priorities are most crucial in terms of transparency, accountability, participation, ecological legitimacy, and data stewardship. There must be transparency so that stakeholders can know how data are being utilized, how models produce results and how decisions are reached. Accountability means that the institutions are held accountable towards restoration outcomes and not the algorithms solely. It must be participatory to ensure that local communities, ecological practitioners and impacted groups can influence priorities and management (Malik, 2025). The ecological legitimacy demands that the restoration objectives and indicators should be based on the real needs of the environment and not only on the most readily measurable digitally. Data stewardship also provides an opportunity to maintain the responsible governance of information throughout the collection, access, sharing, and protection. Such priorities are important since restoration is not a technical optimization problem. It is a governance process that entails the trust of the people, ecological responsibility and long-term stewardship. Easy-to-understand governance priorities thus render AI more believable, context-specific, and regenerative-oriented.

9.2 Implementation Choices

The decisions of implementation will define the future role of AI-supported restoration as an effective governance tool or a theoretical technological project that does not have much practical value. The choices regarding system design, data sources, the extent of deployment, the complexity of the model, the role of the stakeholders, and the performance indicators are all determinants of real-world outcomes. In most instances, the more basic and context-friendly systems can work better than the very complicated systems that require huge infrastructure or technical know-how. It should be implemented on the basis of need ecologically, data availability, operational capacity and readiness of governance as opposed to technological ambition. Options will also have to consider the interaction between AI and field assessments, local institutions, and pre-existing restoration practices (Lebrasse et al., 2022). Pilot based deployment, gradual scaling, and ongoing review can be used to minimize risk and enhance learning in the implementation phases. Above all, implementation decisions must be made so that they are flexible to allow feedback and rectification in the long run. Careful decisions at this point will make AI fortify the practice of restoration in a manner that is viable, decipherable and aligned to institutional and ecological realities.

9.3 Responsible Use

Responsible use can be described as the circumstances in which AI can assist in restoring ecologies in a way that does not weaken the ecological integrity, social legitimacy or sustainability objectives. It demands that AI should be considered as a supportive part of the restoration infrastructure and not an independent power. The first step to responsible use is definite ecological intent, in that the digital systems are created to enhance restoration performance as opposed to merely optimizing data processing or predictive capabilities. It should also be monitored by humans, as the decisions regarding ecology are judgmental, uncertain, and local, and cannot be decided by automated results completely (Programme, 2023). Further, responsible use requires inclusive governance, model application transparency, plurality of knowledge respect and the understanding of the costs of lifecycle environmental costs. These are the factors that can help mitigate technocratic overreach to ensure the possibility of AI becoming a panacea is minimized. Finally, responsible use implies using AI solely when it will provide obvious ecological and governance benefits, and in a manner that enhances accountability, resilience and long-term restorative capacity.

10. CONCLUSION

As stated in this chapter, artificial intelligence can help to make a significant contribution to climate action, but only when it is framed outside of a close-minded mitigation logic and as a part of a more radical regenerative systems lens. The main argument is that AI is not to be utilized as a technical tool of carbon tracking, prediction of environmental changes or efficiency in monitoring, but rather as an element of a restoration-based governance infrastructure that connects ecological restoration, adaptive intelligence, institutional responsibility, and social-economic resilience. The chapter has demonstrated how the regenerative intelligence is most useful when the environmental sensing, analytical capability, decision support, and governance safety are crafted as interdependent parts of ecological restoration systems through the proposed AERSI framework. This potential has also been made out of the discussion to be conditional and not automatic. There is the ability to enhance restoration through AI by enhancing monitoring, predictive planning, and adaptive response, but this hinges on the quality of data, ecological validity, the involvement of locals, institutional capacity, and lifecycle sustainability. The chapter has also highlighted that responsible regenerative AI should incorporate tangible governance policies, cautious use of evidence, acknowledgment of the limitations of digital twins, and significance of incorporating Indigenous and local knowledge into the data design, validation, and decision-making process. Instead of portraying planetary restoration as a mere technological option, the analysis has highlighted the importance to put the scale, feasibility and legitimacy into view with regard to context. The key lesson is that this can only be accurate: AI can be a restoration infrastructure when ecological objectives, governance safeguards and long-term resilience are created in concert as opposed to as silos. In this regard, AI has a future in climate action not in technological growth by itself, but in its ability to aid ecologically based, socially responsible, and institutionally sustainable regeneration courses.

REFERENCES

Aarthi, S., & Ravikumar, R. N. (2026). From regulation to transformation-economic and institutional pathways of environmental policy. In *The Ripple Effects of Environmental Policies* (pp. 49–80). IGI Global Scientific Publishing., DOI: 10.4018/979-8-3373-8998-1.ch002

Abubakar, A. M., Zakarya, I. A., Hasnain, M., Sarkinbaka, Z. M., Mukwana, K. C., & Abdo, A. (2024). Potential breakthroughs in environmental monitoring and management. In *Advances in Geospatial Technologies* (pp. 239–282). IGI Global. DOI: 10.4018/979-8-3693-8104-5.ch011

Afuwape, K. (2025). Legal and regulatory frameworks for blue carbon credit markets. *Journal of Foreign Legislation and Comparative Law*, *21*(3), 88. DOI: 10.61205/S199132220034876-9

Aragani, V. M., & Thirunagalingam, A. (2025). Leveraging advanced analytics for sustainable success. In *Advances in Business Strategy and Competitive Advantage* (pp. 229–248). IGI Global. DOI: 10.4018/979-8-3693-9750-3.ch012

Asorose, E. (2025). Integrating digital twins and AI-augmented predictive analytics for resilient, demand-driven global supply chain orchestration under volatility. *International Journal of Science and Research Archive*, *16*(2), 971–992. DOI: 10.30574/ijsra.2025.16.2.2430

Bains, R., Singh, N., & Singh, N. P. (2025). Artificial intelligence in predictive, personalized, and regenerative endodontics: A narrative review of emerging paradigms. *Journal of Healthcare Research and Education*, *1*(5), 5. DOI: 10.25259/JHRE_9_2025

Bongarts Lebbe, T., Rey-Valette, H., Chaumillon, É., Camus, G., Almar, R., Cazenave, A., Claudet, J., Rocle, N., Meur-Férec, C., Viard, F., Mercier, D., Dupuy, C., Ménard, F., Rossel, B. A., Mullineaux, L., Sicre, M.-A., Zivian, A., Gaill, F., & Euzen, A. (2021). Designing coastal adaptation strategies to tackle sea level rise. *Frontiers in Marine Science*, *8*(740602), 740602. Advance online publication. DOI: 10.3389/fmars.2021.740602

Elliott, C. P., Commander, L. E., Merino-Martín, L., Golos, P. J., Stevens, J., & Miller, B. P. (2022). An approach to defining and achieving restoration targets for a threatened plant community. *Ecological Applications : A Publication of the Ecological Society of America*, *32*(6), e2613. DOI: 10.1002/eap.2613 PMID: 35366034

Esiri, A. E., Sofoluwe, O. O., & Ukato, A. (2024). Digital twin technology in oil and gas infrastructure: Policy requirements and implementation strategies. *Engineering Science & Technology Journal*, *5*(6), 2039–2049. DOI: 10.51594/estj.v5i6.1221

Feng, C., Ye, G., Zeng, J., Zeng, J., Jiang, Q., He, L., Zhang, Y., & Xu, Z. (2023). Sustainably developing global blue carbon for climate change mitigation and economic benefits through international cooperation. *Nature Communications*, *14*(1), 6144. DOI: 10.1038/s41467-023-41870-x PMID: 37783692

Frazier, A. E., & Song, L. (2024). Artificial intelligence in landscape ecology: Recent advances, perspectives, and opportunities. *Current Landscape Ecology Reports*, *10*(1), 1. Advance online publication. DOI: 10.1007/s40823-024-00103-7

Greiner, R. (2025). Conservation fencing for coastal wetland restoration: Technical requirements and financial viability as a nature-based climate solution. *Sustainability (Basel)*, *17*(16), 7295. DOI: 10.3390/su17167295

Hamdan, A., Ibekwe, K. I., Etukudoh, E. A., Umoh, A. A., & Ilojianya, V. I. (2024). AI and machine learning in climate change research: A review of predictive models and environmental impact. *World Journal of Advanced Research and Reviews*, *21*(1), 1999–2008. DOI: 10.30574/wjarr.2024.21.1.0257

Kaymaz Mühling, Ş. M. (2023). Utilizing artificial intelligence (AI) for the identification and management of marine protected areas (MPAs): A review. *Journal of Geoscience and Environment Protection*, *11*(09), 118–132. DOI: 10.4236/gep.2023.119008

Kotwal, V. S.Vasifa S. Kotwal. (2025). A Digital Twin framework for intelligent water treatment, quality monitoring, and autonomous control. *International Journal of Applied Mathematics*, *38*(7s), 565–580. DOI: 10.12732/ijam.v38i7s.500

Kumar, P., & Leonardi, N. (2025). Predicting morphological changes along a macrotidal coastline using a two-stage machine learning model. *Water Resources Research*, *61*(4), e2024WR037523. Advance online publication. DOI: 10.1029/2024WR037523

Kuppusamy, P. (2025). Artificial Intelligence-powered digital twin predictive analytics model for smart healthcare system. In *Advances in Computational Intelligence and Robotics* (pp. 271–324). IGI Global. DOI: 10.4018/979-8-3373-0538-7.ch009

Lebrasse, M. C., Schaeffer, B. A., Coffer, M. M., Whitman, P. J., Zimmerman, R. C., Hill, V. J., Islam, K. A., Li, J., & Osburn, C. L. (2022). Temporal stability of seagrass extent, leaf area, and carbon storage in St. Joseph Bay, Florida: A semi-automated remote sensing analysis. *Estuaries and Coasts*, *45*(7), 2082–2101. DOI: 10.1007/s12237-022-01050-4 PMID: 37009415

Lestari, S. L., Mayaguezz, H., & Febryano, I. G. (2025). Artificial intelligence in mangrove management policy in Indonesia: A systematic literature review. *Jurnal Perikanan Unram*, *15*(6), 3560–3569. DOI: 10.29303/jp.v15i6.2035

Maddipatla, S. (2026). Green AI for sustainable Big Data platforms. *Journal of Information Systems Engineering & Management*, *11*(1s), 228–241. DOI: 10.52783/jisem.v11i1s.14065

Malik, F. (2025). Carbon-conscious intelligence: Life Cycle Assessment and Green standards for Generative AI. *International Journal of Innovative Science and Research Technology*, (IJISRT25AUG585), 1030–1037. DOI: 10.38124/ijisrt/25aug585

Nasirahmadi, A., & Hensel, O. (2022). Toward the next generation of digitalization in agriculture based on digital twin paradigm. *Sensors (Basel)*, *22*(2), 498. DOI: 10.3390/s22020498 PMID: 35062459

Nasruddin, B., Baderan, D. W. K., Rahim, S., Rauf, A., & Hamidun, M. S. (2025). Toward regenerative coastal seascapes: Integrating blue carbon, restoration ecology, and spatial planning across marine and coastal ecosystems. *International Journal of Research and Review*, *401*, 401. Advance online publication. DOI: 10.52403/ijrr.20251245

Oloruntoba, O. (2024). Green cloud computing: AI for sustainable database management. *World Journal of Advanced Research and Reviews*, *23*(3), 3242–3257. DOI: 10.30574/wjarr.2024.23.3.2611

Pasupuleti, M. K. (2024). *EcoInformatics: Advancing climate and biodiversity solutions with geospatial data*. National Education Services., DOI: 10.62311/nesx/rb978-81-971242-6-6

Programme, U. N. E. (2023). *Strengthening transparency of non-state actors: How national experiences and new digital technologies can strengthen the Transparency efforts of non-state actors*. United Nations Environment Programme., DOI: 10.59117/20.500.11822/43573

Rane, N. L., Rane, J., & Paramesha, M. (2024). Artificial Intelligence and business intelligence to enhance Environmental, Social, and Governance (ESG) strategies: Internet of things, machine learning, and big data analytics in financial services and investment sectors. In *Trustworthy Artificial Intelligence in Industry and Society*. Deep Science Publishing., DOI: 10.70593/978-81-981367-4-9_3

Rathinavel, P., Ravikumar, R. N., Aarthi, S., Habib, M., Salaeva, N., & Pardaev, J. (2025). Augmenting human capabilities. In *Advances in Computational Intelligence and Robotics* (pp. 1–30). IGI Global Scientific Publishing. DOI: 10.4018/979-8-3373-2372-5.ch001

Ravikumar, A. S., Kurbanova, N., Sevdim, K. E., Zaynitdinovna, B. D., & Jumaniyazova, N. (2026). Bridging knowledge and policy. In *Advances in Computational Intelligence and Robotics* (pp. 119–150). IGI Global Scientific Publishing. DOI: 10.4018/979-8-3373-7620-2.ch005

Ravikumar, R. N., & Aarthi, S. (2025). Climate data without borders. In *Cross-Border Cooperation in the Age of Climate Change* (pp. 61–90). IGI Global Scientific Publishing., DOI: 10.4018/979-8-3373-3972-6.ch003

Ravikumar, R. N., & Aarthi, S. (2026). Energy-aware Federated Learning for climate monitoring. In *Advances in Computational Intelligence and Robotics* (pp. 113–146). IGI Global Scientific Publishing. DOI: 10.4018/979-8-3373-7554-0.ch005

Ravikumar, R. N., Aarthi, S., Nadarajan, S., Masharipova, G., Mamadiyarov, Z., & Tamilselvi, S. (2025). Big data and cloud computing for climate-oriented remote sensing. In *Advances in Geospatial Technologies* (pp. 165–196). IGI Global Scientific Publishing. DOI: 10.4018/979-8-3373-3501-8.ch006

Sang, L. T., Ravikumar, R. N., Aarthi, S., & Ben, T. L. (2025). Optimizing sales and inventory performance using AI and business intelligence in POS environments. In *Advances in Computational Intelligence and Robotics* (pp. 259–290). IGI Global Scientific Publishing. DOI: 10.4018/979-8-3373-4392-1.ch011

Selvakumar, A., & Natarajan, R. (2026). Neuro-environmental AI orchestration for climate-adaptive cross-border logistics systems. In *Advances in Computational Intelligence and Robotics* (pp. 267–298). IGI Global Scientific Publishing. DOI: 10.4018/979-8-3373-7847-3.ch010

Sharma, K., Verma, N., Batra, N., & Bhatia, S. (2025). A strategic management framework for sustainable machine learning with Eco-AI and digital innovation. In *Enhancing Operational Efficiency and Predictive Maintenance Through Digital Innovation* (pp. 1–14). IGI Global Scientific Publishing., DOI: 10.4018/979-8-3373-2474-6.ch001

Shreha, S. A., Ravikumar, R. N., & Aarthi, S. (2025). Neuro-symbolic AI and Behavioral Intelligence. In *Advances in Computational Intelligence and Robotics* (pp. 121–152). IGI Global Scientific Publishing. DOI: 10.4018/979-8-3373-5012-7.ch005

Siirila-Woodburn, E., Gerlein-Safdi, C., Mueller, J., Dennedy-Frank, J., Negron-Juarez, R., & (OSTI), O. of S. and T. I. (2021). *AI predicted shifts in watershed hydrodynamics driven by extreme weather and fire*. Atmospheric Radiation Measurement (ARM) Archive. DOI: 10.2172/1769660

Tian, Y., Yu, J., & Gu, J. (2024). Overview of digital twins in mining equipment. In *Advances in Transdisciplinary Engineering*. IOS Press., DOI: 10.3233/ATDE240242

Upadhyay, A., Yadav, A., Aggarwal, G., & Maan, V. (2024). Comparative analysis of sustainable AI techniques used to reduce the Carbon Footprints in the healthcare sector. *International Journal of Scientific Research in Engineering and Management*, *08*(11), 1–7. DOI: 10.55041/IJSREM38583

Vemana, V. C. (2025). Digital twins in remote monitoring and telehealth: Bridging the gap with Ai. In *Digital Twins: Building Trust and Security in the Digital Age (Unlocking Innovation in Education, Humanities, Management, and Sciences)* (pp. 20–34). Iterative International Publishers (IIP), Selfypage Developers Pvt Ltd. DOI: 10.58532/nbennurDTBCSW2

Wang, R., Sun, Y., Zong, J., Wang, Y., Cao, X., Wang, Y., Cheng, X., & Zhang, W. (2024). Remote sensing application in ecological restoration monitoring: A systematic review. *Remote Sensing (Basel)*, *16*(12), 2204. DOI: 10.3390/rs16122204

Wang, Y., Qi, Y., Li, J., Huan, L., Li, Y., Xie, B., & Wang, Y. (2023). The wind and photovoltaic power forecasting method based on digital twins. *Applied Sciences (Basel, Switzerland)*, *13*(14), 8374. DOI: 10.3390/app13148374

Wong, J., Hoong, P., Teo, E., & Lin, A. (2022). Digital twin: A conceptualization of the task-technology fit for individual users in the building maintenance sector. *IOP Conference Series. Earth and Environmental Science*, *1101*(9), 092041. DOI: 10.1088/1755-1315/1101/9/092041

Yue, Y., Wang, L., Brandt, M., Zhang, X., & Wang, K. (2024). A social-ecological framework to enhance sustainable reforestation under geological constraints. *Earth's Future*, *12*(5), e2023EF004335. Advance online publication. DOI: 10.1029/2023EF004335

KEY TERMS AND DEFINITIONS

Adaptive Decision-Making: A flexible approach in which restoration actions are revised over time based on new data, feedback, and changing conditions.

AERSI Framework: A conceptual framework proposed in the chapter to explain how AI can support regenerative systems through monitoring, planning, decision-making, and governance.

Carbon Neutrality: A state in which carbon emissions released are balanced by carbon removal or offsetting measures.

Climate Action: Efforts taken to reduce climate risks through mitigation, adaptation, resilience building, and environmental recovery.

Data Governance: Rules and processes for collecting, managing, validating, sharing, and protecting data responsibly.

Digital Twin: A digital model of a real-world ecological system used to simulate conditions, test scenarios, and support planning.

Ecological Legitimacy: The extent to which environmental actions are aligned with real ecosystem needs and responsible stewardship.

Ecological Regeneration: The process of restoring damaged ecosystems so that they recover health, balance, and long-term functioning.

Environmental Monitoring: The continuous observation of ecological conditions such as land, water, biodiversity, and vegetation change.

Governance: The structures, rules, and processes through which restoration decisions are guided, implemented, and made accountable.

Indigenous Knowledge: Traditional, place-based ecological knowledge developed by Indigenous communities over generations.

Institutional Capacity: The ability of organizations and governance systems to manage, support, and sustain AI-enabled restoration efforts.

Lifecycle Sustainability: Consideration of the environmental impact of AI systems across design, deployment, operation, and maintenance.

Local Participation: The involvement of communities and local stakeholders in planning, monitoring, and decision-making processes.

Planetary Restoration: Large-scale ecological recovery aimed at improving environmental health across interconnected natural systems.

Predictive Planning: The use of data and AI models to anticipate ecological change and support better restoration decisions.

Regenerative Intelligence: The use of AI and data systems to support ecological recovery, adaptive planning, and restoration decisions.

Restoration Infrastructure: The combined ecological, digital, and governance systems that support long-term restoration and recovery.

Chapter 5 AI-Enabled Circular Economy Models and Sustainable Supply Chains

Amit Verma
http://orcid.org/0000-0002-9283-7873
Kalicharan PG College, Lucknow, India

Rajiv U. Kalebar
http://orcid.org/0009-0000-2788-1512
School of Business and Management, Christ University, India

U. Divya
http://orcid.org/0000-0003-0506-7063
Alliance Ascent College, Alliance University, India

D. Jogish
http://orcid.org/0000-0003-2991-794X
Sai Vidya Institute of Technology, India

ABSTRACT

The transition toward a circular economy requires intelligent systems capable of redesigning traditional linear supply chains into regenerative value networks. This chapter explores AI-enabled circular economy models that integrate advanced analytics, machine learning, IoT, and predictive optimization to enhance resource efficiency, waste minimization, and sustainable value creation. By embedding artificial intelligence across procurement, production, distribution, consumption, and reverse logistics, organizations can achieve real-time visibility, predictive mainte-

DOI: 10.4018/979-8-3373-9978-2.ch005

nance, demand forecasting, and automated material recovery. The study develops a conceptual framework linking AI capabilities with circular business strategies, sustainable supply chain performance, and long-term competitive advantage. It also examines governance, ethical, and technological challenges associated with AI-driven sustainability transformation.

1. INTRODUCTION

1.1 Problem Context

Allergists have shaped the global industrial system from ancient times. It is based on cycles in which raw materials are extracted from nature, industrially processed, and ultimately discarded as trash. Over the past century, this method has led to industrial and economic expansion, but it has also generated an environmental crisis: resources are running out, waste is rising, and CO2 emissions are rising. Due to their inefficiency in using product resources, materials, and energy flows, linear throughput supply chains are problematic for sustainability. A circular economy is an option. To prolong life, reduce environmental effect, and save money, minerals and other resources should be reused, repaired, restored, or remanufactured. Circularity seems wonderful in principle, but it doesn't necessarily help the environment or climate. Many circular programs focus on recycling or waste reduction rather than changing how we create and utilize things. Even while circular approaches may improve operations, they may still be applied at unsustainable resource extraction and consumption (Toorajipour et al., 2021).

1.2 Gap in Existing Discussions

A lot of research talks about how AI may make things more efficient, cut down on waste, and make procedures easier, but they don't necessarily make a clear distinction between little and big changes. People occasionally say that AI efficiency improvements are circular or sustainable, even when they don't save resources or affect the environment. Better demand forecasts may help reduce waste in inventories, but if production goes up, the benefits for the environment may be limited. Optimized logistics could save transportation costs and emissions, but if delivery frequency or markets grow, "rebound effects" might cancel out these savings. People who talk about AI-enabled circular supply chains don't talk about how they are rarely used with climate change in mind. A lot of people talk about "sustainability," but not about how to deal with climate change, environmental shocks, or long-term plans for change. Climate change is changing the world, and little improvements in efficiency won't be enough to make an impact quickly (Sharma et al., 2022).

A lot of research has not looked at the organizational, institutional, and governance frameworks that are needed to make AI work in circular processes. Data quality, coordination among Sisley chain trade partners, regulatory frameworks, and the capacity of AI-driven interventions to effectuate genuine transformation are essential. AI projects may not work or be coordinated if these prerequisites aren't met.

1.3 Chapter Objective and Scholarly Contribution

This chapter aims to present a more complete and unified understanding of how AI may assist in developing circular, environmentally friendly supply chains. This chapter clarifies the difference between efficiency-driven digital optimization and regenerative supply chain transformation. It also emphasizes that technological skills require organizational preparedness, company collaboration, and institutional support. Recent literature has neglected this. This chapter tackles a significant gap in the literature, which often views AI as a standalone solution rather as part of a socio-technical system. The chapter further argues that AI-supported circular initiatives should be assessed not just on their technological sophistication and efficiency but also on their environmental and climate impacts. This argument dominates the chapter. The chapter begins by reexamining supply chain sustainability and circularity. This establishes a solid theoretical foundation and integrated conceptual framework. It then addresses design techniques, analytical capabilities, management procedures, and critical implementation difficulties. Finally, the chapter summarizes the important ideas and offers new research, practice, and policy to improve the supply chain for the environment and economy (Pournader et al., 2021).

2. HISTORICAL EVOLUTION REFRAMED: WHY EARLIER SUSTAINABILITY MODELS WERE INSUFFICIENT FOR REGENERATIVE AND CLIMATE GOALS

The advancement of sustainable inter-enterprise practice is far from finished; it is now situated between reactive environmental management and systems that are past the point of effectiveness. But prior models, although useful in their time, can't currently lead to conclusions that are regenerative and have a big beneficial effect on the global climate. Understanding its trajectory is beneficial not just as a historical descriptive endeavor, but also as an analytical basis for elucidating why contemporary enterprises increasingly necessitate the implementation of intelligent, AI-enhanced systems (Cannas et al., 2024).

2.1 From Pollution Control to Green Supply Chains

At the beginning of the movement, the focus was on cleaning up industrial systems. In the 1980s, a company made some or all its money by ignoring environmental rules. This is why sustainable development means pollution. Environmental awareness that grew in the second half of the 20th century has led to more environmental laws and made it necessary for businesses to implement end-of-pipe steps. This meant that filters, scrubbers, and other end-of-pipe solutions were needed to clean up exhaust gases or waste. Products from a production method did help the environment, but they didn't function very well. Even though they were required, most of these activities were reactive and didn't impact on the way the linear production model worked. Extraction, production, usage and disposal of resources were essentially one-way. Environmental problems were seen as costs of obeying regulations, not parts of design. The next step in "green" supply chain management was a lot bigger. Companies divided their businesses into networks that took into consideration environmental concerns in the delivery, manufacturing, distribution, and transportation of products. From here, there came the eco-design program and green buying. Environmental managers spend more time on advance logistics and the effects on the environment after sales. Even though the green supply chain was getting bigger, most of the approaches were one-liners. The changes to the new system were usually small and were aimed at cutting down on waste, energy usage, or emissions here and there. That didn't accomplish anything to change the system back. It was hard to keep track of these activities since there weren't any real-time monitoring tools, it was hard to share information across partners, and they didn't know what was missing or where (Riahi et al., 2021).

2.2 From Circular Economy to Intelligent Circular Systems

The circular economy paradigm works in a capitalist corporate context. It is a true transformation in capitalism, unlike the linear "remove-make-consume" loop. Reuse and remanufacture were initially talked about by Circularity. It also made it feasible to pace its goods so that they could recover when they were done. This approach is timely in the center, not in a straight line; it makes closed loops. Alice Cheng noted that early models likewise had problems getting dependable output when they were scaled up. Without real-time data, it would be very hard to keep these flows going. In a complex chain of manufacturing, coordination between supply networks and tracking of resources has become impossible. The Internet of Things, which includes digital technology and artificial intelligence, will dramatically improve the odds of making circular systems work. These technologies can permanently enhance predictive maintenance and demand forecasting. You may also see where things are in real time. AI can help businesses switch from rigid plans to flexible systems that let them learn and react to changes in supply, demand, and the environment. In the meanwhile, machine learning algorithms can find the exact times when reverse logistics may be used to their fullest potential or when resources can be recovered. Predictive analytics may also guess when waste is likely to happen. Making circular systems smarter is a big improvement in how they operate in the circular economy (Shahzadi et al., 2024).

2.3 Analytical Transition

This event is important in history because it raises the level of transactions. This means coordinating the flow of materials and setting up real-time pipelined environmental management, which employs technology that is based on data to make it simpler to manage a lot of resources. Older models faced issues since they didn't have enough information, judgments were made in parts, and the goal was to reduce harm instead of resolving it. On the other hand, smart circular systems employ data to move items. With this method, you can examine, sync, and improve all the supply zones and interfaces at once (Hangl et al., 2022).

But the statement makes it apparent what the remedy to this standoff is: the climate objectives of today don't match with the sustainability models of the past. Reactive pollution control doesn't function to block the flow of pollutants that are incorporated into supply chains. The little gains in efficiency that result from making industrial processes more eco-friendly don't make up for the entire carbon reductions. Also, non-digital circular models don't have the accuracy and scalability needed to manage big networks that cross several nations. We need systems that are both circular in design and clever in how they are used to generate excellent results for the environment or the climate: Being willing to learn more about what it needs at all times For systems that are healthy for the environment, ecological optimization is such a crucial aspect that digital platforms need to be employed as well. They can link suppliers and service providers, which gives us a complete view of how resources are used and how materials move about in **Table 1**.

Table 1. Evolution of Circular Economy Approaches

Stage	Key Focus	Characteristics
Waste Management	Pollution control	Waste treatment systems
Resource Efficiency	Energy optimization	Reduced material consumption
Green Supply Chains	Sustainable procurement	Environmental integration
Circular Economy	Closed-loop systems	Recycling and reuse
AI-Enabled Circular Economy	Intelligent optimization	Data-driven sustainability

In summary, building a circular economy and green supply chain management is part of the gradual process to achieve sustainable production systems.

3. THEORETICAL AND CONCEPTUAL FOUNDATION: LINKING AI, CIRCULARITY, REGENERATION, AND CLIMATE OUTCOMES

3.1 Systems Theory as an Interdependence Lens

Systems theory can help us understand how supply networks operate right now Using AI in circular supply chains can make them more resilient by allowing resources to flow in many ways, giving them several ways to recover, and letting them make decisions that are best for the situation. For this resilience to work, all participants need to be able to share data and work together all the time. Systems theory presents a whole way to look at the parts of the supply chain and how they work together. It shows how AI may help integrated circular transformation move forward while keeping balance and strength (Bhattacharya et al., 2024).

3.2 Sustainability Theory Beyond Generic Balance

Most people claim sustainability theory has economic, environmental, and social goals. This broad perspective is useful in principle, but it lacks the analytical precision needed to evaluate AI-powered circular supply chains. We must think more subtly about the multiple environmental effects and their meanings. So, we must be careful not to confuse sustainability or its goals every time we talk about it. Some adjustments are needed. Efficiency enhancement includes resource reduction and waste elimination from linear systems. However, sustainability transformation must be more advanced. This is because altering capital may mean something different than simplifying or reengineering a process.

Making transportation routes better may progressively reduce the resources needed to get products from C to D. Another example is industrial energy efficiency, which is unclear. Incremental improvements may reduce resource usage, but they do not affect supply chain structure. The performance of circular flowing materials has advanced. It reduces waste and preserves system materials. This entails producing recyclable, reusable, repairable, and manufacturable products. Currently, AI increases material flow tracking, monitoring, and optimization. Recirculating does not guarantee sustainability if the processes use a lot of energy or do not lessen environmental impact. Regenerative results indicate a shift from circularity to systems that actively replenish life's ecology. This implies the system must find trash-removal strategies. Instead, we should develop systems that are less harmful, help the environment recover, and last. Through accurate resource management, mechanical self-adjusting maintenance, and system flexibility, AI may improve results (Farshadfar et al., 2024).

These talents enable more effective resource utilization and less environmental impact. Climate change impacts sustainability in new ways. reduced greenhouse gas emissions, reduced new material consumption, and higher climate change resilience are involved. Not all efficiency and reuse measures assist the climate. Energy-intensive recycling methods may negate their environmental benefits. Thus, AI-enabled supply chains must be examined from several angles. We must determine if they are efficient and assist us achieve circularity, regeneration, and climatic sustainability. This contradicts the belief of many digitization supporters that digitalization is efficient and sustainable. This way of examining sustainability theory allows us to objectively evaluate AI-driven supply chain changes and identify instances where they fail to improve the environment and climate.

3.3 Digital Transformation Theory as Capability Transformation

Digital transformation theory considers AI in supply chains as a capacity change, not a technology transition. This is vital. This method indicates that AI only helps corporate networks, procedures, and decision-making systems. Businesses must develop data-driven capabilities to monitor, assess, and adapt to environmental changes as part of their digital transformation. Analytics, smart decision-making tools, and data infrastructure are needed in supply chains. AI is needed for this transformation because it can forecast, act, and help people make better decisions. AI works best when many things are ready. Strong data is needed for good decisions and forecasts. Information flows more smoothly between supply chain nodes with system interoperability. Competent individuals and a culture that supports AI help a company design and implement AI (Abyaneh et al., 2025).

The digital revolution will change governance and collaboration. Circular supply networks need teamwork. Easy-to-use digital platforms and data-sharing technologies enable this cooperation and decision-making. These ideas may not assist if privacy, trust, and competition are concerns. Another key aspect of digital transformation is real-time decision-making and response. AI-powered systems can monitor and adapt operations, but traditional supply chains must plan and make changes often. Circular supply chains require fast management adaptation and response due to unpredictable material flow and recovery. Ethics and obligation are also challenged by the digital revolution. Given their potential influence on the environment and society, governance frameworks must ensure AI-driven choices are egalitarian, transparent, and sustainable. According to digital transformation theory, AI influences manufacturing. The effects depend on how well the technological, organizational, and institutional parts work together (Chauhan et al., 2022).

3.4 Proposed Integrated Conceptual Framework

This arrangement makes A linked, typable, and performable but unsustainable upgradeable. Even when applications are lengthier, no one has contemplated leap cutting completely past bristle familiarity. AI and flexible sensors send data to central processing engines, which forecast physical device performance. People can't handle it. You may utilize predictive methods to manage a product's life cycle. Sensing gives you real-time data on product use and environmental impact. Automation streamlines operations, while optimization maximizes resources. However, automatons push limitations. The supply chain must shift for these ideas to function. Suppliers should leverage digital platforms and data communication to improve modularity and recycling. The supply chain employs coordinated circular processes.AI, circular techniques, and procedural changes produced various results. Remove waste, extend life, recover resources, and expose the environment. This improves supply chain efficiency. These early insights have helped firms reduce carbon emissions, use fewer raw materials, last longer, and make steady progress toward a decarbonized future that benefits everyone. Results on AI-enabled circular supply chains need to be revised. They include system management standards, data source accuracy and quality, industrial waste disposal laws, SMB company demands, and legal and commercial difficulties. AI integration isn't only about big goals; everyone must benefit. A pipeline with the right capabilities and restrictions produces varied outputs. Our approach focuses on the many elements that impact on how a technological solution fits into operational contexts and organizations, not broad, general ideals. This interpretive framework summarizes how environmental changes may renew AI-enhanced circular supply networks. This combines theory and practice. It states that talents, processes, and outcomes only work together as a whole(Wynn & Jones, 2022).

Figure 1 shows the overall architecture of AI-enabled circular economic systems. It shows the flow of resources through production and consumption, then back to the economy.

Figure 1. Conceptual Framework of AI-Enabled Circular Economy

These frameworks, when combined, allow businesses to create intelligent supply networks that enhance productivity and safeguard the environment.

4. DIGITAL ARCHITECTURE OF INTELLIGENT CIRCULAR SUPPLY CHAIN NETWORKS

Most individuals don't completely understand how these businesses are evolving toward digital infrastructures and combining tools for analysis. That's only part of the tale, though. In a lot of situations, these plans also employ firmware to communicate instructions from a business to its items. Using such quick changes in demand, it's hard to picture how to achieve this using the old supply structure. Verisk Robotics combines technology and smart systems into production lines from the very beginning of a product's life cycle. These include sequence tools that save time and energy. This way, customers get advantages at many levels, from a better product design to cheaper prices. These technologies, along with AI, make it easy for companies to set up circular supply chains. They may follow the path of materials through the life of a product. If you want to map those processes that are related to each other, you need information on where resources are, how products are utilized, and when they'll need to be recycled or reused. This is what digital information-processing infrastructures do: they for example, many processing businesses that still rely on fossil fuels are stuck trying to figure out what a spreadsheet means. As the years go by, things stay the same: company stays connected to the supply chain and may keep an eye on activities to improve its supply chain. These ongoing duties should help people make decisions based on evidence to reach their goals of strength and prosperity (Romagnoli et al., 2023).

4.1 Data Infrastructure and Real-Time Visibility in Supply Chain

Data infrastructure is the initial layer of intelligent circular supply chain architecture. It connects all the nodes in the network and gathers information from them all in one location. Setting up IoT (Internet of Things) for Data2 Modern businesses get a lot of operational data from "their corporate applications, transport programs, equipment manufacturing, and connected hardware." Along with putting all this data together, Internet of Things (IoT) devices can also tell where things are and how they are doing. IoT sensors in factories, transport vehicles, and warehouses can very accurately describe how a product is used, what resources it requires, and what the environment around it is like. Records from these kinds of systems could keep track of things like the temperature of items as they were being sent, the flow of commodities between warehouses, and even how manufacturers made things. We acquired a lot of factual data from these kinds of sources to learn more about how the supply chain truly works.

A business system's logistics database or inventory management system can help your data architecture. They oversee buying things, talking to suppliers, and getting them to the end customers. Now, when the organization's business apps start delivering data to them and their linked IoT devices bring in real-time input, the system can seamlessly combine this information with the data it already has. To make a circular supply chain, where items move through their life cycle, you need to be able to see so much of your product chain. Organizations may utilize excellent data in smart ways to keep track of the life cycle of their goods, not just when they are at the end of their life, but also when they are being reused, repaired, or recycled (Bressanelli et al., 2022).

4.2 AI Analytics with Smart Decision-Making Support

You will need to do something with your dataset after you have put it together to gain useful information from it. This is happening in the analytics part of digital architecture. Also, data is a stage of analysis that has a lot of volume and complexity, which may make it very difficult to understand at times. This is where the wonders of artificial intelligence come in, because it can quickly scan huge datasets and find trends that human analysts would miss. Machine learning models use historical and real-time supply chain data to find patterns in how people behave when they want something, how they use resources, and how well they run their business. For instance, forecast models can tell you where there will be demand for a product or where there might be traffic jams in the transportation networks. Businesses may utilize this information to find potential supply difficulties and to make things operate more smoothly. AI analytics may help companies discover where their operations are repeated throughout the supply chain. In other circumstances, analytic systems could discover that some transport lines are always late or that some manufacturing processes use a lot of energy. Companies may then utilize these kinds of information to change how they do things so that they consume fewer resources and have less impact on the environment. AI algorithms are going through the circular supply chain to find ways to get resources back. Predictive algorithms may try to figure out when things will come back, maybe for recycling, and what parts could be taken from things that look like trash. These insights help build reverse logistics systems, which "pull" things back into manufacturing (Cagno et al., 2021).

4.3 Circular Value Generation through Optimization

If we can make the systems more accurate and better, we can make them more valuable or even optimize them. After that, it coordinates supply chain actions based on the insights gained from AI-enabled analytics. This makes operations run more smoothly. In the whole industrial sector, all optimization systems were designed to facilitate seamless production operations by enhancing operational inputs such as transportation routes, resource allocations, demand projections, and similar factors. Finally, if managers applied these kinds of analytical tools, logistics planning systems could be able to find faster and more energy-efficient ways to move goods. For instance, in manufacturing systems, production schedules may be changed to fit expected demand patterns, which would save the company money. All these best ways to combine inputs not only make more money, but they also make the world a better place. This layer also has another tremendous benefit: it makes cyclic activity easier. Circular supply chains depend on reverse logistics that pick-up machinery and send it back to production systems to be recycled or remanufactured. Optimization Systems help businesses manage these tasks, such as when to recover items for recycling and how to put spent resources back into the production process. **Figure 2** shows the physical configuration for this digital architecture data collecting, analytic processing, and optimization processes, together with all the connections that support origin properly. **Table 2** gives a review of the main technical underpinnings and shows how this design brings together different types of technologies to create smart circular supply chains.

Table 2. Core Components of AI-Enabled Circular Supply Chains

Component	Role
Data Infrastructure	Collect operational information
AI Analytics	Generate predictive insights
Optimization Systems	Improve logistics coordination
Circular Operations	Enable recycling and remanufacturing

Figure 2. AI-Driven Circular Supply Chain Framework for Sustainable Value Creation

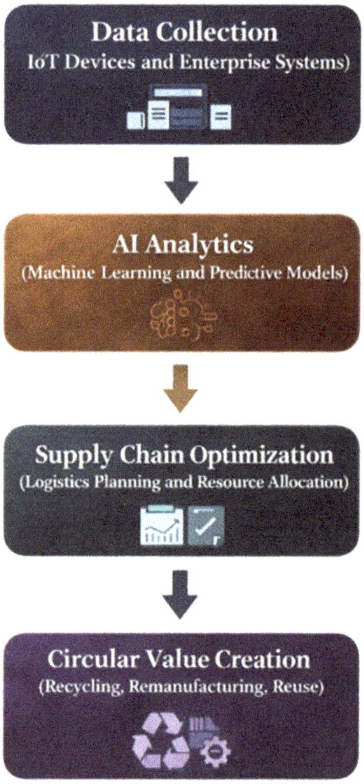

Thus, these digital infrastructures are critical facilitators of the transition to sustainable and regenerative economic systems.

5. AI-DRIVEN MECHANISMS OF REGENERATIVE VALUE CREATION

For the shift from conventional sustainability practices to authentic regenerative systems, it is essential that the descriptive components of these innovative techniques commence their operation as well. Artificial intelligence (AI) is both a facilitator and a system agent for this type of development. However, we can better understand the role of AI by looking at how it works in practice, such as how it affects the flow of materials from start to finish, how it adds years to the lives of products that have already been made or are being reinstalled in a new form with new functions, and how this affects the environment. This section explains how AI-driven processes may provide regenerative value, how connections can lead to success, and what the boundaries are for reaching that goal (Pagoropoulos et al., 2017).

5.1 Predictive Analytics and Resource Efficiency

Predictive analysis is one of the most popular AI applications. In supply chain systems, it is mostly used for planning production, optimizing inventory, and anticipating demand. Using machine learning models that blend prior sales data with market signals, weather trends, and historical customer activities can help the organization perform better. This makes it much easier to forecast demand. It can help cut down on overproduction, keep extra stock to a minimum, and lower the chance that items that are still in stock after they reach the end of their shelf life will become trash. Resource allocation algorithms then make things even more efficient by making the best choices about when to buy things, when to make things, and how to get them to customers for instance, demand estimates need to be linked to sourcing decisions that put secondary materials first. Inventory management should also think about the complete life cycle of a product and make sure that things that are made too much of are reused, reconsidered, or recycled instead of merely thrown away. In other words, predictive analytics is a crucial part of regeneration. It makes it easier to make regenerative decisions, but it can't do it on its own (Reuter, 2016).

5.2 AI in Resource Recovery, Sorting, and Remanufacturing

AI garbage recovery solutions are a great approach to provide value. Complex waste streams have made recycling less effective and remanufacturing less successful in rehabilitation for years. Due to their mix of materials, pollutants, and quality levels, waste streams have historically complicated recycling and remanufacturing. Computer vision and machine learning AI have reversed this issue (Awan et al., 2021).

Computer vision systems with robotic sorting technologies can reliably recognize and separate rubbish stream constituents. Operations are less prone to face uncertainty and judgments are more correct. For instance, usable items can be refurbished and useless ones recycled. AI also helps people determine how to get things back by ensuring returned products and equipment transit seamlessly between collection stations, sorting factories, and remanufacturing facilities. These alternatives consider transportation costs, traffic, environmental concerns, and recovery value. AI improves economic and environmental recovery. These advances won't help AI-supported resource recovery unless the infrastructure, regulations, and secondary material market demand are good. Without proper collecting infrastructure, regular material flows, and material recovery rules, AI-driven recovery solutions are restricted.

5.3 Reverse Logistics Under Real Constraints

In the recycling economy and renewable energy system, logistics in a reversed system are very important. It is more complicated and unpredictable than forward logistics. Respect the flows are still unexpected, and this makes reverse logistics much more problematic than typical supply chains. There are also concerns with inconsistent collections and flow changes caused by bad traffic systems. You can utilize AI to look at past data on returns, customer behavior, and how products are being used to anticipate flows. This doesn't make the uncertainty any less. With this foresight, companies may use more flexible ways to collect data and make smart decisions about how to use their limited resources. Returns in reverse logistics are even more complicated when there are sorting complications.AI tools can solve a jigsaw puzzle made up of different types of materials, goods with different amounts of damage, and items that aren't complete (Lopes de Sousa Jabbour et al., 2018).

5.4 Predictive Maintenance and Life Extension

Another keyway to give AI Regenerative Value is through predictive maintenance. Sensor data, machine learning algorithms, and live monitoring equipment let businesses predict machine failures and prepare for collapse. This reduces maintenance planning time, prevents unplanned downtime, and extends product life. This saves Mother Earth a lot of money. For resources, longer-lasting gear requires fewer replacements. That also means less new materials and resources must be imported and extracted. Longer-lasting items help the environment. Because these activities take longer, creating, transporting, and discarding away the product will cost less environmentally. Saving power is another benefit of predictive maintenance. A well-maintained machine consumes less electricity and pollutes less. AI-powered monitors can discover issues, modify system settings, and bring cars within range. This improves national health and reduces carbon emissions. Extended-life technologies facilitate circular processes like remanufacturing and refurbishing. If companies care about their products, they sell their old stuff when new ones come out better. This helps closed-company models conserve resources and renew. Environmentalists are increasingly emphasizing AI's role in circularizing the economy. Because things are cyclical, our economy and environment may stay the same for a while (Yuan & Pan, 2023).

The predictive aspect of AI is quite restricted without dependable sources of information and monitoring systems. What it can deliver is significantly less than what was hoped for in real life. To set up and run AI-driven maintenance systems, it's also important for an organization to be ready for change and have technical expertise.

5.5 Regenerative Value Creation as a Distinct Outcome

An efficiency plan aims to reduce resource consumption and waste, whereas regeneration aims to bring nature and industrial systems back to life and make them stronger. At its heart, regenerative value generation is made up of several different parts that are all connected. First, it wants to get materials back into production cycles so that value may be gained from waste resources. This involves recycling old metal parts or other things that have been worn out. You may get these things back and then make new ones out of them. Second, the focus is on making things last longer. Regenerative systems cut down on the demand for new production and the extraction of resources by extending the life of existing systems through maintenance, repair, and refurbishment. Third, the goal of regenerative value generation is to reduce the need for new resource extraction. This is done by using more secondary materials and designing systems that encourage the recovery and reuse of materials. Fourth, it means lowering the environmental impact along the whole supply chain. This goes from making sure that as little waste as possible is made during production to making packaging materials that are less harmful to the environment by design. **Table 3** provides a summary of prominent artificial intelligence technologies that facilitate a circular economy and their essential functions in green supply chain ecosystems (Srivastava et al., 2026).

Table 3. AI Technologies Supporting Circular Economy

Technology	Application
Machine Learning	Demand forecasting
Computer Vision	Automated waste sorting
IoT Analytics	Product lifecycle monitoring
Natural Language Processing	Sustainability data analysis

In summary, artificial intelligence technologies provide robust instruments to facilitate the shift towards circular economic systems. AI systems facilitate the redesign of supply chains by allowing predictive resource management, enhancing recycling operations, and assisting sustainability analysis, so fostering long-term environmental and economic sustainability.

6. DESIGN PRINCIPLES FOR SMART AND CIRCULAR SUPPLY CHAIN ECOSYSTEMS

The main goal of conventional supply chains has always been to deliver goods quickly and easily from the manufacturer to the customer. When the product is no longer useful, it is thrown away as trash, and the supply chain comes to an end. In contrast, circular supply chains take a different approach. Instead of throwing away things for good, circular models try to keep materials in use for as long as possible by reusing, repairing, and remanufacturing them. For this to happen, we need to rethink how supply chains are set up, how products are made, and how organizations work together. Because of this, circular supply chains don't just have items flowing forward; they also contain echo flaps. Forward supply chains are about getting items from producers to customers using Huawei Channel Networks in a normal way. On the other hand, the reverse supply chain mostly picks up dead items and sends them back to be recycled. Even while dealing with two transposed directions at once is a logistical nightmare, it does help companies get back valuable resources and lower their carbon impact (Roy et al., 2022).

6.1 – Bridging Forward and Reverse Supply Chain Flows

Returning and Reverse Logistics must be part of circular supply networks. This notion applies to all ways of getting back products, such as fixing them or reprocessing them. Forward logistics is used by traditional supply chain systems to get goods to customers through transportation, warehouses, and stores. In reverse logistics, there is generally a distinct department that takes care of returning items and getting rid of trash. So, reverse logistics, such recycling centers and industrial parks, is very important to the circular supply chain. Organizations that need to return products build supply networks. There are several ways to get rid of things. Some can be fixed and sold again, while others can be taken apart and utilized with fresh parts. Think about this plan; it could be too much. Analytics models predict that a specific number of things will need to be collected at any given time. This data assist businesses figure out how to recycle and make new things. This kind of AI system also looks at data on transportation and logistics, such as the typical location of old consumer goods across the country. Then they may transform it into cheap operational solutions that get discarded items to recovery facilities in a timely manner. Companies can develop by setting up supply channels that merchant's control. It doesn't simply make people happy; it also sells things. As they ink, reverse logistics processes make things that will be used in future manufacturing cycles (Shekarian et al., 2021).

6.2 Product Design for Circularity

Decisions on how to develop a product might make supply chain circularity harder or easier. No matter how good logistics or AI systems are, products that are hard to take apart, fix, or add to make recovery and reuse harder. To make things circular, you need to think about the whole life cycle of a product early in the creation process. Key design ideas include repairability, which lowers the amount of material that goes through the device and makes it last longer, and modularity and disassembly, which make it easier to get materials back.

By looking at data on how well materials operate, how they fail, and how they are really used, artificial intelligence may aid with better circular product design. For example, predictive analytics could uncover parts that are likely to break down often or go wrong in various ways. This provides you with the chance to redesign them so that they last longer or to mend them more quickly. Most of these returns are based on the quality of lifetime data that is shared by numerous companies. You can't make fresh product designs with AI help if you don't have data services integration. The way things are made has an impact on both the environment and the wallet. We can cut down on the need for new materials and carbon emissions by making products last longer and encouraging people to use them again. The rewards depend on how you get back what you lost and use it again. Making a product recyclable doesn't imply it will be recycled in places that don't have enough recovery facilities. Products need to deal with the real recovery abilities of the environment (Govindan & Soleimani, 2017).

6.3 Sector Specificity and Limits of Generalization

Manufacturing circularity recovers resources, reuses parts, and creates new products. AI-powered technologies can monitor and sort to ensure quality and compliance. However, the product's complexity and various materials to fix may cause businesses problems. Retail circular methods focus on inventory management and reducing output and waste. AI might improve supply-demand matching, reducing unsold inventory. Materials are harder to return than in manufacturing. Instead of internal rehabilitation, sales exist.

Logistics systems simplify and complicate circularity. Logistics networks pollute the environment but must move goods quickly and easily. AI can aid with routing and load consolidation, but the fuel, infrastructure, and networks will determine how environmentally benign this is. In circular energy systems, maximizing resources, increasing energy sources, and improving the system are more important than receiving materials back. AI may improve and reduce energy utilization in energy systems. Material-heavy companies differ from this one. Making systems less resource-intensive and robust is more important than closing the materials loop. These sectoral differences make AI-enabled circular supply chains unreliable. Strategies must meet company operational, technological, and regulatory needs (Mishra et al., 2023).

6.4 SMEs versus Large Firms

Companies find it tougher to use predictive analytics, improve reverse logistics, or manufacture based on manufacturing data. Cost is another issue.AI, digital platforms, and reverse logistics infrastructure are too expensive for SMEs. SMEs have little influence in the supply chain, making it challenging to agree on how to operate, who brings stuff back, and how to get everyone together. Only with the correct legal entity forms and platforms that allow partners to collaborate can SMEs develop circular ecosystems. Sharing transportation, a digital marketplace and platform for secondary materials, and data systems may help SMEs join the circular value chain. Companies that use AI may acquire information from many sources to provide smaller firms with knowledge they never had before. Large firms must focus on cooperation and staff management, not individual skills. All components of a large circular supply chain must function together, including outside partners and government laws. Old habits and organizational inertia can hinder growth, and new technologies can't fix that. Even with new technologies, large companies must consider how to make their global operations more effective, cheaper, and greener (Ben-Daya et al., 2019).

This design paradigm cuts down on waste since it lets you fix and remake things. You can utilize things that work again, and you can recycle things that break to get resources back. This cuts down on mining for raw materials and makes things last longer, which is better for the environment. **Figure 3** shows how long circular supply chains last. In these systems, natural and industrial production processes recycle or remanufacture materials before they are taken out.

Figure 3. Circular Supply Chain Lifecycle

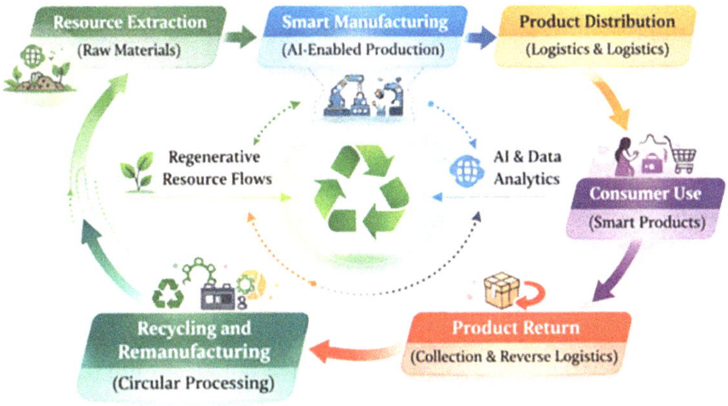

To enable this through intelligent and circular supply chain ecosystems, enterprises must reimagine their products to goods, resources and logistical systems interaction over their entire lifetime. Integrating forward and reverse supply chain flows using AI-enabled logistics coordination allows organizations to create renewable resource streams that increase economic efficiency by converting used products into useful inputs for new production cycles and use product designs such as standard modular components to extend the lifespan of products being manufactured. These design ideas serve as underpinning for intelligence circular supply chains that can reduce wastages, save resources and foster sustainable development to last (Yang et al., 2021).

7. SUSTAINABILITY INTELLIGENCE AND CLIMATE-OUTCOME ANALYTICS

7.1 What Sustainability Intelligence Actually Means

Sustainability intelligence is the capacity of a company to collect, analyze, comprehend, and use data about the environment and circularity to help make decisions about the supply chain. Environmental knowledge is part of both operational and strategic operations. It goes beyond only tracking compliance and sustainability reporting. Sustainability intelligence incorporates sustainability into decisions on procurement, manufacturing, logistics, and recovery.

It's very important to know the distinction between data availability and decision relevance. A lot of companies can get huge volumes of environmental data via sensors, business systems, and digital platforms. Third, decision-integration capabilities add these signals to systems for procurement, logistics, product design, and strategic planning. Without it, sustained intelligence is just a description, not a breakthrough (Voulgaridis et al., 2022).

To have sustainable intelligence, you need to know about the uncertainties and limits of environmental data. To figure out emissions, material flows, and recovery rates, people employ guesses, assumptions, and bad data. Organizations need to come up with ways to deal with uncertainty instead of assuming that all indications are accurate and similar. Governance frameworks need to define standards for data quality and explain how to use unclear information when making decisions.

7.2 Environmental Indicator Selection and Validation

To be sustainable, you need to be able to find and check environmental indicators. Not all signs are helpful, therefore picking the incorrect ones might have bad effects. So, companies need to use indicators that help them reach their goals for climate and sustainability as well as their operations. Carbon emissions are still used to measure how much carbon has been cut down. Carbon measurements don't entirely indicate how well the environment is doing. When energy sources have different amounts of carbon, you need to think about how much energy is used and how much carbon is released. Material intensity looks at how well industrial resources are used by comparing new and recycled materials. For circular supply chains, it's very important to know how much trash is diverted and collected. These figures show that people are more likely to reuse things than throw them away. High recovery rates don't mean that the quality is high. Recycling items that have broken down may waste resources. Pick an indicator based on how useful and good the content you got back is (Upadhyay et al., 2021).

7.3 Translating Indicators into Decisions

When procurement includes sustainability information, companies may evaluate vendors on cost, quality, and environmental performance. Carbon intensity, resource efficiency, and recovery strategies impact supplier selection and contracting. Environmentally friendly options may cost more or be harder to source, therefore procurement decisions usually include trade-offs. Thus, sustainability intelligence must combine environmental, economic, and operational variables to make multi-criteria judgments. Planning logistics and route is crucial. Optimizing routes may cut prices and hurt the environment despite significant transportation emissions. Sustainability intelligence helps organizations to integrate emissions data in routing algorithms for greener logistics. However, routing solutions must include infrastructure constraints, delivery periods, and demand variations (Kirchherr et al., 2017).

Recovery must be prioritized in circular supply chains. Businesses must prioritize repairs, remanufacturing, and recycling, as not all returned products may be reused. Sustainability intelligence helps individuals choose recovery strategies by analyzing their economic and environmental impacts. While recycling lower-value materials, remanufacturing high-value parts with high embodied energy is achievable. Supplier assessment includes initial purchasing decisions and ongoing performance monitoring. Sustainability intelligence continuously assesses supplier conduct to discover environmental performance abnormalities and assist corrective measures. Complex supply chains with varied supplier behaviors require dynamic review. Product redesign employs sustainability intelligence strategically. Environmental indicators can help you create materials that are easy to recycle, disassemble, and utilize. Design modifications may be needed due to excess material or low recovery rates. However, redesigning often needs engineering, marketing, and operational collaboration and long development cycles. Organizational incentives, decision-making processes, and data systems must collaborate to integrate sustainability intelligence in all these decision domains. Environmental indicators won't alter behavior unless they're employed to assess performance or make judgments. Businesses must include sustainability into governance, awards, and KPIs.

7.4 Climate Metrics and Long-Term Strategy

Environmental indicators provide you with information about how things are going right now, but climate measurements let you plan for the long future. These steps connect everyday choices to bigger transition pathways by focusing on how to reduce, adapt to, and bounce back from climate change in organizations. In this case, getting rid of carbon is quite important. Companies may use sustainability intelligence to figure out how much pollution their supply chain is causing and how to reduce it down. But initiatives to cut down on carbon emissions must also consider emissions that happen when materials are bought, moved, and used. This necessitates a comprehensive view of system-wide effects and several indicators. Another important method is to replace materials. To have less impact on the environment, you can substitute products that use a lot of carbon or resources with ones that are better for the environment. Sustainability intelligence looks at how resources are made, used, and disposed of. In addition to unexpected impacts, choices for substitutes must consider performance, cost, and availability (Ghisellini et al., 2016).

Because the weather is so unpredictable, it's even more crucial to plan for resilience. Changes in the law, a lack of resources, and bad weather can all affect supply routes. Sustainable intelligence may make people more resilient by helping them recognize hazards and use flexible techniques. Localized production, diverse sourcing, and adaptable logistics help mitigate disruptions. But techniques for resilience may need a thorough look at how cost-effective they are. For climate-oriented decision-making, it is very important to make sure that compliance expectations and transition strategies are in line with each other. Organizations must deal with changing rules and laws around climate performance and sustainability, as well as the expectations of stakeholders and industry standards. Sustainability intelligence helps you analyze how well you're following the rules, meeting your goals, and making progress. Compliance should be a part of a strategy that encourages taking care of the environment and making long-term value, not something that comes first (Silva & Sehnem, 2022).

Lastly, businesses need to change the way they think about performance to use climate-outcome analytics and sustainability intelligence. Companies need to see the links between economic, environmental, and social effects instead of thinking of environmental measures as less important than financial ones. This transition necessitates modifications in stakeholder engagement, leadership focus, decision-making processes, as well as technological and cultural dimensions. Sustainability information in both operational and strategic domains might help companies go beyond reactive compliance to proactive and informed decision-making (Morseletto, 2020).

As shown in **Table 4**, the main methods of value co-created from regenerative AI are outlined below which display how varying schemes enable environmental sustainability while also leading to different economic performances.

Table 4. Regenerative Value Creation Mechanisms

Mechanism	Impact
Recycling optimization	Reduced material waste
Remanufacturing systems	Extended product lifecycle
Resource recovery	Efficient material reuse
Predictive maintenance	Reduced equipment failure

8. GOVERNANCE, RESPONSIBLE AI, AND INSTITUTIONAL CONDITIONS FOR IMPLEMENTATION

AI makes circular supply chains possible. For this shift to happen, organizations need to be ready. "Mine" means that rules and policies need to work effectively along with how governance works. In both methods, Artificial Intelligence (AI) gives the supply chain improved visibility, the capacity to foresee what will happen, and the ability to optimize; nevertheless, these functions only work within a larger socio-technical environment that is regulated by laws and requires trust and responsibility. AI-driven circular systems might make things less efficient, add new forms of opacity, and go against the same sustainability goals they were meant to help if they aren't properly governed. Governance is not simply a side issue; it is important for figuring out if AI-enabled circularity truly does influence the environment and climate (Bocken et al., 2016).

8.1 Why Governance Matters in AI-Enabled Circular Systems

A circular AI system requires enterprises to share data. Information goes down the manufacturing chain from manufacturers to suppliers to transportation and logistics to waste recyclers to clients. System must constantly state what materials go in, how they are used, and what comes out. They must also disclose their lifetime pollution. These links may help or hurt. Sharing knowledge aids recycling, demand prediction, and resource allocation. It also raises problems regarding data accuracy, utilization, and stakeholder access. Data governance collects, verifies, shares, and handles emergencies. AI system transparency allows users to know how they function and if they generate or dispose of garbage. Accountability shows who is responsible for resource or operation failures. Auditability lets firms and governments follow their rationale and facts. This is needed to prove sustainability and keep up with environmental protection methods. These properties are required for AI-based circular systems. Is the discomfort of a full-process disruption more relevant as it hinders system integration? Which is worse? Governance reduces risks and enables circular systems (Yadav et al., 2020).

8.2 Governance Mechanisms

How AI-powered circular systems function firm data governance standards tell the firm where to obtain, store, and handle data. These are vital too. These guidelines should focus on privacy, data standards, security, and interoperability. Standardization is crucial in circular systems. Collaboration is impossible if the data doesn't satisfy the parties' and platforms' definitions. Third-most critical in AI governance is model validation assurance. To ensure they perform in all conditions, AI models for anticipating demand, classifying rubbish, and maximizing recovery must be regularly tested. Contextual and technological assessments may ensure models function properly in various circumstances. Always alert for straying models When the market changes fast. Openness and accountability require audits and explainability. AI results and rationale should be clear to everyone. Audit trails make data inputs, model parameters, and decision outputs easy to review. Companies must demonstrate that they obey the standards and report to outside bodies that they are green and responsible. So, their methods work. People must monitor the government. Review boards or committees can help with major decisions like choosing a partner, altering materials, or funding recovery infrastructure. They ensure AI-based decisions meet the company's goals, legal obligations, and growth ambitions. Making standards for sharing information and holding suppliers accountable can help you manage a smart, distributed supply chain. Data sharing, functionality, and compliance should be outlined in contracts and agreements. Collaborative systems let users pick who may see their data, making data sharing easier. These tactics help rival groups collaborate and set goals before starting work (Seuring & Müller, 2008).

8.3 Responsible AI Risks

AI systems pose hazards that must be handled. Modeling mistakes, even little ones, can be significant. Unreliable estimations can lead to overproduction, mis categorization, and slow rearrangement. Problems worsen when circular system supply chains progress. This makes an already terrible condition worse or simpler to govern, harming the environment. Disaster might result from supply networks not prepared for environmental and economic issues. Another concern is that AI may make unjust conclusions (Platon et al., 2024). AI models based on old data may reflect data collector biases. Supplier evaluation hurts some areas or enterprises. Traceability issues make system administration difficult. Due to supply chain uncertainty, it's impossible to know where products come from, how they get there, and what occurs next. This requires comprehensive AI data and reporting structure. Without such, it's hard to measure pollution, recover materials, or see how much is recycled. Too many proprietary systems and providers raise risk. Most AI solutions are developed and managed by outside organizations. This might confuse matters and create dependencies that threaten the company's independence. Vendor lock-in increases system transfer costs and disrupts power. Companies must weigh the pros and cons of running their own data operations vs employing new technologies (Pagell & Wu, 2009).

8.4 Policy and Regulatory Context

The rules and processes that govern AI-enabled circular systems affect their performance. Extended Producer Responsibility (EPR) rules make producers accountable for their products' long-term effects. Environmental and economic incentives are combined in EPR systems. They want firms to make durable, recyclable, and reusable products. Environmental disclosure rules make it harder for producers to mislead about emissions, resource consumption, and waste disposal. The reader can understand that even these disclosures will make the whole firm operate ethically and honestly, helping him evaluate his work performance. AI can automate data collection and processing. Without specific measuring criteria to compare with values people currently have and recognize to be true, any AI system will fail (Sarkis et al., 2011).

8.5 Implementation Readiness

For example, making significant investments up front for little return wouldn't work well unless you were previously established and had money to fall back on when you started using new technologies like AI. Setting up and keeping these systems running costs a lot of money and requires regular changes. For enterprises with limited money, the procedures and costs of building supply chains are different. This makes supply chains unequal because certain portions of the chain are moving forward with technology while others are falling behind. Another big challenge is getting all the Value chain suppliers to work together. People with varied aims, talents, and constraints must work together in circular systems. Putting together technologies and goals to make sure that everyone agrees is what coordination entails. People have trouble with fitness since delivery cycles don't always happen on time and data reporting standards aren't the same for everyone. People are not eager to adapt, don't know enough about the benefits of AI-run circular economies, or don't comprehend them, therefore it's hard to convince them to agree. Without all the details on how new technologies work, governments need to be flexible and determined to make sure that progress happens. Organizations might challenge their utilization. Figure 4 depicts the general pattern of sustainability analytics. To do an analysis, you need to have chain after chain and a circular supply environment (Morash & Clinton, 1998).

Figure 4. Sustainability Analytics Cycle

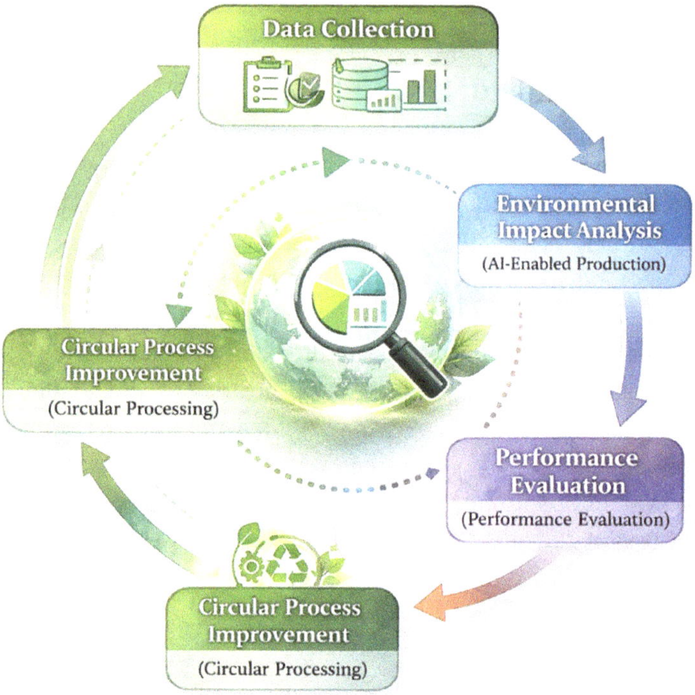

The primary environmental indicators used in sustainability intelligence systems are summarized in **Table 5**, which highlights the metrics organizations monitor when evaluating circular supply chain performance.

Table 5. Sustainability Performance Indicators

Indicator	Description
Carbon emissions	Environmental impact
Resource utilization	Efficiency of material use
Recycling rate	Percentage of recovered materials
Energy consumption	Operational sustainability

Hence, the application of such data-driven methodologies supports organizations to make their supply chain broad aligned with larger sustainability goals while enabling a constant evolution of regenerative economic systems (Xia et al., 2024).

9. CRITICAL CHALLENGES, TRADE-OFFS, AND BOUNDARY CONDITIONS

9.1 Rebound Effects and Over-Optimization

One of the biggest problems for AI-driven sustainability solutions is rebound effects. The benefit is that AI systems can make operations more efficient in ways that no human could ever hope to match. For example, they can cut down on waste, fine-tune inventory levels, improve routing, and cut down on idle capacity. The result is better company performance without putting too much strain on resources. It also makes feasible completely circular manufacturing methods that were never possible before. But these benefits don't always reach the degree of effect that each person in that larger supply chain feels. In many circumstances, making things more efficient lowers operating costs, which leads to more output and consumption. However, this supply chain no longer "looks" like what it should be called "sustainable development. "It's strange that in a circumstance like this, the systems are always running yet still use business models that aren't good for the long term. For example, greater demand forecasting may lead to less excess inventory being retained because of false predictions about demand. However, if the overall amount of output goes up each year, then net material throughput and emissions are also going up (Roberts et al., 2024).

AI can speed up logistics, but this might reduce the life of products instead of lengthening them. One negative aspect of over-optimization is that individuals may get so focused on quantifiable efficiency that they miss bigger changes in the system and long-term complaints. Most of the time, AI systems are set up to reach short-term goals like lowering costs, speeding up deliveries, or increasing inventory turnover. They are not set up to reach long-term regenerative goals. So, businesses may get to a point of "efficient linearity," where their operations are very well-tuned yet depend on constantly getting and getting rid of resources. If there is no such alignment, efficiency increases might make unsustainable patterns of production and consumption even stronger instead of changing them (Chi & Liu, 2023).

9.2 Data Quality and Infrastructure Dependence

AI-enabled circular systems can only realize their full potential if the data is high quality, easy to get, and easy to mix. To figure out how much environmental benefit an item may have, circular supply chains need highly detailed information on the ingredients that make up, how it is used, and how long it can be returned for recovery. Right now, though, this kind of data is often incomplete, inaccurate, or just not available to everyone in a supply network (Duan et al., 2019).

Bad data can cause bad decisions. If you don't prepare for reverse logistics, it may be difficult when item return rates are lower than expected. Incorrect information regarding materials may stop recycling or slow remanufacturing. Also, if incorrect environmental indicators make supply chain techniques appear greener than they are. AI development requires IoT sensor installations, supply chain monitoring systems, clouds, platforms, and data exchange protocols. Supply networks without common data formats or weak system linkages make it harder for AI to coordinate the same mental process. Many smaller suppliers, remote workers, and informal sector businesses don't have trustworthy data or share it. These are black holes in circular systems. AI requires data and infrastructure to operate, but it can't replace absent or erroneous data. Its performance depends on the data environment's development. Thus, data governance and interoperability standards must be funded as complimentary endeavors (Laskurain-Iturbe et al., 2021).

9.3 Digital Costs and Unequal Access

Technology-intensive AI takes time and money. People rarely consider this complete perspective while discussing such issues. Research and development-related fat spending includes personnel and data infrastructure software platforms. Even when functioning, AI systems require expensive maintenance.AI requires huge algorithms and dynamic feedback optimization systems with fixed hypothesis space points, making power expensive. Some of the energy saved by becoming digital keeps the old structure running. “This adoption pattern will cause company or regional growth differences.” Companies with multiple branches and good financial and technical performance may construct superior AI-driven circular models. Due to a lack of money, knowledge, and market, SMEs may struggle to innovate. These costs affect how technology is employed in various regions and with different technologies. Digitalization, governmental backing for AI, and skilled personnel all affect whether AI may be employed in a circular system. Players who already have better digital skills and digital ecosystems will get more resources for themselves than others. This results in a scenario where the advantages of smart circularity, contingent upon resources, infrastructure, or institutional backing, establish a boundary condition (Dolci et al., 2024).

9.4 Profitability versus Sustainability Tensions

In AI-enabled circular supply chains, what is lost becomes the shortest-term competition company and the long-term sustainability. People typically utilize AI systems to save expenses and make work more efficient, which may not be what is needed for circularity. For instance, a business may utilize AI to find the best ways to recover resources, such high-value ones or ones that are easy to recycle. It will, on the other hand, disregard waste streams that are less valued but nonetheless significant for the ecosystem. A choice made based on cost and efficiency could also focus on incremental modifications instead of more drastic ones. For instance, restructuring items and substituting their inputs with more environmentally friendly ones might make this happen on a much bigger scale: changing how people use things so that less waste is made before anything gets to consumers. Circular measures that help the environment in the long run, such extending product lifecycles, developing repair infrastructure, or creating closed-loop systems, may cost more in the near term and take longer to pay off. AI-based efficiency increases, on the other hand, usually pay off right away. From a management point of view, this makes them much more appealing (Bag et al., 2025).

9.5 Technology Dependence and Platform Control

AI regulates and relies on Supply Chain as we use it more. Most AI capabilities are only available from vendor platforms, niche providers, or clouds. Technology is only accessible to a few. This dependency makes planning and management difficult. Companies may be limited to one platform, limiting negotiation and activity. When each supplier uses its own data system and there's no standard procedure, things might get complicated (Arab Momeni et al., 2024). Some individuals don't know how they choose or whether carbon footprint and other pollution claims are true. Switching providers may result in less secure data, incompatible technology, or long-term technological issues. Platform owners determine supply chain setup. Successful technical platform suppliers may standardize protocols, data accessibility, and decision-making. This may affect circular ecosystems and human interaction. Small and medium-sized enterprises can't adjust the device's regulations and system, and Wall Holding makes this worse. AI is a dark box. This makes socializing difficult. If the decision-making process is unknown, transparency cannot be used to hold someone accountable for mistakes, prejudice, or unexpected outcomes. This may help determine if circular habitats are detrimental or beneficial. Platform restrictions complicate AI-enabled circularity with digital technologies. Openness and compatibility are crucial to maintaining supplier trust and preventing power concentration. Operations cycle in a circular governance structure. Automation must incorporate sustainable policies into decision-making and daily living (Morshedi et al., 2026).

When necessary, there is human control. This governance structure goes back to **Figure 5**. It shows how AI-driven processes may be monitored by making constant changes based on policy frameworks. This is a clear illustration of how to use AI responsibly across the circular supply chain. **Table 6** lists its main governance concepts.

Table 6. Governance Principles for Circular AI Systems

Principle	Purpose
Transparency	Explain algorithmic decisions
Accountability	Human supervision
Fairness	Prevent algorithmic bias
Compliance	Follow environmental regulations

Figure 5. Governance Framework for AI-Enabled Circular Systems

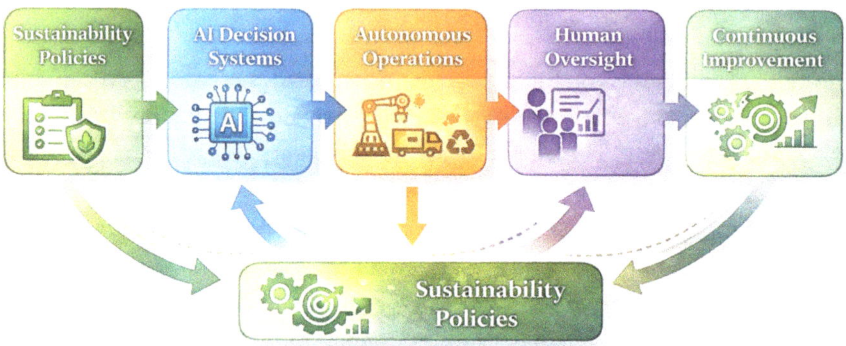

The practical and efficient management of circular supply chains is only possible if ethical AI practices and governance are adopted. Organizations can, indeed, leverage sustainable-based AI technology, maintaining their commitment to sustainability by complying with regulatory standards and avoiding AI black boxes that do not involve human supervision and decision in the loop. As a result, the process of responsible implementation enables organizations to design supply chains that are not only effective but also consequential with their larger environmental and social goals (George, 2023).

10. CONCLUSION

Each chapter of the book gives a clear and honest point of view. It uses AI perception, modeling, optimization, visibility, and environmental intelligence to provide a whole analytical framework for circular business strategies. This will help make the supply chain more environmentally friendly and help future generations. The chapter looks at how social space may become regenerative value by bringing together different parts of interconnected supply networks. This study demonstrates that AI does not inherently foster a regenerative or climate-positive society. AI can help with efficiency, forecasting, and making the most use of resources, but it can't help with sustainability. Overconsumption, consumerism, and virgin resource extraction can make things more efficient, but only intentional circular design methods that focus on extending the life of products, reducing waste by recovering materials, and having a low environmental impact can create regenerative value with AI. Governance institutions facilitate this convergence by transforming environmental measurement techniques from a weak disciplinary framework to one that accurately reflects societal performance and the interests of its members. When companies work together in supply chains using digitalization systems that are transparent, accountable, and responsible, they can stay in business and flourish. In some circumstances, AI may promote linear forms.

First, the results demonstrate that managers shouldn't think of adopting AI as a technology investment choice. Instead, businesses should set measurable goals for the environment and the circular economy, such cutting down on emissions, recovering materials at the end of their life, and diverting trash. These goals should help you choose and use AI tools. Governance systems must make sure that models are correct, that data is of high quality, and that everyone in the business is responsible. Companies will also have to pay for coordinating with suppliers, logistics service providers, recyclers, and other partners in reverse flow and recovery. To make AI-driven circular projects work, these groups need to be brought together in shared data ecosystems and connected processes. Instead of learning new algorithms, managers now must deal with sophisticated socio-technical systems. "Success is more about governance than new formulae."

The policy implications of this chapter highlight how the adoption and performance of AI-driven circular supply chains are affected by the laws and institutions that are in place. Companies may be more likely to establish circular systems for materials, technology, and digital monitoring if they have a lot of different products and good waste management. Okay? For successful supply chain management information sharing, this policy framework needs uniform reporting standards, data set compatibility, and traceability requirements. AI's smart sustainability is restricted without them. Policy decides how much operations will cost and what rules must be followed. Both affect long-term success and profitability. The institutional framework, not the individual business, primarily determines how to switch to an AI-enabled regeneration system.

There are numerous things that this framework hasn't explored. Sector-specific empirical research is necessary to assess AI-enabled circular approaches, given the disparities in data availability regarding recovery logic and value-retention mechanisms across manufacturing, retail, logistics, and energy systems. It is yet unknown where small and medium-sized businesses may use these solutions within their financial and technical restrictions for data infrastructure. According to longitudinal research, AI-driven circular projects may offer enduring regeneration and climate benefits instead of immediate efficiency. Better governance audit frameworks might look at how transparent, accountable, and ethical AI-enabled decision-making systems are. Lastly, advances in efficiency may have effects that come back to them. This increases consumption and takes away the advantage of the environment.

REFERENCES

Abyaneh, A. G., Ghanbari, H., Mohammadi, E., Amirsahami, A., & Khakbazan, M. (2025). An analytical review of artificial intelligence applications in sustainable supply chains. *Supply Chain Analytics, 12*, 100173. DOI: 10.1016/j.sca.2025.100173

Arab Momeni, M., Jain, V., & Bagheri, M. (2024). A multi-objective model for designing a sustainable closed-loop supply chain logistics network. *Logistics (Basel), 8*(1), 29. DOI: 10.3390/logistics8010029

Awan, U., Sroufe, R., & Shahbaz, M. (2021). Industry 4.0 and the circular economy: A literature review and recommendations for future research. *Business Strategy and the Environment, 30*(4), 2038–2060. DOI: 10.1002/bse.2731

Bag, S., Routray, S., Rahman, M. S., & Gupta, S. (2025). Digital innovation for circular supply chain sustainability and resilience for achieving carbon neutrality: An empirical study. *Journal of Environmental Management, 386*, 125665. DOI: 10.1016/j.jenvman.2025.125665 PMID: 40359861

Ben-Daya, M., Hassini, E., & Bahroun, Z. (2019). Internet of things and supply chain management: A literature review. *International Journal of Production Research, 57*(15–16), 4719–4742. DOI: 10.1080/00207543.2017.1402140

Bhattacharya, S., Govindan, K., Ghosh Dastidar, S., & Sharma, P. (2024). Applications of artificial intelligence in closed-loop supply chains: Systematic literature review and future research agenda. *Transportation Research Part E, Logistics and Transportation Review, 184*, 103455. DOI: 10.1016/j.tre.2024.103455

Bocken, N. M. P., de Pauw, I., Bakker, C., & van der Grinten, B. (2016). Product design and business model strategies for a circular economy. *Journal of Industrial and Production Engineering, 33*(5), 308–320. DOI: 10.1080/21681015.2016.1172124

Bressanelli, G., Adrodegari, F., Pigosso, D. C. A., & Parida, V. (2022). Towards the smart circular economy paradigm: A definition, conceptualization, and research agenda. *Sustainability (Basel), 14*(9), 4960. DOI: 10.3390/su14094960

Cagno, E., Neri, A., Negri, M., Bassani, C. A., & Lampertico, T. (2021). The role of digital technologies in operationalizing the circular economy transition: A systematic literature review. *Applied Sciences (Basel, Switzerland), 11*(8), 3328. DOI: 10.3390/app11083328

Cannas, V. G., Ciano, M. P., Saltalamacchia, M., & Secchi, R. (2024). Artificial intelligence in supply chain and operations management: A multiple case study research. *International Journal of Production Research*, *62*(9), 3333–3360. DOI: 10.1080/00207543.2023.2232050

Chauhan, C., Parida, V., & Dhir, A. (2022). Linking circular economy and digitalisation technologies: A systematic literature review of past achievements and future promises. *Technological Forecasting and Social Change*, *177*, 121508. DOI: 10.1016/j.techfore.2022.121508

Chi, Z., & Liu, Z. (2023). Driving circular economy through digital technologies: A systematic literature review. *Sustainability*, *15*(24), 16608. DOI: 10.3390/su152416608

Dolci, V., Bigliardi, B., Petroni, A., Pini, B., Filippelli, S., & Tagliente, L. (2024). Integrating Industry 4.0 and circular economy: A conceptual framework for sustainable manufacturing. *Procedia Computer Science*, *232*, 1711–1720. DOI: 10.1016/j.procs.2024.01.169

Duan, C., Xiu, G., & Yao, F. (2019). Multi-period E-closed-loop supply chain network considering consumers' preference for products and AI-push. *Sustainability (Basel)*, *11*(17), 4571. DOI: 10.3390/su11174571

Farshadfar, Z., Mucha, T., & Tanskanen, K. (2024). Leveraging machine learning for advancing circular supply chains: A systematic literature review. *Logistics (Basel)*, *8*(4), 108. DOI: 10.3390/logistics8040108

George, B. (2023). Artificial intelligence for Circular Economy 2.0: Opportunities, challenges, and outlook. *Clean Technologies and Recycling*, *3*(3), 166–171. DOI: 10.3934/ctr.2023011

Ghisellini, P., Cialani, C., & Ulgiati, S. (2016). A review on circular economy: The expected transition to a balanced interplay of environmental and economic systems. *Journal of Cleaner Production*, *114*, 11–32. DOI: 10.1016/j.jclepro.2015.09.007

Govindan, K., & Soleimani, H. (2017). A review of reverse logistics and closed-loop supply chains: A Journal of Cleaner Production focus. *Journal of Cleaner Production*, *142*(Part 1), 371–384. DOI: 10.1016/j.jclepro.2016.03.126

Hangl, J., Behrens, V. J., & Krause, S. (2022). Barriers, drivers, and social considerations for AI adoption in supply chain management: A tertiary study. *Logistics (Basel)*, *6*(3), 63. DOI: 10.3390/logistics6030063

Kirchherr, J., Reike, D., & Hekkert, M. (2017). Conceptualizing the circular economy: An analysis of 114 definitions. *Resources, Conservation and Recycling*, *127*, 221–232. DOI: 10.1016/j.resconrec.2017.09.005

Laskurain-Iturbe, I., Arana-Landín, G., Landeta-Manzano, B., & Uriarte-Gallastegi, N. (2021). Exploring the influence of Industry 4.0 technologies on the circular economy. *Journal of Cleaner Production*, *321*, 128944. DOI: 10.1016/j.jclepro.2021.128944

Lopes de Sousa Jabbour, A. B., Jabbour, C. J. C., Godinho Filho, M., & Roubaud, D. (2018). Industry 4.0 and the circular economy: A proposed research agenda and original roadmap for sustainable operations. *Annals of Operations Research*, *270*(1), 273–286. DOI: 10.1007/s10479-018-2772-8

Mishra, A., Dutta, P., Jayasankar, S., Jain, P., & Mathiyazhagan, K. (2023). A review of reverse logistics and closed-loop supply chains in the perspective of circular economy. *Benchmarking*, *30*(3), 975–1020. DOI: 10.1108/BIJ-11-2021-0669

Morash, E. A., & Clinton, S. R. (1998). Supply chain integration: Customer value through collaborative closeness versus operational excellence. *Journal of Marketing Theory and Practice*, *6*(4), 104–120. DOI: 10.1080/10696679.1998.11501814

Morseletto, P. (2020). Targets for a circular economy. *Resources, Conservation and Recycling*, *153*, 104553. DOI: 10.1016/j.resconrec.2019.104553

Morshedi, M., Hargaden, V., Papakostas, N., & Ghadimi, P. (2026). The adoption of digital technologies in circular supply chains: From theoretical developments to practical applications. *Logistics (Basel)*, *10*(1), 18. DOI: 10.3390/logistics10010018

Pagell, M., & Wu, Z. (2009). Building a more complete theory of sustainable supply chain management using case studies of 10 exemplars. *The Journal of Supply Chain Management*, *45*(2), 37–56. DOI: 10.1111/j.1745-493X.2009.03162.x

Pagoropoulos, A., Pigosso, D. C. A., & McAloone, T. C. (2017). The emergent role of digital technologies in the circular economy: A review. *Procedia CIRP*, *64*, 19–24. DOI: 10.1016/j.procir.2017.02.047

Platon, V., Pavelescu, F.-M., Antonescu, D., Constantinescu, A., Frone, S., Surugiu, M., Mazilescu, R., & Popa, F. (2024). New evidence about artificial intelligence and eco-investment as boosters of the circular economy. *Environmental Technology & Innovation*, *35*, 103685. DOI: 10.1016/j.eti.2024.103685

Pournader, M., Ghaderi, H., Hassanzadegan, A., & Fahimnia, B. (2021). Artificial intelligence applications in supply chain management. *International Journal of Production Economics*, *241*, 108250. DOI: 10.1016/j.ijpe.2021.108250

Riahi, Y., Saikouk, T., Gunasekaran, A., & Badraoui, I. (2021). Artificial intelligence applications in supply chain: A descriptive bibliometric analysis and future research directions. *Expert Systems with Applications*, *173*, 114702. DOI: 10.1016/j.eswa.2021.114702

Roberts, H., Zhang, J., Bariach, B., Cowls, J., Gilburt, B., Juneja, P., Tsamados, A., Ziosi, M., Taddeo, M., & Floridi, L. (2024). Artificial intelligence in support of the circular economy: Ethical considerations and a path forward. *AI & Society*, *39*(3), 1405–1418. DOI: 10.1007/s00146-022-01596-8

Roy, T., Garza-Reyes, J. A., Kumar, V., Kumar, A., & Agrawal, R. (2022). Redesigning traditional linear supply chains into circular supply chains: A study into its challenges. *Sustainable Production and Consumption*, *31*, 113–126. DOI: 10.1016/j.spc.2022.02.004

Sarkis, J., Zhu, Q., & Lai, K. (2011). An organizational theoretic review of green supply chain management literature. *International Journal of Production Economics*, *130*(1), 1–15. DOI: 10.1016/j.ijpe.2010.11.010

Seuring, S., & Müller, M. (2008). From a literature review to a conceptual framework for sustainable supply chain management. *Journal of Cleaner Production*, *16*(15), 1699–1710. DOI: 10.1016/j.jclepro.2008.04.020

Shahzadi, G., Jia, F., Chen, L., & John, A. (2024). AI adoption in supply chain management: A systematic literature review. *Journal of Manufacturing Technology Management*, *35*(6), 1125–1150. DOI: 10.1108/JMTM-09-2023-0431

Sharma, R., Shishodia, A., Gunasekaran, A., Min, H., & Munim, Z. H. (2022). The role of artificial intelligence in supply chain management: Mapping the territory. *International Journal of Production Research*, *60*(24), 7527–7550. DOI: 10.1080/00207543.2022.2029611

Shekarian, E., Das, S., & Krishnan, R. (2021). Analyzing the structure of closed-loop supply chains: A game theoretic approach. *Sustainability (Basel)*, *13*(3), 1397. DOI: 10.3390/su13031397

Silva, T. H. H., & Sehnem, S. (2022). Industry 4.0 and the circular economy: Integration opportunities generated by startups. *Logistics (Basel)*, *6*(1), 14. DOI: 10.3390/logistics6010014

Srivastava, S., Shardeo, V., Dwivedi, A., & Paul, S. K. (2026). Analyzing enablers of artificial intelligence for decarbonization: Implications for circular supply chains. *Annals of Operations Research*, *359*(1), 953–980. DOI: 10.1007/s10479-025-06843-x

Toorajipour, R., Sohrabpour, V., Nazarpour, A., Oghazi, P., & Fischl, M. (2021). Artificial intelligence in supply chain management. *Journal of Business Research, 122*, 502–517. DOI: 10.1016/j.jbusres.2020.09.009

Upadhyay, A., Mukhuty, S., Kumar, V., & Kazancoglu, Y. (2021). Blockchain technology and the circular economy: Implications for sustainability and social responsibility. *Journal of Cleaner Production, 293*, 126130. DOI: 10.1016/j.jclepro.2021.126130

Voulgaridis, K., Lagkas, T., Angelopoulos, C. M., & Nikoletseas, S. E. (2022). IoT and digital circular economy: Principles, applications, and challenges. *Computer Networks, 219*, 109456. DOI: 10.1016/j.comnet.2022.109456

Wynn, M., & Jones, P. (2022). Digital technology deployment and the circular economy. *Sustainability (Basel), 14*(15), 9077. DOI: 10.3390/su14159077

Xia, H., Chen, Z., Milisavljevic-Syed, J., & Salonitis, K. (2024). Uncertain programming model for designing multi-objective reverse logistics networks. *Cleaner Logistics and Supply Chain, 12*, 100155. DOI: 10.1016/j.clscn.2024.100155

Yadav, G., Mangla, S. K., Bhattacharya, A., & Luthra, S. (2020). Exploring indicators of circular economy adoption framework through a hybrid decision support approach. *Journal of Cleaner Production, 277*, 124186. DOI: 10.1016/j.jclepro.2020.124186

Yang, M., Fu, M., & Zhang, Z. (2021). The adoption of digital technologies in supply chains: Drivers, process and impact. *Technological Forecasting and Social Change, 169*, 120795. DOI: 10.1016/j.techfore.2021.120795

Yuan, S., & Pan, X. (2023). The effects of digital technology application and supply chain management on corporate circular economy: A dynamic capability view. *Journal of Environmental Management, 341*, 118082. DOI: 10.1016/j.jenvman.2023.118082 PMID: 37141715

Chapter 6

AI for Climate-Smart Agriculture: Pathways for Resilience in Developing Countries

Monica C.
http://orcid.org/0009-0006-6611-1960
Dayananda Sagar University, India

Nisha Goyal
http://orcid.org/0000-0002-7897-0635
Dayananda Sagar University, India

Navita Vijay
http://orcid.org/0000-0002-4572-3654
Dayananda Sagar University, India

ABSTRACT

The study focuses on the use of artificial intelligence (AI) as a as a tool to advance Climate-Smart Agriculture (CSA) within developing nations, considering smallholder farming systems. The systematic review of 35 peer-reviewed articles (2015-2025) and thematic synthesis is used to assess the contribution of various AI applications to CSA objectives. It also identifies the socio-economic, institutional, and technological factors that drive their adoption. The results indicate that AI applications can enhance productivity and use of resources more efficiently but the benefits to climate resilience is conditional. The effectiveness relies on the system-level integration, access, and adaptation to the local socio-economic and institutional contexts. The discussion indicates that there has been a consistent gap between

DOI: 10.4018/979-8-3373-9978-2.ch006

technical feasibility and practice. This is primarily due to infrastructural issues, low digital literacy, lack of institutional support, and risks based on data asymmetry, platform dependency, and marginalisation. The key contribution of the study is the establishment of a system-level analytical framework that characterizes the difference between the high-tech and scalable AI-based solutions. It correlates the technological potential with conditions of adoption in the smallholder environment. Using this framework, the study suggests evidence-based avenues to enable inclusive adoption of AI. AI has the potential to enable climate-resilient agriculture not individually as a technological solution, but rather as a part of accessible systems, robust institutional intermediaries, and inclusive governance frameworks. This ensures that the economic and data-related benefits are more evenly distributed between farmers and technology providers.

1. INTRODUCTION

In developing countries, like India, Pakistan, and Ethiopia, agriculture continues to be one of the pillars of economic development, food security and livelihoods in rural areas. In these areas, smallholder farming systems form a significant segment of the population and constitute a source of income and subsistence (Pawlak & Kołodziejczak, 2020). These systems are however more susceptible to climate change, environmental degradation and socio-economic limitations. Increasing temperatures, unpredictable rains, extended droughts, floods, the spread of pests and diseases are affecting the productivity of agriculture and increasing rural susceptibility (Ali et al., 2025; Dhanaraj et al., 2025). These issues reveal the shortcomings of traditional farming methods, which are dependent on past climate conditions and fixed decision-making processes.

In contrast, Climate-Smart Agriculture (CSA) has emerged as a planning tool that tackles productivity, resilience, and environmental concerns. CSA is goal-focused, and emphasizes long-term sustainability and adaptive capacity in climate uncertainties (Ma & Rahut, 2024). Although CSA clearly defines what needs to be done, it offers minimal suggestions on how the objectives can be achieved under resource-based limitations. Simultaneously, Smart Agriculture has developed into a technology-driven paradigm. It utilizes technology, including Artificial Intelligence (AI), the Internet of Things (IoT), and remote sensing to enhance efficiency and decision-making (Wolfert et al., 2017; Liakos et al., 2018). Smart Agriculture and CSA are two different models that are used interchangeably. Smart Agriculture deals with technological potential and optimization, but CSA focuses on outcomes of resilience, sustainability, and food security (De Avila & Barbosa, 2025). This contrast is crucial: technologies are not always climate-smart, and their applicability is determined by their contribution to the CSA objectives.

The major issue does not lie in the creation of the technology, but to align with the limitations and requirements of smallholder agricultural systems. Figure 1 presents a conceptual flow of AI integration in CSA. It emphasizes the connection between the outcome of technological innovation and the outcomes of CSA in smallholder contexts. AI has gained a lot of attention because of its capability to handle large and complicated datasets. Crop classification, yield prediction, irrigation optimization, and disease identification are some of the applications that forecast accurately especially under controlled and data rich conditions (Ferraris et al., 2023; Ramzan et al., 2024). The applications, however, do not equally contribute to CSA objectives.

As an example, yield prediction models can enhance productivity but provide decreasing support to resilience unless applied collectively with adaptable decision-making systems (Van Klompenburg et al., 2020). Conversely, AI-driven systems of early detection of droughts and pest outbreaks are more directly relevant in CSA since they increase adaptive capacity (Uztürk & Büyüközkan, 2024). This contrast has revealed a very important gap in the literature as technical performance is usually valued more than contextual relevance and real-world impact. It is enhanced by the structural bias towards large-scale and capital-intensive agriculture.

Most of the studies focused on individual applications of AI but they did not examine how the tools would be utilized in larger agricultural systems (Mana et al., 2024; Reitano et al., 2025). Therefore, there is a lack of clarity regarding which solutions of AI are most feasible, scalable, and successful in the conditions of resource constraints. In addition, some of the most important socio-economic and institutional drivers such as access to technology, digital literacy, affordability, and data governance are often ignored. Such fragmentation reduces the value of research to policymakers and smallholder farmers.

Figure 1. Conceptual flow of AI Integration in Climate Smart Agriculture

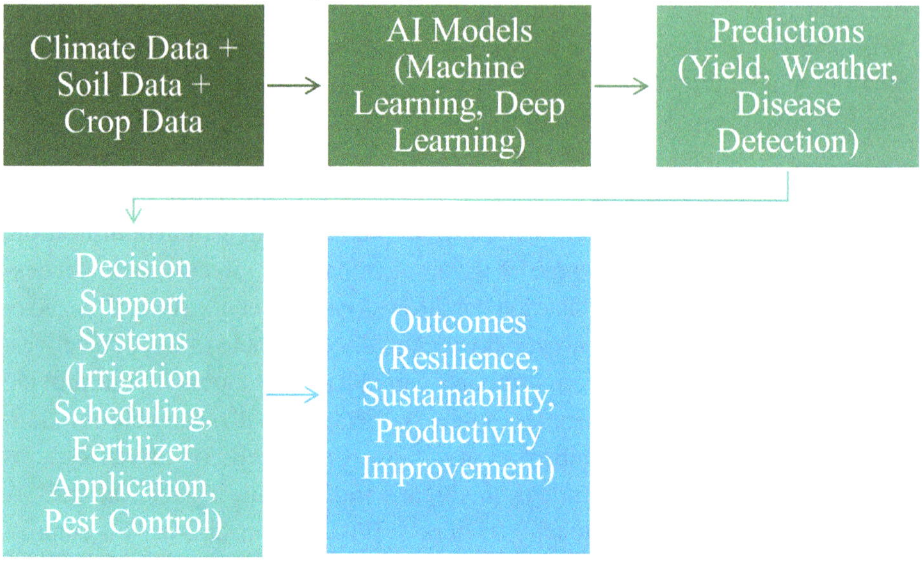

Source: *Authors' own compilation*

In order to overcome these shortcomings, a systematic and context sensitive framework is needed to address the gap between technological potential and CSA relevance. This study incorporates the findings of a PRISMA-based systematic review and thematic synthesis to reflect AI as a part of a broader socio-technical system. Its central contribution is the creation of the analytical framework. It shows that the success of AI relies on the compatibility between the technological capabilities, institutional support, and socio-economic conditions. It examines and prioritizes the applications of AI in a systematic way based on its relevancy, scalability, and adaptation to CSA objectives in smallholder systems. Instead of analysing the technical ability, the study determines the best AI solutions under climatic uncertainty conditions and limited resources. This way, it realigns AI as a context-specific agent of climate resilience instead of a generalized solution.

The study focuses on two objectives.

Objective 1: To evaluate the role of various AI applications in resilience, productivity, and sustainability in CSA systems.

Objective 2: To examine the socio-economic, technological, and institutional constraints that characterise AI adoption by smallholder farmers, where the scalability, inclusiveness, and performance trade-off are emphasised.

By fulfilling these objectives, the study goes beyond descriptive analysis and builds an organized idea of how and in what situations AI can contribute positively to CSA. Based on the framework, this study offers a comparative synthesis of AI implementation in terms of predictive, diagnostic, monitoring, and advisory systems. It also outlines the evidence-based pathways to inclusive adoption. These pathways connect technological potential and the infrastructural, institutional, and governance circumstances needed to scale it through the context of developing countries.

The chapter is structured as follows. Section 2 provides the methodology, Section 3 establishes the thematic analysis, Section 4 assesses the applications of AI in agriculture, Section 5 covers opportunities, Section 6 discusses the challenges, Section 7 suggests avenues of adoption, and Section 8 concludes the study.

2. METHODOLOGY

The PRISMA framework is used in the study because it provides transparency, reproducibility and rigour in terms of methodology in the literature review. PRISMA offers a designed process of identifying, screening and selecting highly relevant studies by minimizing selection bias and increasing the reliability of synthesis results. The review is conducted through four main steps, such as identification, screening, eligibility, and inclusion. Each step is clearly outlined in the study to maintain replication and transparency in the selection process.

2.1 Data collection

The Scopus database was chosen because of its comprehensive coverage of peer-reviewed, interdisciplinary literature in the areas of agriculture, environmental science, and AI. The search was conducted in March 2026 and included articles published between 2015 and 2025. This timeframe was selected based on the aim of identifying the latest developments in the field of AI and its use in CSA. To maintain transparency and uniformity, structured Boolean search strategy was used in titles, abstracts and key words. The search query was generated by combining keywords based on the AI technologies, climate-based agriculture, digital farming systems, and smallholder settings. The keywords used are Machine learning OR artificial intelligence OR deep learning AND climate-smart agriculture OR climate resilience AND precision agriculture OR smart agriculture OR digital agriculture AND smallholder farmers or small farms AND developing countries. The search was conducted as one query to provide consistency of results. Peer-reviewed journal articles in English only were used to ensure quality and comparability. Grey literature, book chapters, and conference papers were excluded. The preliminary search provided 114 articles that were used in the next screening process. All records that were retrieved were exported to a reference management system where the screening and selection process was carried out in a systematic manner. The application of inclusion and exclusion decisions was consistent, determined by set criteria to reduce subjectivity and ensure transparency.

2.2 Selection criteria

The selection of the study was performed in a systematic and reproducible way that included four steps: identification, screening, eligibility and inclusion in the PRISMA framework. Figure 2 shows the process of selecting the studies, in accordance with the PRISMA framework. The database search initially provided 114 articles. Once 4 duplicate records had been removed, 110 articles were left in the screening phase. In the screening phase, the titles and abstracts were examined to determine their relevance in AI applications in agriculture. Research articles were filtered out when that had a non-agricultural application of AI, irrelevant technical areas, or no relationship with climate, agriculture, and smallholder farming systems. This phase led to 54 articles being excluded and the remaining 56 articles were subjected to full-text evaluation.

At the eligibility stage, the full-text articles were considered with predefined inclusion and exclusion criteria. AI or machine learning applications in agriculture studies that fulfilled the following criteria were included. Studies need to be relevant to CSA, smart agriculture, or climate resilience, with a direct emphasis on smallholder farming systems or on developing-country contexts. The eligible studies also had to report the findings of the productivity, resilience, sustainability, or adoption. The studies were not included if they focused on large-scale commercial agriculture without implications to smallholders. It was also excluded if they lacked conceptual or empirical connection to AI in agricultural systems, and were not peer-reviewed journal articles. This process led to the exclusion of 21 studies during the eligibility phase and a final sample of 35 articles to be included in the review.

Figure 2. Flow diagram based on PRISMA for selecting studies addressing AI applications in Climate Smart Agriculture

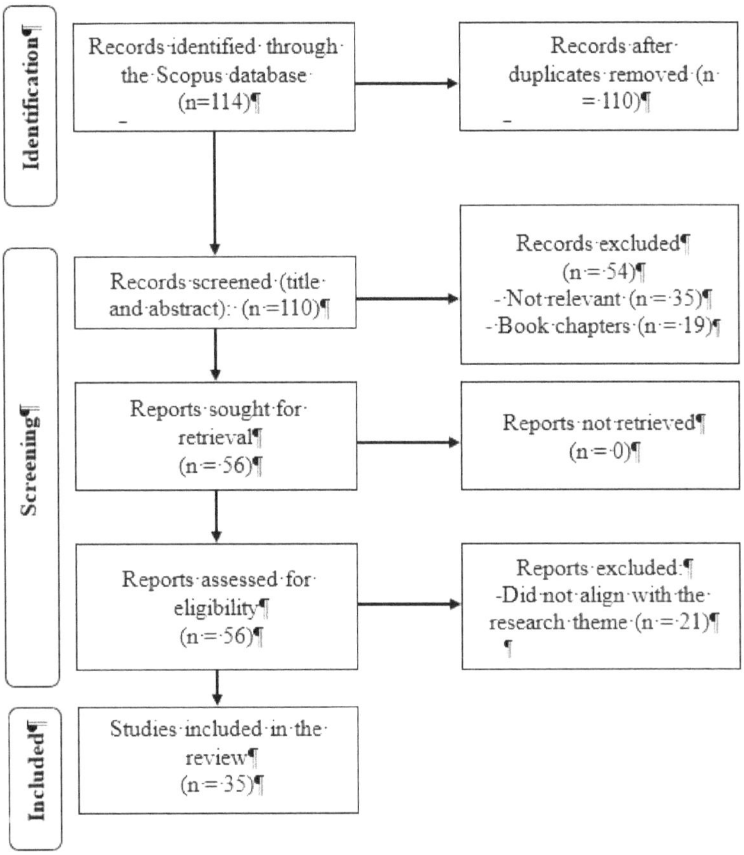

Source: *Adapted from Page MJ et al. The PRISMA 2020 Statement. BMJ 2021.*

The intersection of this research is consistent with the final sample of 35 studies. Similar systematic reviews in newer domains such as digital agriculture and AI-based farming tend to use the corpora of similar size (Erike et al., 2025; Screpnik et al., 2025). This is due to the narrower set of inclusion criteria and the relative novelty of the research topic. This study is not about statistical generalisation, but the depth of the analysis based on thematic synthesis. Here, the narrower yet clearly defined and carefully chosen dataset will be suitable. As it allows a cross-study analysis and defining patterns on the system level that relate technological performance, adoption barriers, and institutional requirements. It is therefore, the conceptual saturation that determines sufficiency of the dataset and not just volume.

2.3 Analysis

Qualitative thematic analysis was employed to analyse, compare and interpret the studies across different research designs and contexts. Instead of using a formal inductive coding scheme, themes were determined through a close reading approach, where the literature was sorted manually. Each of the identified studies was reviewed and analysed based on their research focus, methodology and the outcomes they presented. Key themes were grouped manually, based on recurrent patterns and similarities found within the studies to capture significant areas of intersections in the literature. Themes such as crop prediction and yield optimisation, pest and disease management, integration of AI and IoT, CSA and adoption dynamics, AI to assist policy were included. The creation of themes was done in a cyclical process, and the comparison between the studies is done to narrow the scope and to have internal consistency. Although the process was interpretive rather than code-driven, attempts were made to ensure that there were coherence and reliability by continuously checking findings. It is on the basis of this thematic framework that a comparative and critical analysis of the literature is possible. This goes beyond descriptive summaries to observe larger trends, relationships, and gaps in the research. It helps to comprehend the role of various applications of AI that can support CSA in various fields.

2.4 Limitations

However, in spite of its systematic structure, there are a number of limitations. First, the review was limited to Scopus, peer-reviewed journal articles. This could have resulted in missing out on relevant grey literature, such as policy reports and pilot studies that frequently reflect early-stage developments in AI-based agriculture. Second, differences in study design, datasets, and evaluations metrics are a limitation on direct comparability of studies, and this can impact the consistence of synthesis results. Third, despite the fact that the review concentrates on the smallholder farming systems in the developing countries, there is a lot of contextual difference across regions. Based on this, results cannot be generalisable across contexts. Lastly, findings cannot be empirically validated in real world agricultural environments, due to the lack of primary data collection. However, a synthesis, through the combination of the PRISMA framework with thematic synthesis gives a transparent, structured, and analytically rigorous methodology. This helps in identifying key patterns, gaps, and priorities in AI-enabled CSA research.

3. THEMATIC ANALYSIS

Figure 3. Thematic analysis of AI in Climate Smart Agriculture

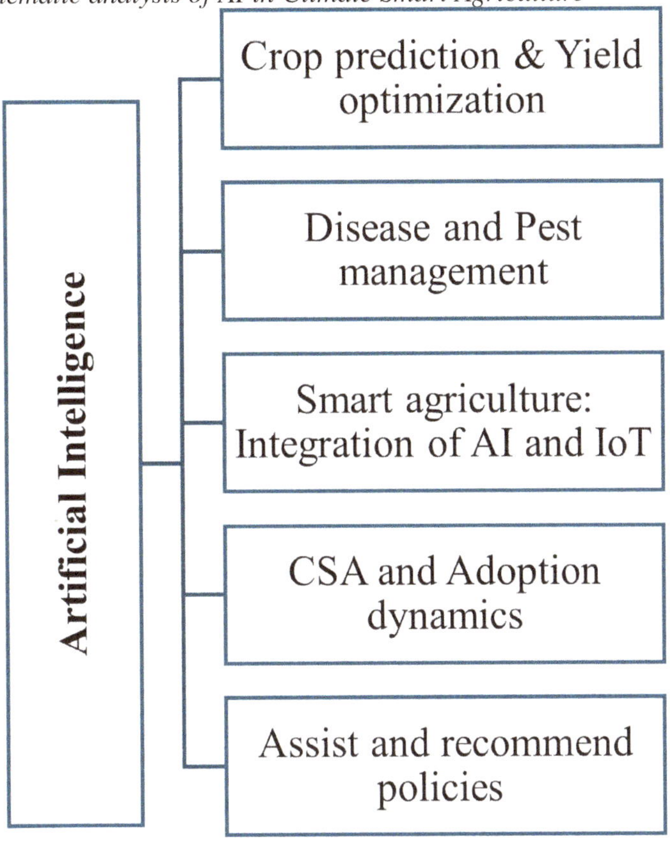

Source: *Authors' own compilation*

The thematic analysis is applied not merely to categorize findings, but to show how various strands of the literature can be linked to a wider CSA context. It is useful in systematic reviews where the studies may vary in methods, technologies, datasets, and contexts, but focus on a common problem. The key question is not whether AI can undertake certain tasks in agriculture, but rather on what circumstances it can contribute to the productivity, resilience, and sustainability of agriculture in the developing countries. Based on the PRISMA framework as described in the methodology, this section generalises the chosen studies into a singular analytical framework. The review does not consider prediction and yield optimisation, pest and disease management, integration of AI and IoT, CSA and adoption dynamics, AI to assist policy as different themes. Instead, it views these themes as a component of a socio-technical system as shown in figure 3.

The main idea is that the usefulness of AI in CSA is not based on the performance of specific models. Instead, it focuses on the ability to integrate models to interact with current data and transform it into practical recommendations. It is also facilitated by the provision of institutional and socio-economic environments. Based on this, AI in agriculture is conceptualised as a multi-layered system that involves 4 connected functions. The functions include data generation, analytical processing, decision translation and socio-economic enablement. IoT and sensing devices produce data; predictive and diagnostic systems convert the data into analysis results. The advisory systems convert results into practical prescriptions; institutions, infrastructure, and capabilities of users decide whether the prescription can be put into action. This view changes the emphasis on the performance of the models, but it also gives an opportunity to determine the most relevant AI applications to CSA in smallholder contexts.

3.1 AI for crop prediction and yield optimization

The oldest field of AI application in agriculture is crop prediction and yield optimisation. In the reviewed studies, it was found that machine learning models are capable of detecting non-linear connections between rainfall, temperature, soils characteristics, and crop performance. These models are able to predict the outcome better than traditional statistical methods (Ferraris et al., 2023; Ramzan et al., 2024). This renders crop forecasting as one of the powerful technical fields within the literature. But the technical performance does not have consistency to be linked to the practical value of decision making. The majority of yield prediction models are created in data-rich or controlled environments and assume that the inputs are stable, production environment is comparably homogeneous, and digital infrastructure is reliable. The assumptions restrict their application to smallholder systems where landholdings are fragmented, the data sets are sparse and the climate conditions are extremely variable. Moreover, other models exclude socio-economic variables like labour supply, input prices, risk aversion by farmers, and access to irrigation. These variables are powerful forces that determine whether the anticipated gains can be realised. This poses a significant constraint to CSA.

Although yield prediction may enhance productivity planning, it does not boost resilience. Even a model which predicts the output well can be of limited use when it fails to guide farmers to act in response to climatic shocks, changing rainfalls or uncertainty in the market. Crop prediction in this context provides only a conditional contribution to CSA. It becomes more valuable as a component of a larger decision-support systems, and not as an analytical instrument on its own. The literature thus indicates that crop prediction is a necessary but not a complete contribution in AI-assisted CSA (Jabed & Murad, 2024). Its usefulness is enhanced with the real-time monitoring, climate-risk analytics, and advisory platforms that turn forecasts into definite management decisions. In the absence of such linkages, predictive systems can only be analytically strong rather than operationally robust. This trend previews a larger observation throughout the review that good model performance does not always become adaptive capacity unless it is a part of an organized system.

3.2 AI in disease and pest management

The second significant area of application is AI-driven disease and pest management systems. The analysed literature indicates that image-based classification systems can effectively diagnose crop diseases and pest damages within a relatively short period of time and with high precision under controlled conditions (Balasubramaniam et al., 2024). They are strong in terms of operational speed, consistency, and can handle big image dataset more effectively than manual diagnosis. But the systems turn out to be reactive than adaptive. Majority of the tools detect symptoms once they occur and fail to predict outbreaks before they cause any damage. This restricts their climate-responsiveness, especially in environments where pest dynamics are shifting with increasing temperatures, rainfall distribution, and uncertainty in their ecological stability. Diagnosis cannot be equated to resilience but it can be achieved through detection alone. The literature also indicates that disease and pest tools are not often well integrated into broader climate or early warning systems (Yang et al., 2026).

In the absence of predictive analytics, weather information and local advisory services, detection technologies provide partial assistance. Farmers are aware of what problem exists, but they do not know how or when to treat, or prevent any problem. Similarly, disease detection systems, have the same structural vulnerability with crop forecasting of high technical ability and low ability to translate into anticipatory decision making. In comparison with crop prediction, disease detection is more directly related to immediate farm-level action since it responds to apparent threats to production. However, it is also narrower and more reactive in comparison to advisory systems. The pest and disease AI is an intermediate component in the agricultural intelligence system. It is useful in the diagnosis, but tactically incomplete, when it is not related to predictive and advisory, and climate-risk capabilities. This confirms the necessity of cross-thematic integration, instead of parallel creation of individual AI instruments.

3.3 Smart agriculture: Integration of AI and IoT

The limitations of predictive and diagnostic systems themselves lead to the significance of the real-time data. In this aspect AI and IoT integration is important. In the studies reviewed, the IoT enabled sensors, remote monitoring devices, and connected field systems are able to provide high-frequency data on soil moisture, crop conditions, and other environmental variables. This allows AI models to operate dynamically and not based on static or past datasets (Dhakal et al., 2026; Nawaz & Babar, 2024). At the system level, the IoT serves as the technical support of the interconnection of conditions in the field with analytical intelligence. It enhances responsiveness of crop prediction models, boosts diseases and pest surveillance, and is helpful in managing inputs more accurately. Ideally, this makes AI and IoT systems very pertinent to the CSA goals of efficient water utilisation, precise fertiliser application, more responsive management in the times of changing climatic conditions.

Nevertheless, the review results in a consistent finding that benefits of AI and IoT systems are offset by significant issues of accessibility. The systems are capital-intensive, maintenance-reliant and frequently require consistent electricity, connectivity, and technical assistance. Consequently, they are far more suitable to commercial or semi-commercial farming systems than low-resource smallholder systems. This reveals a fundamental conflict in the literature. The data-based systems tend to be the least scalable in the environments which are the most vulnerable to climatic risk. This is another theme that enables the connection of the previous sections. Crop prediction models are sensitive to timely information. Due to continuous monitoring, pest and disease systems become more useful as earlier detection is possible. However, both rely on infrastructures that tend to be lacking in the environment most highlighted by CSA scholarship in developing nations. However, AI and IoT integration is technically strong, although socially imbalanced. Its role in CSA is based not only on its accuracy, but also on its ability to address cost, infrastructure, and coordination barriers of deployment models.

3.4 AI in CSA and adoption dynamics

Technologies fail or succeed not on technical basis and the adoption does not happen outside AI systems but it is a part of their efficacy. The social, economic and institutional contexts influence their usefulness. The ability to interpret and act upon AI-generated information is affected by various factors. These factors include infrastructure, affordability, digital literacy, trust, extension support, gender relations and the local organisational capacity. The studies demonstrates that these factors are particularly serious in the smallholder context (Amalan & Aram, 2025; Nxumalo & Chauke, 2025). Farmers might not have smartphones, a connection to the Internet, formal education, or trust towards algorithmic suggestions. Additional barriers to access and use are common amongst women, older farmers as well as marginalised groups. The extension systems can bc weak, and the local institutions do not have the potential to mediate the digital technologies. In this case, even technically successful tools can have a small impact in reality. Due to this reason more complex systems tend to show less performance in practice than those that are simpler.

The tool of advisory created through a mobile platform or enabled by extension systems has a greater chance of being adopted than two complex infrastructures that require capital investment and specialised knowledge. It does not mean that more advanced technologies are not important but rather the access and usability are part of climate-smart effectiveness. A system that cannot be implemented at scale cannot play a significant role in resilience. The issue does not solely lie in the fact that predictive tools require more data or the disease systems require more integration. It is also because the agricultural intelligence system is also filtered with social inequality and institutional weakness. This implies that the socio-economic conditions do not influence the pace of adoption, but transforms the value of all the technological elements in the system. This renders inclusivity as an analytical need and not a policy issue.

3.5 AI to assist and recommend policies

Among the identified themes, advisory systems are the integrative and most practical. As compared to independent prediction or detection systems, AI-powered advisory systems combine various information source. They convert the information sources into context-driven suggestions to farmers, extension services, and in some cases even policymakers (Godara et al., 2024). Their unique value is not the production of raw analytics but conversion of analytics to decisions. This translation capability is essential especially during climate uncertainties. Farmers not only need predictions, but also operational recommendations on what to grow, when to irrigate, how to react to weather variability, and how to handle risk in case the environment changes. Advisory systems are potentially capable of incorporating climate information, soil information, pest warnings and agronomic advice in a single interface or mediated service. By doing so, they bridge the analytical strata of AI to the practicalities of agricultural management.

The advisory systems are more scalable when compared to data intensive infrastructures since advisory systems can use rather low-cost environments like mobile phones, SMS systems, or extension-mediated platforms. This makes them particularly pertinent in developing countries. Simultaneously, their performance is strongly reliant on the quality of data, its representativeness, and trust in institutions. Advisory systems constructed on unrepresentative datasets can give inappropriate or biased recommendations. Failure to understand or trust the system by the farmers can lead to non-compliance of the advice. Even with weak local extension systems, the last phase of the recommendation to action process can fail. With these constraints, there is a particular strength of the advisory systems in the literature since they balance the shortcomings of the other themes. They are capable of not only making predictive tools more actionable, but also disease detection more useful, and partly compensate low technical capacity by translation and mediation. Due to this fact they can be viewed as the best example that illustrates how integration of a system is more important over the performance of isolated models in CSA.

3.6 Cross theme synthesis

The thematic analysis has shown that there is a growing trend. AI in agriculture is growing very fast when it comes to individual tools, but it is much slower when it comes to a system. Predictive models increase the accuracy of forecasting, pest and disease tools assess and diagnose, IoT provides real-time monitoring, and advisory systems turn analysis into an action plan. However, these components tend to grow in parallel but not in coordination. This fragmentation occurs as a result of three cross-cutting gaps. First, analytical systems are weakly connected with anticipatory climate adaptation. Yield prediction and disease diagnostics enhance accuracy. But they do not support uncertainties management by farmers, unless related to early warning systems, scenario-driven thinking, and actionable recommendations. Second, the most advanced technologies are not always available. Systems that are data-intensive require infrastructure, capital and technical capacity that are often lacking in smallholder contexts. Third, the linkages between farm level tools and large-scale institutional systems are not yet well developed. The benefits of AI are limited to localisation without extension support, cooperatives, inclusive governance arrangements or public agencies. One may inquire further to see that these restrictions are not independent of each other, but rather they support one another.

The poor performance of crop prediction among small holders is not only a result of the constraints in data, but also the poor translation of advisories and the institutional mediation. The limited scalability of the IoT-based systems implies the presence of cost obstacles as well as lack of coordinated service models. This implies that the system is incomplete, and failures in one component of the system, including infrastructure or institutional capacity, decrease its overall efficacy. However, the technical constraints in predictive or diagnostic model can be offset by enhancing the integrative aspects, especially advisory systems and institutional intermediaries. This interdependence supports that AI is a contributor to CSA not by individual technological functionality but by the power of relationships among the system components. An intervention based on a single dimension like model accuracy will not deliver substantial results unless it is supported by improvements in the data access, user capacity, and institutional support.

Based on comparative evidence, all applications of AI do not equally contribute to CSA. Productivity is mainly supported by predictive systems, whereas operational diagnostics is offered by disease and pest tools. IoT systems are more precise and responsive, yet difficult to access. Advisory platforms have the greatest contribution since they bridge the gap between analysis and action and can be delivered at low-cost channels. They are not technologically advanced, but rather have practical relevance. One of the main lessons is that there has always been a disconnect between technical feasibility and its practical implementation. Most AI applications can be predicted with great accuracy in controlled conditions, but they fail to be automatically applied to field-scale operations especially with smallholder systems. The infrastructural constraints, affordability, digital literacy and institutional support influence adoption. Thus, impact of technical feasibility is necessary but not sufficient requirement. The success of AI in CSA will depend on the alignment of the technological design with the socio-economic and institutional reality. This gap is to be interpreted as an ongoing aspect of AI-assisted agriculture that justifies the idea that technologies need to be judged not by their performance but by their compatibility.

The dependencies demonstrate that AI in agriculture is not to be viewed as a collection of independent tools. But it is a multi-faceted socio-technical system where the generation of data and data analytics, advisory translation, institutional mediation, and user capability are interdependent. The technical failure in one area tends to be failure in another part of the system, like poor extension support, inadequate infrastructure, or ineffective governance organization. The analytical conclusion is that the climate-smart relevance is not possible to be assessed on an application-by-application basis. This should be evaluated on its ability to combine various parts of AI to facilitate actionable, scalable, and sustainable decision-making in smallholder conditions. In each of these themes, there is a critical contrast to short-term operational benefits and long-term resiliency benefits. Although most AI applications can enhance efficiency, precision, and productivity in the short term, such benefits are not always adaptive capacity in climate uncertainty. In other situations, optimisation-oriented systems can actually make systems more vulnerable as they promote stability, reliance, or standardised decision models. This difference plays a key role in assessing climate-smart relevance by AI and guides the directions facilitated in subsequent sections.

4. THE USES OF AI IN INTELLIGENT FARMING

This section builds on the system level synthesis in Section 3 to explore AI applications as differentiated components of a larger socio-technical agricultural system. It is not what these technologies do, but rather how the technical capacity relates to the structural realities of smallholder farming, and the outcome-based logic of CSA. In this regard, the applicability of AI applications is not assessable in terms of analytical performance. Their climate-sensible value will be based on whether they can be translated into action, implemented in circumstances of resource constraint and maintained with institutional and infrastructural support.

AI applications are categorized into four segments such as advisory systems, data-intensive monitoring, automation, and value-chain AI. These applications are compared in the real-world in regards to precision, scalability, inclusivity, and practical considerations. Such a comparative framework provides a major trend throughout the literature. The technological advancement and climate-smart utility does not always go hand in hand. Most technically advanced systems tend to be the least compatible with the social, economic, and institutional environments that influence agricultural decision-making in developing countries (Wolfert et al., 2017; Liakos et al., 2018; Birner et al., 2021). The problem is not the existence of AI but rather the interaction between the various types of AI and the systems where they are likely to be applied.

4.1 Advisory systems

The advisory systems emerge as the most practically relevant category since their fundamental role is not an analytical, but a translational one. Advisory platforms transform the scattered types of data such as weather signals, crop conditions, market trends, or agronomic expertise into advice that can be implemented by farmers in particular situations. The advice can be delivered via mobile channels or extension-mediated platforms. They are structured in such a way that they do not inherently contradict the infrastructural and economic limits of smallholder agriculture (Qwaid et al., 2026; Liakos et al., 2018). Their significance lies in the mediating role. The strong predictive or diagnostic results do not necessarily produce climate-smart outputs unless they are converted into actionable outcomes. Thus, advisory systems play a central role in the field of AI-enabled CSA due to their ability to bridge the gap between technical intelligence and practice. Their usefulness is reinforced by institutional intermediaries. They include extension systems, cooperatives, local support actors, facilitating the interpretation of recommendations, contextualisation of recommendations to local circumstances, and building trust in their applicability (Birner et al., 2021). This does not turn advisory systems into a mere subset of applications, but an essential mechanism by which the larger AI system is enabled to function.

Advisory tools can tolerate a lower level of precision compared to data-intensive monitoring systems because of its broader coverage. It does not imply a technical trade-off but a structural gain. In environments where connectivity is weak, the amount of capital available is small, and land is fragmented, the ability to provide useful information on a low-cost basis is more decisive. The ability to create highly granular information that cannot be regularly accessed by farmers are less decisive. Likewise, in contrast to automation technologies, advisory systems do not require major restructuring of production systems or high-risk capital expenditure. They can minimize decision uncertainty without creating greater dependence on infrastructure or other forms of external technical control. This puts advisory systems in a climate-smart capacity as the greatest contributors to inclusive adoption. They are effective not merely because they are analytically sophisticated, but because they link information, institutions and farmers in a manner that can be compatible with the realities of smallholder agriculture. Here, the advisory systems illustrate a wider principle throughout the review. Technologies can best serve CSA when they are integrated into readable, interpretable, and institutionally backed processes of decision-making.

4.2 Data-Intensive surveillance

One of the most technologically advanced areas of digital agriculture is the data-intensive monitoring systems such as the IoTs, Unmanned Aerial Vehicle (UAV), and field sensors. They are mostly effective in producing high-frequency and high-resolution data. It has the potential to advance farm management, enhance predictive modelling, and be used in more responsive decision-making processes (Agrawal & Arafat, 2024; Shaikh et al., 2022). Analytically, these systems have a higher precision, timeliness and finer granularity than less advanced advisory models. Nonetheless, there is a recurring pattern of this similarity in the studies reviewed (Shepherd et al., 2018; Ayre et al., 2019). The characteristics contributing to the analytical potential of monitoring systems also contribute to their inability to operate in a smallholder context.

Monitoring technologies rely on good electricity, digital networks, maintenance, and capacity as compared to advisory systems, that run on low-cost mobile device channels. Their technical importance is highly dependent on infrastructure and service ecosystems and are least robust in areas where climate vulnerability is the greatest. Monitoring technologies are better than advisory systems as they enhance data quality, yet accessibility is low. Their advantages are noticed at the information production stage whereas their disadvantages are noticed at the affordability, maintenance and long-term use. Consequently, they tend to work well in pilot projects, controlled environments or semi-commercial farms but far less effectively as inclusive large-scale solutions to smallholder fragmented agriculture. This does not diminish their technical value, but demonstrates that the aspect of being precise is not climate-smart relevant.

Meanwhile, the monitoring systems should not be considered irrelevant. They are more flexible than automation since at times they can be integrated with shared services, institutional platforms, or models supported with extensions instead of complete reshaping of production systems. They are not the best stand-alone solutions but rather enabling elements that has the potential to enhance advisory services, detect earlier risks in production and help to use the resources more efficiently. This cross-sectional evidence indicates the comparative space of data-intensive monitoring between more analytically accurate and less accessible advisory systems on one hand. The more practical but structurally limited automation, on the other. The contribution it makes on climate smart can be considered as technically robust but institutionally conditional. Monitoring systems can only produce significant value when accuracy is converted to actionable decisions by having support systems that are accessible to farmers.

4.3 High capability low relevance automation systems

Automation systems play a significant role due to the fact that they are the future of AI capability in agriculture. The benefits they provide include labour-saving, standardisation, operational efficiency, and in certain situations significant gains in productivity. But in comparison with structural reality of developing-country agriculture, automation turns out to be the least-compatible category. Its prerequisites, such as capital investment, technical specialisation, standardisation of field conditions, and partial amount of land consolidation makes it inaccessible to many smallholder systems (Liakos et al., 2018; Wolfert et al., 2017). The main problem with this is not that automation is normally encountered with the same barriers to ordinary adoption, but rather it is based on types of agricultural organisation. These are fundamentally dissimilar to the type of fragmented and resource-constrained farming systems.

Unlike advisory systems, that operate within the current constraints, and sometimes monitoring technologies, that can be modified by service models. The automation often demands deeper transformation of landholding patterns, investment capacity and technical support systems. This renders it structurally misaligned and not just hard to adopt. This difference is important as it indicates that technology development does not necessarily increase the applicability of AI in CSA. Indeed, automation can depict the reverse. It shows how the innovation agendas shift towards a capital intensive, commercially based approach to agricultural futures that are poorly adapted to the situations in the framework of climate resilience and inclusive development. Automation, in this way, is a limit of the technical skill that AI has, rather than a limit of usefulness of AI in the small-holder agriculture.

Therefore, its features should be viewed critically. Automation is useful as a measure of technological feasibility yet it only adds little to the broadening CSA unless the structural conditions to implement automation are already established. Its purpose in developing-country contexts is not so much as a realistic, but rather as a comparative standard, in which the compatibility of structures is more important than technical sophistication. Automation is the most technically capable and least structurally compatible with smallholder agriculture as compared to both advisory and monitoring systems. This comparison highlights a more general finding of the review. The technological sophistication is not the determinant of climate-smart relevance, but alignment with socio-economic and institutional conditions.

4.4 Value-chain AI

Value-chain AI falls into a different category since it does not work at the level of farm management. But it works at the level of commercial intermediation, coordination across markets, logistics, information exchange, and commercial activities. Apps like price prediction, e-commerce websites, and supply-chain optimisation enable the enhancement of information flows, the minimisation of transaction costs, and market responsiveness (Zhai et al., 2020). In this context, value-chain AI would broaden AI-enabled agriculture beyond production to larger system efficiency. This increased scalability enables value-chain AI to be more scalable compared to other technologies at the farm level. It can spread more easily across dispersed agricultural systems as it does not necessarily rely on physical on-farm infrastructure. However, the synthesis indicates that this scalability is associated with an alternative risk group.

Unlike advisory systems that are largely evaluated based on their capabilities to assist farmers in making decisions. The value-chain systems also have to be evaluated based on their capacity to share control over data, coordination, and economic value. The evidence shows that such systems tend to reinforce the role of platform operators, aggregators and agribusiness actors (Sauvagerd et al., 2024; Fairbairn & Kish, 2023). The farmers continue as consumers within the systems, which they lack control over. Value-chain AI is more viable in the infrastructural constraint conditions compared to monitoring and automation technologies. However, compared to the advisory systems its benefits are not directly correlated with farmer agency.

The increased access to information can decrease uncertainty, but is not guaranteed to change bargaining power, ownership of market information, or the conditions under which farmers engage in digital platforms. This leads to an increase in system efficiency and a retention or even an increase in distributional inequalities. Thus, CSA of value-chain AI is heavily reliant on governance rather than technical functionality. Its advantage is that it could be more coordinated, whereas its hazards are the concentration of value and power. In absence of protection of data utilisation, platform responsibility, and sharing of benefits, such systems can enhance agricultural productivity and strengthen market power asymmetry. This renders value-chain AI a crucial and yet politically sensitive field of AI implementation, in which the inquiry is not just what the technology can effectively enable, but who benefits the most through its growth.

4.5 Synthesis

The comparative analysis indicates that AI applications vary not just in terms of their technical operation but also in the type of agricultural systems. Advisory systems are the most scalable since they are compatible with low-resource contexts and empower decision-making without significant changes to the infrastructural transformations. Monitoring systems will be more precise but the usefulness depends on the support systems that are not well distributed. Automation is the most advanced technical capacity that has the least structural appropriateness to the smallholder realities. Value-chain AI increases the efficiency in the system, but creates more acute concerns regarding governance, value capture, and farmers control. The overall inference is that climate-technology relevance depends more on the compatibility of technology as a framework than on the complexity of technology.

Affordable, interpretable, institutionally mediated, and adaptable to prevailing constraints are more likely to produce meaningful and inclusive results than systems that are optimised to work with analytical or operational performance (Wolfert et al., 2017; Birner et al., 2021). This further supports the existence of the feasibility-adoption gap. AI has a role to play in CSA not when it can maximise technical sophistication, but when it can align itself with socio-economic and institutional situations under which agricultural decisions are formulated. This section demonstrates that the applications of AI in the agro-industry cannot be evaluated with the help of a generalized model of innovation. There are varying trade-offs between the precision, reach, agency and equity in different categories. Thus, the more pertinent question is not about which technology has become most advanced. But what kinds of AI can be integrated into the workings of smallholder systems without causing the growth of dependency or exclusion.

5. OPPORTUNITIES FOR AI ADOPTION (SMART AGRICULTURE) IN DEVELOPING COUNTRIES

Based on the comparative analysis above, this section defines the most significant opportunities of AI in CSA in developing nations. This is not to assume that AI is a homogenous opportunity across systems in agriculture, but aimed at gauging which types of AI are already generating a useable value and which are aspirational, conditional, or disparate. This difference is significant as the literature fails to justify the general account of technological change. Rather, it shows differentiated opportunity structures that are influenced in terms of cost, institutional support, availability of data, and the socio-economic status of the targeted users.

One of the key findings is that the best opportunities lie in applications that lower decision uncertainty and do not demand large amounts of capital, infrastructure or technical reliance. These include mobile advisory system, early warning and weather-based services, less expensive resource optimisation tools, and some digital market platforms. In comparison, the highly automated or sensor-intense systems tend to be limited to pilot projects, semi-commercial farms, or more resource-rich production systems. The opportunities should be seen as a relational concept. AI does not open opportunities in the abstract but under certain structural conditions that define the possibility of transferring technical capability to effective and fair use.

5.1 Smallholder advisory support

The most direct and significant opportunity is on advisory and decision-support system that can be specific to smallholder agriculture. In most developing economies, agricultural production is run by the smallholders. They work under limited capital accumulation, limited access to agronomic services, unreliable input markets and are highly sensitive to climatic shocks (Vijayakumar et al., 2025). In this context, the technologies of the greatest value are those that can be used to make stronger decisions but do not require significant financial or infrastructural change. According to the literature, mobile advisory tools have consistently been one of the most validated scalable types of AI in the smallholder agricultural industry (Cerjak et al., 2025). These tools are significant because they are adapted to the realities of daily agriculture. They are able to assist in making decisions regarding planting, irrigation, responding to pests, and timing in the market which is already more accessible as compared to capital intensive alternatives.

Smallholders use less infrastructure as compared to sensor-heavy systems. The tools are less risky in financial terms than full automation and are not based on land consolidation or uniform production settings. They are also effective when combined with extension systems and local institutional networks that boost trust, understanding and adoption. The significance of this opportunity is not only that advisory systems are simpler to implement, but that it represents a more realistic approach to AI-enabled agriculture in the smallholder context. They do not remove the deeper structural constraints, but are much more compatible with them. Advisory systems cannot be regarded as a second-best alternative to more sophisticated AI, but it is a major channel in which AI will be able to deliver significant CSA outcomes in the scale.

5.2 Climate resilience opportunity

The second big opportunity is in climate-risk management. The farmers in developing nations become more vulnerable to unpredictable rainfall, floods, droughts, heat stress, and changing pest pressures. The AI applications that facilitate anticipation and prompt response become particularly important to CSA. These systems are not just important because they allow prediction. They are important because they will enhance adaptive decision making under uncertainty. The weather forecasting and early warning systems provided by mobile or advisory channels are the most developed applications in this field. They are more established than highly localised climate simulation models, as they can run on a larger scale and still assist in meaningful preventive action. Relative to generic crop prediction tools, early warning systems have a more direct role in resilience. They assist farmers in taking action prior to the occurrence of damage by altering the input use, planting timelines, irrigation timing, harvesting choices, or risk-management behaviours.

Meanwhile, the literature hints that resilience opportunity is not even (Mehrabi et al., 2020). More sophisticated uses, like integrated scenario modelling or climate-adaptive crop planning, are still promising but less put into practice. Their effectiveness depends on the quality of data, calibration in all types of agro-ecological conditions, the institutional backdrop, and how purely the farmers believe in themselves and comprehend the advice they are being provided. Thus, the opportunity does not depend merely on the predictive sophistication but the ability to collapse climate uncertainty into useful advice. This difference is important as it helps to understand the type of resilience AI can and will be able to assist. Perfect forecasting is a challenging task in volatile situations, but making flexible, anticipatory decisions stronger is a better opportunity. AI is most helpful within CSA as it decreases uncertainty by enhancing adaptive capacity instead of augmenting technical prediction accuracy.

5.3 Optimal use of resources

Optimal use of resources are also significant opportunities in AI to enhance water use, fertiliser application, and soil management. Some of the areas where

AI can decrease waste without compromising or expanding productivity include irrigation scheduling, nutrient recommendation systems and soil monitoring tools. These systems tend to yield quantifiable efficiency benefits in controlled settings, as well as in semi-commercial facilities. However, the most important question is not whether optimisation can be technically implemented or not, but under which circumstances it can be used. In the literature, there lies a distinct difference between high-precision optimisation systems and more convenient types of incremental efficiency support (Saiz-Rubio & Rovira-Más, 2020). Models that are sensor-intensive and automation-oriented can be more powerful in terms of technical characteristics, yet demand more infrastructure, maintenance and investment. Simple systems, on the other hand, can be realistic at smallholder conditions and tend to give smaller efficiency gains. Figure 4 shows the relative efficiency gain in the irrigation system using AI-based systems compared to traditional irrigation methods. The data-based irrigation system, which relies on real-time soil moisture and weather patterns, can save water by a large percentage without negatively or positively influencing the crop yield. But these efficiency gains are frequently observed in controlled or semi-commercial settings. They are not applicable to smallholder farming systems without proper infrastructure, cost effectiveness, and institutional implications (Akanbi et al., 2025).

Figure 4. Irrigation Efficiency Comparison (Akanbi et al., 2025)

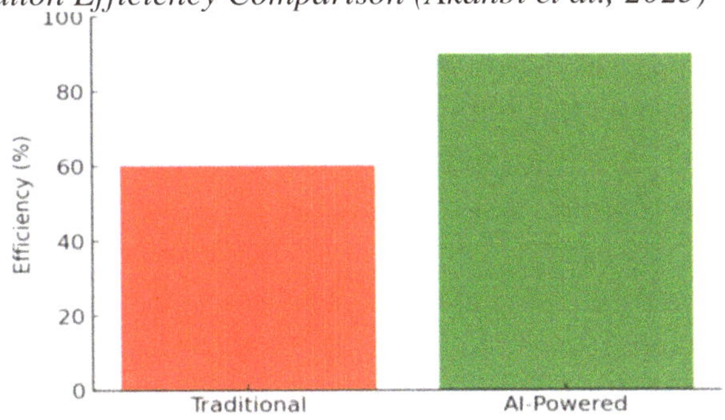

Incremental optimisation presents the best opportunity as compared to full automation in the case of smallholders. Combined solutions like common irrigation, streamlined digital recommendations, community-based water management or decision-support tools are less expensive. They are more in line with the realities of fragmented landholding and reduced investment potential. Their strength is not that they can maximise their precision but enable efficiency improvements without re-creating the costs and dependencies of high-tech systems. This comparison is significant, as it refocuses the emphasis on the requirements of technological maximalism to the availability of impact. The key opportunity is not to turn agriculture into data-intensive but to create small, scalable, and farmer-accessible resource efficiencies. Resource-use optimisation supports a broader conclusion. The climate-smart value is optimal not where AI is most advanced, but where efficiency gains may be realised in the face of real circumstances of institutional and economic scarcity.

5.4 Financial inclusion

Another area that AI can be used productively is financial inclusion. AI-based credit rating, agricultural insurance, and electronic financial services could facilitate the mitigation of traditional obstacles to rural finance and application using alternative data. This includes transaction histories, production records, or remote sensing indexes. These instruments have the potential to lower administration expenses, enhance risk evaluation, and increase access to fund to farmers without formal collateral or conventional credit scores. However, according to the literature the opportunities cannot be separated by deeper governance risks (Birhane, 2021; Frost et al., 2019). In comparison with the conventional financial system, AI-driven models can expand the number of people receiving services. They can also categorize and differentiate farmers in a way that are difficult to challenge.

Individuals with weaker digital presence, such as women, poorer households or socially marginalised groups, will be having higher riskier scores. This is not due to their production capacity, but data environments, which they are depicted in. Therefore, new kinds of exclusion are enabled by opaque algorithmic systems in under the guise of technical neutrality. The opportunity in this case is conditional as opposed to being direct. AI can increase inclusivity, although the systems in which it is implemented is transparent, accountable and vulnerable to mechanisms that prevent biased decision-making. Otherwise, the inclusion can turn into a selective type of access that redistributes financial services unevenly and increases data dependency and farmer vulnerability. Therefore, financial AI is a field where the gains of innovation can be seen, only when governance and fairness are considered as one of the primary design principles and not as a secondary issue.

5.5 Market accessibility

Another significant opportunity provided by AI-enabled marketplaces and market intelligence systems is by combining fragmented and opaque markets in developing-country agriculture. This has remained a major constraint in the agricultural sector. The price information, demand projection, and buyer network tools could mitigate the uncertainty of transactions, enhance coordination, and further strengthen the involvement of farmers in the market. Such systems can generate short-term practical value in the contexts where there is a limited bargaining capacity due to intermediary dominance and price transparency. These are some of the successful real-world applications of AI-enabled agriculture as they address a visible coordination issue without over-relying on infrastructure on the farm. They are more easily diffused and can introduce wider system efficiencies compared to high-capital production technologies.

Nevertheless, it is also evident in the literature that better information does not necessarily lead to better welfare of farmers (Mossie, 2022). Platform operators, data aggregators, and other larger players in agribusiness are likely to take a larger share of the value created. The farmers do not have so much control over the pricing mechanisms, the contract terms, or the data that is created during the use of platforms. Thus, market AI is both a real and unequal opportunity. Such systems have the potential to enhance information accessibility and minimize some types of uncertainty. But controlling the platform architecture, data utilization, and the institutional environment can determine whether such information access can be transformed into bargaining power by farmers. This depicts a larger trend throughout the review. The opportunities generated by AI are not just determined by technical functionality but by regulation and allocation.

5.6 Synthesis of opportunities

The cross-domain comparison gives a consistent conclusion. The most valuable uses of AI in agriculture in developing countries are those that are already in action, and where the conditions are already constrained. The mobile advisory systems, early warning tools, less expensive resource optimisation models and certain market platforms are not technologically advanced. But they are the most structurally adapted to the realities of the smallholders. In comparison, the capital-heavy automation, highly sensorised infrastructures and advanced simulation-directed systems are significantly lower in terms of inclusive adoption. In most cases their promise is not a falsehood but it is exaggerated when spoken in isolation of cost, institutional weaknesses, system of service and skewed infrastructure.

Thus, the literature substantiates a differentiated over a universal view of opportunity (Carrer et al., 2022). Low-cost, decision-oriented and institutionally mediated opportunities are proven. Aspirational opportunities are more data-driven, infrastructure-based and business-oriented. These opportunities are important but not separate areas of innovation. They are influenced by the same structural conditions that restrict adoption. Infrastructure, institutional capacity, access to data and governance arrangements play a role as both the facilitators and filters of the realisation. This implies that AI-enabled agriculture opportunity is non-uniform. The technologies that have most significant potential effects tend to be those which minimize the uncertainty of decisions without adding dependency, complexity, or exclusion. This gives the necessary transition to the next section. The opportunities are not enough but the systemic barriers to its further realisation have to be addressed.

6. CHALLENGES AND BARRIERS TO AI ADOPTION (SMART AGRICULTURE) IN DEVELOPING COUNTRIES

Expanding upon the opportunities that have been discussed in the previous section, this section investigates the institutional challenges that inhibit the implementation of AI in CSA. Although some AI applications have explicit technical and practical potential. They are limited by interdependent infrastructural, economic, institutional, and socio-cultural determinants. These barriers are not only important to limit access to technology but also influence the understanding of AI, its adoption, and to whom the use of AI ultimately benefits. The barriers cannot be viewed as the only standalone obstacles as they create a network where limitations in one area will increase the limitation on others. Poor infrastructure leads to higher costs of use, poor institutions diminish trust and capacity to translate, low digital literacy exacerbates risks of exclusion, poor governance contributes to uneven control of data and economic value.

The outcome is that there has been a persistent gap between technical feasibility and the adoption at the field level, especially when it comes to the smallholder farming systems. Even though AI has the potential to enhance productivity, resource usage, and climate resilience, its adoption in developing countries is slow and disproportionate, especially among smallholder systems. It is not merely a matter of access, but also of disparate sharing of the benefits and restricted inclusiveness. As farmers produce the necessary data and are involved in implementation, Agri-tech companies, the platform providers, and inputs suppliers often absorb a disproportionate portion of economic value. This can lead to new power imbalances (Shuliang et al., 2025). Thus, these challenges cannot be solved by diffusion of technology alone. It involves focusing on the institutional and political-economic circumstances by which AI can be made accessible, controllable, and fair.

6.1 Farmer's perception

The perception of the farmer has been a key determinant of adoption since it acts as an intermediating variable between technical systems and application. The notion of perception is usually perceived in simple terms of awareness or willingness; however, the literature has shown that it as a multifaceted situation (Ricart et al., 2023). The farmers who have been exposed to AI technologies previously are more likely to use it. This tendency depends not only on the information but also on the trust, usability, the cultural suitability, and perceived economical value (Shuliang et al., 2025). Adoption is filtered by judgment regarding the understandability, credibility, and riskiness of a technology. In developing countries such judgements are highly mediated by disparate access to digital literacy and training. Women, older farmers, and marginalised populations tend to experience more obstacles in using AI-based tools. This enhances the risk that technological innovation will become socially selective instead of being widespread in empowerment. In that case, the reluctance to AI cannot be viewed merely as a form of resistance to innovation. It tends to be a logical reaction to uncertainty, unclear decision processes, lack of clear payoffs, risk of disruption of livelihood. This removes the perception as a problem of individual behaviour to socio-technical.

Confidence in AI does not arise out of awareness, but through the existence of institutions and mechanisms of interaction through which farmers can test, question and learn new systems. Thus, the importance of sensitisation programmes, demonstration farms, mobile tutorials, peer learning and participatory experiments is not due to the persuasive effect they have on farmers. It is due to the minimisation of uncertainty and making technologies more comprehensible in actual farming settings. The process can be further enhanced by implementing AI literacy into extension programs and agricultural education, which introduces digital learning into the established institutional frameworks. The perception of farmers indicates that adoption is not a factor of technical supply alone. It depends on whether AI gets into agricultural systems via mediation forms that render it socially plausible and practically significant. Thus, perception is not a peripheral factor in adoption, it is part of the processes by which feasibility is converted or not converted into use.

6.2 Infrastructural availability

The outer limit of feasibility of AI adoption is influenced by infrastructural constraints. The lack of access to electricity, internet connectivity, digital devices, education, and interoperable systems restricts the ability of farmers to adopt technologies of smart agriculture (Mallinger et al., 2024). In rural areas infrastructure is not a background phenomenon but a filtering mechanism that excludes which types of AI can be implemented outside a demonstration environment. This barrier is not a matter of access. Poor infrastructure further complicates the operations by making it more reliant on trained operators, technical troubleshooting and maintenance of the systems. Consequently, despite the existence of AI tools, they might still be unusable or unstable enough to be adopted in the long term. The absence of interoperability complicates the situation further by establishing fragmented digital landscapes where technologies fail to integrate effectively. This results in compromising trust, as well as the increment in switching costs. It is thus argued in the literature that infrastructure can be considered as both material and systemic (Rotz et al., 2019).

Good internet and electricity connections are important. Along with low-bandwidth system design, local language interfaces, simplified user environments, and technical support structures to enable farmers to use the tools without having specialised knowledge. Training programs organized by the government and non-governmental organizations can alleviate some of these pressures. But infrastructural development is not enough until it is supplemented by institutional frameworks that maintain their daily usability. Community-based AI service centres are significant as they fill a gap that cannot be addressed alone by infrastructure. It is the gap between access to devices and the ability to make them operational in practice (Mallinger et al., 2024). This causes infrastructure to be both a structural, yet comparative, obstacle. It not only defines whether AI can be operationalised; it also defines the types of AI that are prioritised, the individuals who can interact with them, and the stability of technical systems to produce long-term value.

6.3 High cost

One of the most apparent and long-term obstacles to the adoption of AI in developing-country agriculture is economic cost. The high cost of hardware, sensors, UAVs, recurrent costs associated with maintenance, upgrades, and technical training make many AI systems unaffordable to smallholder farmers (Agrawal & Arafat, 2024). To farmers with small profit margins and small amounts of savings, it is not just a matter of whether a technology functions, but whether the risk of investing in a technology can be rationalized at varying returns. One of the studies reveals that there are multiple levels through which cost works (Bahmutsky et al., 2024; Munz et al., 2024). One is the initial expense of equipping them, the recurrent expense of maintaining and upgrading, and there is the indirect expense of depending on specialised expertise or external services. These strata are significant as they explain why affordability cannot be narrowed down to price. Subscription-based or subsidised systems can continue to be a burden even when they create a dependency on proprietary systems or when their long-term cost-effectiveness is uncertain.

Another important matter in this aspect is that of temporal asymmetry. Expenses are direct, measurable, and observable, but the benefits are uncertain and delayed on other conditions like infrastructure, climatic situation and institutional goodwill. This imbalance can assist in understanding why not all the technologies can be diffused. Farmers are being requested to take on current risk to get future returns that cannot be assured. This is particularly an issue in highly risky farming systems, where a failed investment can directly impact livelihoods. The risk of the technological lock-in is also pointed out in the literature (Hackfort, 2023). By relying on proprietary platforms to access their data, receive recommendations, or maintain service continuity, farmers can lose their autonomy and be less successful in bargaining. To this extent, even high cost is an entry barrier and a mechanism by which dependency can be re-created once adopted. Subsidies, microfinance, and subscription can enhance affordability, without solving the problem. They can only postpone but not remove the economic risk that comes with adoption unless they are created with safeguards against exploitative pricing and reliance on service providers (Agrawal & Arafat, 2024).

6.4 Small holders

The structural obstacles to smallholder farmers go beyond costs or infrastructural inequalities. Numerous technologies in AI are designed with large-scale commercial farming in mind, rather than the fragmented, heterogeneous, and resource-constrained environment of smallholder agriculture (Vijayakumar et al., 2025). This poses a greater issue of mismatch in design. Technologies are sometimes technically transferable in theory but strategically misaligned in reality. The adoption of standardised AI solutions is complicated by limited capital, poor institutional support, fragmented landholdings and varied agro-ecological conditions. The literature identifies various mechanisms. These include microfinance, cooperative ownership, modular tools, low-cost sensor, open-source software, and mobile advisory platform as pathways to bring AI into the world of smallholder realities (Sreeram et al., 2025; Poenaru et al., 2025). Such mechanisms are important as they minimize barriers by spreading costs and lowering access thresholds.

Nevertheless, the study also points out that adaptation alone is not sufficient (Barroso-Barroso et al., 2025). Most of these solutions find it difficult to scale since they rely on collective action, coordination of services, or institutional mediation, which are weak in most rural contexts. This implies that it is not only necessary to develop smaller or less expensive technologies, but to create surrounding systems which can support their use. In its absence, it will be localised and difficult to maintain. The problem of data asymmetry is a more serious concern that arises out of the literature (Chamorro-Padial et al., 2024). When the smallholders engage in digital systems they can produce valuable agricultural data, but tend to lack control over how these data are stored, used, or monetised. This distributes the economic benefit of participation more toward technology providers and less toward farmers and contributes to sustaining inequity instead of conquering it (Vijayakumar et al., 2025). The barrier here is not under-inclusion but exploitative inclusion. Farmers can be integrated into AI systems on conditions that fail to improve their agency of control of value. With this, the smallholder question becomes the main focus of the review. It is not merely the category of users in contrast to others, but the pivotal point where the conflict between the technical promise and the fair relevance is most apparent.

6.5 Cultural perceptions

In the transition to AI, cultural values and traditional farming practices are important elements that can be underestimated in technological-oriented discussions. In most developing-country contexts, lived experience, community norms, and intergenerational learning are the sources of farming knowledge. In such circumstances, the implementation of AI does not merely result in a new instrument. It can also interfere with the conventional methods of risk perception, power, and decision making. The literature indicates that resistance to adopt AI in these situations is often rational and not conservative (Hunecke et al., 2017). In areas where farmers do not have locally relevant success stories or evidence of long-term reliability, it is usually logical to maintain the status quo, which will minimize livelihood risk. This is particularly evident when the recommendations of AI are perceived to be abstract, externally imposed, or not adequately rooted in local agro-ecological realities.

This implies that cultural opposition cannot be viewed as a separate issue independent of how technology is designed. It represents a relative lack of balance between the implementation of technologies and the establishment of trust within the agricultural community. Participatory methods, co-design processes and long-term engagement are significant. It is not because it enhances acceptance, but because it changes the conditions, under which technology is analysed. They enable the farmers to compare new systems with local knowledge, check on the limitations and evaluate whether the benefits they promise can be trusted. Clear communication of the advantages and disadvantages is particularly vital in this case and exaggerating the capabilities of AI can negatively affect trust. This way, cultural perception will bring a new aspect to the feasibility-adoption gap. The technologies that are seen as technically promising will not be socially compelling until they are mediated into culturally relevant forms of practice.

6.6 Ethics, data governance and inclusion

Ethics and data governance are not incidental issues but are predominant factors that determine the path of AI adoption as either inclusive or extractive. In developing-country agriculture, the issues of data ownership, privacy, algorithmic biases, and accountability are of particular concern. This is because the regulatory frameworks tend to be weak, and farmers have limited ability to challenge the opaque decisions. Such risks are realised in various forms. The credit scoring systems developed through AI based on digital histories of transactions, mobile activities, or remote-sensed metrics can profitably discriminate against farmers who have a smaller digital footprint. Algorithms can also fail to recognize risk in a highly heterogeneous or data-sparse settings. Therefore, it can exclude price inappropriately, or deny assistance to groups that are already vulnerable. These types of bias are not always deliberate but are structurally generated using uneven representations of data and assumptions in models. In the absence of transparency and contestability mechanisms, exclusion can be replicated in a way that is perceived to be technical in nature. The issue of governance goes beyond bias to the issue of ownership and value capture. Farmers are often not sure how their data are being collected, stored, shared and monetised. This would present a scenario where involvement in AI systems could bring value to others without actual control or advantage to the data creators.

The outcome is the creation of digital asymmetry where farmers are users and providers of raw data but decisions and monetisation are vested in other places. There is a trend in the literature that indicates concrete response to governance (Haque et al., 2022). The importance of explainable AI models, independent audit systems, transparency requirements in use of data, and regulatory oversight is that they establish mechanisms by which technical systems can be audited as opposed to mere trust. Farmer data cooperatives and platform cooperatives advance this by trying to reallocate the power and control of digital infrastructure and shared governance models. Public/hybrid data trusts also provide means of harmonizing data governance with public interests instead of merely private extraction. The use of AI is also a political-economic process. It is not merely a question of the functioning of systems, but rather how the systems bring together rights, responsibilities and value. In the absence of governance systems that render accountability and farmer agency, AI-based inclusion can be biased, partial, or exploitative. Thus, ethical design is not an optional feature that is implemented after innovation. It constitutes a component of institutional architecture that AI needs so that it can be used to add value to CSA.

6.7 Bridging the gap between feasibility and adoption

The gap between technological possibilities and the actualisation is not just a barrier amongst others, it is the collective consequence of all the barriers mentioned above. Most AI systems work well when operated in controlled environments and pilot projects but fail to scale to field level applications and it is not because farmers are not willing to be innovative. It is because the technical feasibility is insufficient to address the infrastructural, economic, institutional, cultural, and governance set of conditions necessary to sustain the adoption period. The failure to scale pilot projects is a most noticeable gap. Inadequate long-term funding, poor institutional coordination, inadequate indulgence with the current farming practices, and absence of trust are the reasons that the promising technologies are either local or transitory.

The literature presents that the issue of adoption is in fact an issue of alignment of the system (Daum et al., 2025). The success of technologies does not occur through innovation, but when the ecology surrounding them enable them to be usable and sustainable. To overcome this gap, a change in perspective is needed. Instead of asking how to diffuse more AI, the more pertinent question is how to create and sustain AI systems that are self-sustaining in the local realities. It involves intergovernmental, interagency, intercompany and agricultural community cooperation. The better focus on context specificity, institutional preparedness, and inclusive design is needed (Shuliang et al., 2025; Mallinger et al., 2024; Agrawal & Arafat, 2024; Vijayakumar et al., 2025).

6.8 Synthesis of barriers

The study outlines three fundamental failures of the systems as a whole that account for the gap between the potential of technology and real-life implementations. First, AI outputs were poorly translated to actionable decisions at a farm level and could not be used in practice. Second, infrastructure deficits, high costs, and affordability barriers limit access to AI technologies. Third, the inequalities that are associated with governance, especially in terms of data ownership, platform dependency, and value capture, influence the beneficiaries of AI adoption. These intertwined constraints provide the analytical foundation of the pathways in the subsequent section. In the studies reviewed, the benefits of use of AI are generally unevenly distributed, with agri-tech companies and platform operators. The providers of inputs are enjoying greater benefits than farmers, especially when ownership of data and governance of the platform is centralized.

This distributional disparity strengthens the current power disparities and minimizes the scope of AI services to inclusive agricultural growth. This gap in the potential of technology and its actual implementation must be perceived as structural, and not transitional. The absence of AI in agriculture is not only limited by the inaccessibility, but also by the overall organisation of infrastructure, institutions, incentives, and rights. To fill this gap, it needs directional evidence-based pathways that are responsive to the identified system failures within this section. The barriers do not exist independently but include the circumstances in which AI can possibly operate. The next section translates these structural limitations into specific avenues towards inclusive adoption.

7. PATHWAYS FOR INCLUSIVE AND SUSTAINABLE AI ADOPTION IN CSA

The challenges highlighted reveal that the low development of AI in CSA is not merely a technological challenge, but the result of interrelated infrastructural, financial, institutional, and governance barriers. In this study, three common patterns can be identified as to why technically viable AI systems tend to fail to move into practice in the case of smallholder contexts. There is a gap between the outputs of analysis and actionable decision making because of poor advisory translation and ineffective institutional mediation. Second, there are barriers of access to data-intensive and capital-intensive systems in terms of infrastructural constraints and affordability. Third, in terms of governance, issues such as data asymmetry, dependency on platforms, and value disproportionately affect those who enjoy the benefits of AI adoption. The directions taken in this part are the result of these patterns.

Instead of broad policy guidelines, the pathways are designed to address the limitations that have been found in the literature. Each of the pathways deals with one of the system-level failures such as translation, access, or governance. But they all contribute to the alignment of technological capability, socio-economic and institutional conditions. One of the key assumptions is that climate-smart effectiveness is not about technological advancement, but compatibility of systems. In this regard, it is necessary to understand the difference between the high-technology and capital-intensive AI solutions. They are mostly applicable to commercial agriculture and low-cost scalable systems and are better placed in the context of smallholder environments. Thus, the pathways are more focused on accessibility, adaptability, and institutional integration as opposed to technological maximalism.

7.1 Basic infrastructure and availability of technology

One of the key requirements that appear throughout the studies reviewed is access to underlying digital infrastructure (Frija, 2024; Kumar et al., 2024). Inadequate reliability of electricity, low connectivity, and access to digital devices are major limitations to adoption. These limitations have a direct impact on the usability of data intensive AI systems and support the feasibility-adoption gap. The infrastructure constraint remains a central factor limiting the applications of data-intensive AI systems. Thus, inclusive AI adoption not only enhances technological innovation but also access conditions. These involve spending on rural connection, decentralised energy connection, community shared digital networks and inexpensive devices. The studies reveals that one of the most predictable factors in non-adoption among smallholders is related to infrastructure gaps (Dixit et al., 2023; Choruma et al., 2024). Nonetheless, the growth of infrastructure is not sufficient in itself unless it is accompanied by technology design encompassing local limitations. Based on this low-cost, low-bandwidth, and mobile-based systems should be prioritized as they can operate with a limited connectivity. This can be in the form of SMS-based advisory platforms, offline-enabled applications, and easier user interfaces. The advisory systems were found to be more scalable due to its minimum infrastructure requirements specifically.

7.2 User-centred design and capacity building

The studies indicate that the adoption of AI devices is not solely dependent on the access to technology, but on how the users can interpret and respond to the output provided by AI (Rajbongshi et al., 2026; Mohan et al., 2025).The low digital literacy, mistrust towards the algorithmic systems, and a lack of training minimize its effective application, especially among women, older farmers, and disadvantaged populations. Digital literacy and trust are not peripheral, but core factors to determine whether AI-generated output is converted to real farm-level decisions. Meanwhile, the thematic synthesis shows that even the effective analytical systems cannot be of much use until their results are converted into usable knowledge. This implies that user capability forms a part of overall effectiveness. This pathway will focus on capacity building through training programmes, digital literacy, and participatory extension strategies. There is consistent evidence in the literature that adoption becomes better when users are engaged in technology design and implementation processes (Lan & Ban, 2025). Integrating AI tools with farmers enhances relevance, usability, and trust. Instead of considering AI as an external intervention, it is now framed as a co-produced decision-support system and integrated into local knowledge systems. This is in accordance with evidence that adoption is enhanced when technologies are tailored to user contexts and facilitated by institutional learning processes (Abdul-Majid et al., 2024; Santos et al., 2025).

7.3 Institutional readiness and intermediary systems

One of the most significant findings among the studies is that AI systems can hardly work alone as technologies (Dhanaraju et al., 2022; Duguma & Bai, 2024). The advisory systems as well as the institutional mediators are essential in the translation of the analytical work into the actionable decisions. The technically sophisticated tools have little practical effect when these systems are weak or not present. The poor institutional capacity, especially in extension services and local organisations, is a significant limiting factor to adoption. This supports the finding that institutional preparedness can serve as a scaling driver of AI in agriculture (Araújo et al., 2021). The pathway focuses on empowering intermediate systems, such as extension services, cooperatives, producer organisations and local non-governmental organizations. These actors are very crucial in decoding AI outputs, contextualising recommendations and assisting farmers in decision-making. The institutional intermediaries execute various essential roles that define the usability of AI systems in practice. These consist of converting complex analytical outputs to local recommendations, aggregating farmer demand to facilitate access to digital services, enabling human contact to build trust, and helping iterative learning through feedback. These actors make AI an operational system by helping to bridge the gap between aggregated outputs of algorithms and the decisions made at the farm level. Various studies have revealed that institutional mediation is among the best predictors of effective adoption and continued use (Osrof et al., 2023; Ayim et al., 2022). This supports the thematic result that advisory translation and institutional mediation are the key gap between analytical potential and practical effects.

7.4 Governance, data systems and equity

The governance and data-related concerns are the focal points of the long-term implications of AI adoption. Farmers are likely to produce valuable data without having control over its use, which creates data asymmetry, dependency on the platform and unequal value appropriation. These governance challenges are consistently linked with unequal value capture and insufficient control of data by farmers. These issues suggest that the implementation of AI is not just a technical change, but also a political-economic one. The studies indicate that value capture is often limited to technology providers and platform developers, but not to farmers (Duncan et al., 2026; Ciliberti et al., 2024). This pathway revolves around the implementation of clear and open data governance systems. The main aspects are the farmer-centric data ownership models, interoperability standards, equitable contractual agreements, regulatory oversight to promote accountability and equitable distribution of benefits. New models, including data cooperatives, platform cooperatives, and public data trusts, provide potential ways to redistribute control of agricultural data. These strategies directly address the governance gaps and have been more closely highlighted in recent CSA and digital agriculture literature (Romera et al., 2024).

7.5 Resilience and adaptive AI systems

One of the primary differences that can be drawn based on the thematic synthesis is between short-term operational efficiency and long-term climate resilience. Most AI applications enhance efficiency by optimising irrigation or pest detection. However, this increase in efficiency will not directly lead to an increase in adaptive capacity in climate uncertainty. Predictive models depend on past data and are concerned with relatively stable environments. This poses a structural constraint. Machine learning systems that are trained on historical rainfall pattern, temperature range, or crop response will make inaccurate suggestions when the climatic conditions change beyond the past pattern. AI systems should progress beyond fixed prediction, to adaptive and probabilistic decision-support, in order to be effective under high climate uncertainty. These consist of scenario-based recommendations, probabilistic weather forecasts, and dynamic advisory systems, which can update guidance to the changing conditions. These methods enable farmers to provide flexible responses to variability instead of fixed optimisation plans. This direction, therefore, prioritises the advancement of AI-based systems that facilitate adaptable decision-making over fixed optimisation. Here, the idea of climate-smart value is not to maximise predictive accuracy in a stable environment, but rather the ability to respond flexibly in risk-sensitive ways under uncertainty.

7.6 Context-specific and scalable solutions

The mismatch between most AI applications and the realities of smallholder farms seems to be a structural issue. The high cost, infrastructure-based solutions frequently do not match fragmented landholdings, limited capital and the diverse agro-ecological environments. This approach underscores the necessity of place-specific and scalable solutions that can respond to local socio-economic and institutional factors. These encompass modular technologies, shared service designs, cooperative based access designs, and low-cost digital tools that are capable of working under constraint conditions. Instead of seeking technological homogeneity, successful use of AI should be based on diversity in the agricultural contexts. This indicates that compatibility with local systems as opposed to technological sophistication is the most important factor in determining scalable and inclusive outcomes. Figure 5 presents an integrative framework that conceptualises AI as a multi-layered socio-technical system involving data, analytics, decision-making, and institutional support to climate-smart outcomes.

Figure 5. Integrative framework for sustainable agriculture

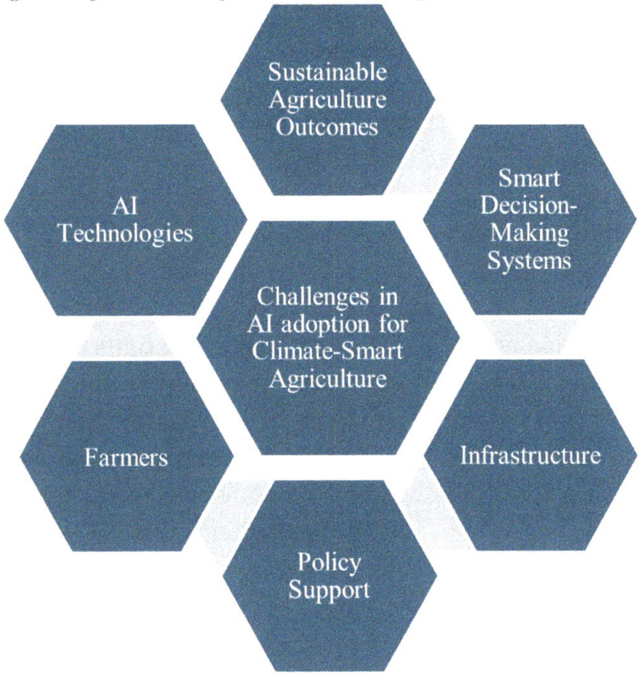

Source: *Authors' own compilation*

Instead of viewing AI applications as independent instruments, the framework emphasizes the role of effectiveness in relationship among five mutually reliant elements. The most basic level is data generation which comprises field-level data like climate data, soil data, crop health data, market data and data submitted by farmers. These are gathered by sensors, remote sensing technology, or by digital action. These inputs are then fed into analytical processing in which AI and machine learning models transform raw data to predictive, diagnostic or optimisation outputs. The outputs only have a practical value when they are translated into decisions. The third element is decision translation which describes the task of advisory systems and user interfaces. It translates analytical results into actionable, situation-specific recommendations to farmers and intermediaries. This layer plays a vital role in eliminating the gap between the technical capability and real usability. The fourth element, institutional mediation, indicates the contribution of extension services, cooperatives, public agencies and local organisations towards mediation. These actors decode AI readings, create trust, contextualize recommendations, and enable implementation processes by providing training and feedback. The power of this layer is that it determines how successfully AI systems work in real-world scenarios. Lastly, the framework concludes with climate-smart outcomes such as productivity, efficiency, climate resilience and sustainability. These outputs, however, are not the direct outputs of AI systems they are the outcomes of convergence of all the previous elements.

The model also points out the two cross-cutting dimensions. First, governance systems and data systems mediate the ways data are owned, shared and monetised that determine equity, transparency and value distribution issues. Second, socio-economic and infrastructural factors, such as connectedness, cost, and digital-capability, are enabling or limiting factors at each layer of the system. Notably, the framework supports a dynamic of feedback, where the results guide further data creation and modification of the system. This indicates the necessity of trial-and-error learning and ongoing adaptation within climate uncertainty. The framework can be applied as an analytical tool to identify the gaps at the system level including weak advisory translation, inadequate institutional support, or imbalance in governance. It serves as a roadmap to create AI systems and then find ways to make them accessible, scalable, and reflective of the reality of smallholder agriculture. The framework supports the key finding of this study by highlighting the importance of integration over individual performance. The climate-smart value of AI lies in system compatibility and not technological sophistication.

8. CONCLUSION

This study aims to analyse the role of AI in enhancing CSA in developing nations and specifically its application in smallholder farming systems. The study was based on the PRISMA framework of 35 peer-reviewed articles and a thematic analysis. The analysis emphasises to create a system-level perception of how and in which conditions AI can be applied to make organizations productive, resilient, and sustainable. The main findings of the study is that AI cannot be an independent technological solution to climate-related agricultural problems. Although most applications exhibit good technical capabilities especially in crop prediction, pest detection, and resource optimisation, their impact on CSA outcomes is conditional. Their effectiveness relies on their implementation in wider socio-technical frameworks comprising of data access, advisory translation, institutional mediation and governance systems. The technological potential and the adoption gaps that have continually occurred throughout the literature are indicative of system alignment failure and not a shortage of technological capacity.

One of the main contributions of this study is the formulation of an integrative analytical framework. This framework conceptualizes AI-enabled agriculture as a multi-layered system that connects data generation, the analytical processing, decision translation, and institutional support. It underscores the fact that climate-wise outcomes are not direct functions of AI models, but rather, the sequence of actions between technological elements and enablers. This way, the analysis does not look at the individual applications in isolation but the compatibility at the system-level with the realities of smallholder. This evidence also proves that not every AI application has the same contribution to CSA. The most scalable and most effective systems are low-cost, advisory-based systems that are delivered via mobile platforms and assisted by extension services. They are feasible in developing-country contexts. These systems minimize uncertainty in decision making, they operate within infrastructural limitations and can be incorporated to current agricultural practice.

In comparison, the technically advanced data-intensive monitoring systems and automation technologies are not very wide in their usage. It is because of their high costs, demanding infrastructure, and lack of fit with fragmented landholding systems. This difference strengthens the necessity to distinguish between high-tech solutions that are applicable to commercial agriculture and context-specific solutions that are applicable to smallholder systems. Simultaneously, this study also reveals that the adoption of AI is not a purely technical process, but it is neutral. The allocation of economic benefit, ownership of data and access to digital systems is asymmetrical. The review has shown that the disproportionate distribution of value most often occurs in agri-tech companies, platform operators, and suppliers of inputs. It is due to the reason that farmers take on risks associated with data asymmetric, platform dependence, and low bargaining power. The adoption of AI without proper governance structures can also increase inequalities rather than decreasing them. The digital literacy, gender and marginalisation problems also emphasize that inclusivity cannot be implied but needs to be consciously implemented into AI systems.

These results are converted into a series of pathways by translating them into structured reactions to three system-level constraints. They are weak translation of AI outputs into actionable decisions, limited access as a result of infrastructural and cost barriers, and governance-related disparities in the system in capturing data and value. In this regard, inclusive AI adoption needs the following. First, investment in digital infrastructure that is accessible and affordable. Second, capacity building and user-friendly design to enhance interpretation and trust. Third, strong institutional to mediate the use of technology. Fourth, clear and equitable data governance infrastructures. Fifth, adaptive AI systems capable of undergoing decision-making under climate uncertainty.

One of the key differences that arise as a result of the analysis is the difference between short-term operational gains and long-term resilience outcomes. Most AI applications enhance efficiency and productivity, but these benefits do not necessarily develop into adaptive capacity. In other instances, optimisation-driven systems can also become more vulnerable when they rely on stability or become dependent on external platforms and inputs. Thus, Climate-smart relevance is reliant on AI systems that improve flexibility, risk management, and capacity to respond to uncertain and changing environmental conditions. AI can make a valuable contribution towards CSA only when it is developed and deployed as a component of inclusive, institutionally-assisted systems. It resonates with the socio-economic realities of smallholder farming. Capital-intensive and high-tech solutions are not likely to provide broad capacity resilience or sustainability in developing country contexts.

Future studies can further enhance this framework by focusing on the scientific field-level validation, regional institutional model, and long-term effects of AI adoption on livelihoods, resilience, and equity. Specifically, more emphasis should be placed on the agricultural data governance and the architecture of participatory AI systems in creating equitable and responsible digital agriculture ecosystems. AI has a high potential to contribute to CSA, yet, not all possibilities are distributed equally and automatically. It adds value not with respect to technological development, but the circumstances within which it is located, managed, and mediated. To maintain resilience, sustainable, and inclusive agricultural systems, a balance between technology, institutions, and social equity is necessary.

REFERENCES

Abdul-Majid, M., Zahari, S. A., Othman, N., & Nadzri, S. (2024). Influence of technology adoption on farmers' well-being: Systematic literature review and bibliometric analysis. *Heliyon*, *10*(2), e24316. DOI: 10.1016/j.heliyon.2024.e24316 PMID: 38312653

Agrawal, J., & Arafat, M. Y. (2024). Transforming Farming: A Review of AI-Powered UAV Technologies in Precision Agriculture. *Drones (Basel)*, *8*(11), 664. DOI: 10.3390/drones8110664

Akanbi, M. B., Adedotun, K. J., Raji, A. K., & Banjoko, I. K. (2025). Optimizing water resource management in agriculture using ai-powered solar irrigation systems. *Journal of Science Innovation and Technology Research*. https://doi.org/DOI: 10.1016/j.sintl.2024.100292

Ali, Z., Muhammad, A., Lee, N., Waqar, M., & Lee, S. W. (2025). Artificial Intelligence for Sustainable Agriculture: A Comprehensive Review of AI-Driven Technologies in Crop Production. *Sustainability (Basel)*, *17*(5), 2281. DOI: 10.3390/su17052281

Amalan, A. A., & Aram, I. A. (2025). Artificial Intelligence adoption in Non-Chemical agriculture: An Integrated Mechanism for Sustainable Practices. *Sustainability (Basel)*, *17*(19), 8865. DOI: 10.3390/su17198865

Araújo, S. O., Peres, R. S., Barata, J., Lidon, F., & Ramalho, J. C. (2021). Characterising the Agriculture 4.0 Landscape—Emerging Trends, challenges and opportunities. *Agronomy (Basel)*, *11*(4), 667. DOI: 10.3390/agronomy11040667

Ayim, C., Kassahun, A., Addison, C., & Tekinerdogan, B. (2022). Adoption of ICT innovations in the agriculture sector in Africa: A review of the literature. *Agriculture & Food Security*, *11*(1), 22. Advance online publication. DOI: 10.1186/s40066-022-00364-7

Ayre, M., Collum, V. M., Waters, W., Samson, P., Curro, A., Nettle, R., Paschen, J., King, B., & Reichelt, N. (2019). Supporting and practising digital innovation with advisers in smart farming. *NJAS Wageningen Journal of Life Sciences*, *90–91*(1), 1–12. DOI: 10.1016/j.njas.2019.05.001

Bahmutsky, S., Grassauer, F., Arulnathan, V., & Pelletier, N. (2024). A review of life cycle impacts and costs of precision agriculture for cultivation of field crops. *Sustainable Production and Consumption*, *52*, 347–362. DOI: 10.1016/j.spc.2024.11.010

Balasubramaniam, S., Nelson, S. G., Arishma, M., Rajan, A. S., & Satheesh Kumar, K. (2024). Machine Learning based Disease and Pest detection in Agricultural Crops. *EAI Endorsed Transactions on Internet of Things, 10*. Advance online publication. DOI: 10.4108/eetiot.5049

Barroso-Barroso, C., Vega-Muñoz, A., Maradiaga-López, J., Salazar-Sepúlveda, G., & Carabantes-Silva, R. (2025). Smart Farming and the SDGs: Emerging research patterns and sustainability implications. *Agriculture, 16*(1), 81. DOI: 10.3390/agriculture16010081

Birhane, A. (2021). Algorithmic injustice: A relational ethics approach. *Patterns (New York, N.Y.), 2*(2), 100205. DOI: 10.1016/j.patter.2021.100205 PMID: 33659914

Birner, R., Daum, T., & Pray, C. (2021). Who drives the digital revolution in agriculture? A review of supply-side trends, players and challenges. *Applied Economic Perspectives and Policy, 43*(4), 1260–1285. DOI: 10.1002/aepp.13145

Carrer, M. J., De Souza Filho, H. M., De Mello Brandão Vinholis, M., & Mozambani, C. I. (2022). Precision agriculture adoption and technical efficiency: An analysis of sugarcane farms in Brazil. *Technological Forecasting and Social Change, 177*, 121510. DOI: 10.1016/j.techfore.2022.121510

Cerjak, M., Medici, M., Faletar, I., Sundeep, J. V., & Canavari, M. (2025). Adoption of mobile-based agricultural extension services: Evidence from South India. *Journal of Rural Studies, 120*, 103851. DOI: 10.1016/j.jrurstud.2025.103851

Chamorro-Padial, J., García, R., & Gil, R. (2024). A systematic review of open data in agriculture. *Computers and Electronics in Agriculture, 219*, 108775. DOI: 10.1016/j.compag.2024.108775

Choruma, D. J., Dirwai, T. L., Mutenje, M. J., Mustafa, M., Chimonyo, V. G. P., Jacobs-Mata, I., & Mabhaudhi, T. (2024). Digitalisation in agriculture: A scoping review of technologies in practice, challenges, and opportunities for smallholder farmers in sub-saharan africa. *Journal of Agriculture and Food Research, 18*, 101286. DOI: 10.1016/j.jafr.2024.101286

Ciliberti, S., Frascarelli, A., Polenzani, B., Brunori, G., & Martino, G. (2024). Digitalisation strategies in the agri-food system: The case of PDO Parmigiano Reggiano. *Agricultural Systems, 218*, 103996. DOI: 10.1016/j.agsy.2024.103996

Daum, T., Scheiterle, L., Yameogo, V., Adegbola, Y. P., Mulinge, W., Kergna, A. O., Daudu, C., Angara, U. A., Zossou, R. C., Nientao, A., Fatunbi, O., Isuyi, L., & Birner, R. (2025). Moving beyond the productivity paradigm: Agricultural innovation systems and sustainable transformation in Africa. *Agricultural Systems*, *229*, 104445. DOI: 10.1016/j.agsy.2025.104445

De Avila, F. R., & Barbosa, J. L. V. (2025). Smart environments in digital agriculture: A systematic review and taxonomy. *Computers and Electronics in Agriculture*, *236*, 110393. DOI: 10.1016/j.compag.2025.110393

Dhakal, M., Banerjee, M., Pandey, K. K., & Seniya, C. (2026). Ai and IoT are no longer limited to high-tech labs but are improving nutrition-focused agricultural farming. *Discover Food*, *6*(1), 69. Advance online publication. DOI: 10.1007/s44187-025-00777-w

Dhanaraj, R. K., Maragatharajan, M., Sureshkumar, A., & Balakannan, S. P. (2025). On-device AI for climate-resilient farming with intelligent crop yield prediction using lightweight models on smart agricultural devices. *Scientific Reports*, *15*(1), 31195. DOI: 10.1038/s41598-025-16014-4 PMID: 40854942

Dhanaraju, M., Chenniappan, P., Ramalingam, K., Pazhanivelan, S., & Kaliaperumal, R. (2022). Smart Farming: Internet of Things (IoT)-Based Sustainable agriculture. *Agriculture*, *12*(10), 1745. DOI: 10.3390/agriculture12101745

Dixit, K., Aashish, K., & Dwivedi, A. K. (2023). Antecedents of smart farming adoption to mitigate the digital divide – extended innovation diffusion model. *Technology in Society*, *75*, 102348. DOI: 10.1016/j.techsoc.2023.102348

Duguma, A. L., & Bai, X. (2024). How the internet of things technology improves agricultural efficiency. *Artificial Intelligence Review*, *58*(2), 63. Advance online publication. DOI: 10.1007/s10462-024-11046-0

Duncan, E., Ruder, S., Kc, K. B., Wittman, H., & Fraser, E. (2026). The political economy of agricultural data governance: Farmer perspectives on trust in data sharing. *Journal of Rural Studies*, *123*, 104076. DOI: 10.1016/j.jrurstud.2026.104076

Erike, A., Ikerionwu, C., Azubogu, A., & Obodoagwu, V. (2025). Is AI for illiterate farmers? A systematic literature review of AI and machine learning applications and challenges for precision agriculture. *Discover Artificial Intelligence*, *5*(1), 204. Advance online publication. DOI: 10.1007/s44163-025-00457-9

Fairbairn, M., & Kish, Z. (2023). Setting data free: The politics of open data for food and agriculture. *New Media & Society*, *25*(8), 1935–1959. DOI: 10.1177/14614448231174520

Ferraris, S., Meo, R., Pinardi, S., Salis, M., & Sartor, G. (2023). Machine learning as a strategic tool for helping cocoa farmers in Côte D'Ivoire. *Sensors (Basel)*, *23*(17), 7632. DOI: 10.3390/s23177632 PMID: 37688090

Frija, A. (2024). A system readiness approach to support the packaging and scaling of innovation bundles for farming systems transformation. *Agricultural Systems*, *221*, 104148. DOI: 10.1016/j.agsy.2024.104148

Frost, J., Gambacorta, L., Huang, Y., Shin, H. S., & Zbinden, P. (2019). BigTech and the changing structure of financial intermediation. *Economic Policy*, *34*(100), 761–799. DOI: 10.1093/epolic/eiaa003

Godara, S., Begam, S., Bana, R. S., Bedi, J., Jain, R., Haque, M. A., Parsad, R., Marwaha, S., Patial, M., Shirzad, S., & Nirmal, R. (2024). TPTC: Topic-wise problems' trend clusters for smart agricultural insights extraction and forecasting of farmer's information demand. *Scientific Reports*, *14*(1), 29272. DOI: 10.1038/s41598-024-80488-x PMID: 39587214

Hackfort, S. (2023). Unlocking sustainability? The power of corporate lock-ins and how they shape digital agriculture in Germany. *Journal of Rural Studies*, *101*, 103065. DOI: 10.1016/j.jrurstud.2023.103065

Haque, A. B., Islam, A. N., & Mikalef, P. (2022). Explainable Artificial Intelligence (XAI) from a user perspective: A synthesis of prior literature and problematizing avenues for future research. *Technological Forecasting and Social Change*, *186*, 122120. DOI: 10.1016/j.techfore.2022.122120

Hunecke, C., Engler, A., Jara-Rojas, R., & Poortvliet, P. M. (2017). Understanding the role of social capital in adoption decisions: An application to irrigation technology. *Agricultural Systems*, *153*, 221–231. DOI: 10.1016/j.agsy.2017.02.002

Jabed, M. A., & Murad, M. A. (2024). Crop yield prediction in agriculture: A comprehensive review of machine learning and deep learning approaches, with insights for future research and sustainability. *Heliyon*, *10*(24), e40836. DOI: 10.1016/j.heliyon.2024.e40836 PMID: 39720079

Kumar, V., Sharma, K. V., Kedam, N., Patel, A., Kate, T. R., & Rathnayake, U. (2024). A comprehensive review on smart and sustainable agriculture using IoT technologies. *Smart Agricultural Technology*, *8*, 100487. DOI: 10.1016/j.atech.2024.100487

Lan, J., & Ban, Q. (2025). The Farm-Level Economic and Environmental Benefits of Precision Agriculture Technology Adoption: A Meta-Analysis of Global Evidence. *Sustainability (Basel)*, *17*(24), 11223. DOI: 10.3390/su172411223

Liakos, K. G., Busato, P., Moshou, D., Pearson, S., & Bochtis, D. (2018). Machine Learning in Agriculture: A Review. *Sensors (Basel), 18*(8), 2674. DOI: 10.3390/s18082674 PMID: 30110960

Ma, W., & Rahut, D. B. (2024). Climate-smart agriculture: Adoption, impacts, and implications for sustainable development. *Mitigation and Adaptation Strategies for Global Change, 29*(5), 44. Advance online publication. DOI: 10.1007/s11027-024-10139-z

Mallinger, K., Corpaci, L., Neubauer, T., Tikasz, I. E., Goldenits, G., & Banhazi, T. (2024). Breaking the barriers of technology adoption: Explainable AI for requirement analysis and technology design in smart farming. *Smart Agricultural Technology, 9*, 100658. DOI: 10.1016/j.atech.2024.100658

Mana, A., Allouhi, A., Hamrani, A., Rehman, S., Jamaoui, I. E., & Jayachandran, K. (2024). Sustainable AI-based production agriculture: Exploring AI applications and implications in agricultural practices. *Smart Agricultural Technology, 7*, 100416. DOI: 10.1016/j.atech.2024.100416

Mehrabi, Z., McDowell, M. J., Ricciardi, V., Levers, C., Martinez, J. D., Mehrabi, N., Wittman, H., Ramankutty, N., & Jarvis, A. (2020). The global divide in data-driven farming. *Nature Sustainability, 4*(2), 154–160. DOI: 10.1038/s41893-020-00631-0

Mohan, R. N. V. J., Rayanoothala, P. S., & Sree, R. P. (2025). Next-gen agriculture: Integrating AI and XAI for precision crop yield predictions. *Frontiers in Plant Science, 15*, 1451607. DOI: 10.3389/fpls.2024.1451607 PMID: 39845494

Mossie, W. A. (2022). The Impact of Climate-Smart Agriculture Technology on Productivity: Does Row Planting Matter? Evidence from Southern Ethiopia. *TheScientificWorldJournal, 2022*, 1–11. DOI: 10.1155/2022/3218287 PMID: 35991088

Munz, J., Maurmann, I., Schuele, H., & Doluschitz, R. (2024). Digital transformation at what cost? A case study from Germany estimating the adoption potential of precision farming technologies under different scenarios. *Smart Agricultural Technology, 9*, 100585. DOI: 10.1016/j.atech.2024.100585

Nawaz, M., & Babar, M. I. K. (2024). IoT and AI: A panacea for climate change-resilient smart agriculture. *Discover Applied Sciences, 6*(10), 517. Advance online publication. DOI: 10.1007/s42452-024-06228-y

Nxumalo, G. S., & Chauke, H. (2025). Challenges and opportunities in smallholder agriculture digitization in South Africa. *Frontiers in Sustainable Food Systems, 9*, 1583224. Advance online publication. DOI: 10.3389/fsufs.2025.1583224

Osrof, H. Y., Tan, C. L., Angappa, G., Yeo, S. F., & Tan, K. H. (2023). Adoption of smart farming technologies in field operations: A systematic review and future research agenda. *Technology in Society*, *75*, 102400. DOI: 10.1016/j.techsoc.2023.102400

Page, M. J., McKenzie, J. E., Bossuyt, P. M., Boutron, I., Hoffmann, T. C., Mulrow, C. D., Shamseer, L., Tetzlaff, J. M., Akl, E. A., Brennan, S. E., & Chou, R. (2021). The PRISMA 2020 statement: An updated guideline. *BMJ (Clinical Research Ed.)*, •••, 372. PMID: 33782057

Pawlak, K., & Kołodziejczak, M. (2020). The role of agriculture in ensuring food security in developing countries: Considerations in the context of the problem of sustainable food production. *Sustainability (Basel)*, *12*(13), 5488. DOI: 10.3390/su12135488

Poenaru, M. M., Manta, L. F., Gher escu, C., & Manta, A. G. (2025). Shaping the future of horticulture: Innovative technologies, artificial intelligence, and robotic automation through a bibliometric lens. *Horticulturae*, *11*(5), 449. DOI: 10.3390/horticulturae11050449

Qwaid, M. A., Sarker, M. T., Shawon, S. M., & Zubair, H. (2026). AI-enabled smart farming framework for sustainable date palm cultivation in arid regions using machine learning and IoT integration. *Scientific Reports*, *16*(1), 5125. DOI: 10.1038/s41598-026-36106-z PMID: 41526481

Rajbongshi, A., Johora, F. T., Hossain, A., Sarker, M. S., Rahman, M. H., Rahman, M. W., Alotaibi, F. T., & Moni, M. A. (2026). Leveraging explainable AI for sustainable agriculture: A comprehensive review of recent advances. *Artificial Intelligence Review*, *59*(3), 105. Advance online publication. DOI: 10.1007/s10462-025-11459-5

Ramzan, S., Ali, B., Raza, A., Hussain, I., Fitriyani, N. L., Gu, Y., & Syafrudin, M. (2024). An innovative artificial neural network model for smart crop prediction using sensory network-based soil data. *PeerJ. Computer Science*, *10*, e2478. DOI: 10.7717/peerj-cs.2478 PMID: 39650399

Reitano, M., Segovia, M. S., & Nayga, R. M.Jr. (2025). A systematic review on the impact of Artificial Intelligence in the agri-food supply chain. *Food Policy*, *137*, 102983. DOI: 10.1016/j.foodpol.2025.102983

Ricart, S., Gandolfi, C., & Castelletti, A. (2023). Climate change awareness, perceived impacts, and adaptation from farmers' experience and behavior: A triple-loop review. *Regional Environmental Change*, *23*(3), 82. Advance online publication. DOI: 10.1007/s10113-023-02078-3

Romera, A., Sharifi, M., & Charters, S. (2024). Digitalization in agriculture. Towards an integrative approach. *Computers and Electronics in Agriculture*, *219*, 108817. DOI: 10.1016/j.compag.2024.108817

Rotz, S., Gravely, E., Mosby, I., Duncan, E., Finnis, E., Horgan, M., LeBlanc, J., Martin, R., Neufeld, H. T., Nixon, A., Pant, L., Shalla, V., & Fraser, E. (2019). Automated pastures and the digital divide: How agricultural technologies are shaping labour and rural communities. *Journal of Rural Studies*, *68*, 112–122. DOI: 10.1016/j.jrurstud.2019.01.023

Saiz-Rubio, V., & Rovira-Más, F. (2020). From Smart Farming towards Agriculture 5.0: A Review on Crop Data Management. *Agronomy (Basel)*, *10*(2), 207. DOI: 10.3390/agronomy10020207

Santos, M. R. C., Carvalho, L. C., & Francisco, E. (2025). A Capability-Based framework for Knowledge-Driven AI innovation and sustainability. *Information (Basel)*, *16*(11), 987. DOI: 10.3390/info16110987

Sauvagerd, M., Mayer, M., & Hartmann, M. (2024). Digital platforms in the agricultural sector: Dynamics of oligopolistic platformisation. *Big Data & Society*, *11*(4), 20539517241306365. Advance online publication. DOI: 10.1177/20539517241306365

Screpnik, C. R., Zamudio, E., & Gimenez, L. I. (2025). Artificial Intelligence in Agriculture: A Systematic Review of crop yield Prediction and Optimization. *IEEE Access : Practical Innovations, Open Solutions*, *13*, 70691–70697. DOI: 10.1109/ACCESS.2025.3560631

Shaikh, T. A., Mir, W. A., Rasool, T., & Sofi, S. (2022). Machine Learning for smart agriculture and Precision Farming: Towards making the fields talk. *Archives of Computational Methods in Engineering*, *29*(7), 4557–4597. DOI: 10.1007/s11831-022-09761-4

Shepherd, M., Turner, J. A., Small, B., & Wheeler, D. (2018). Priorities for science to overcome hurdles thwarting the full promise of the ‘digital agriculture’ revolution. *Journal of the Science of Food and Agriculture*, *100*(14), 5083–5092. DOI: 10.1002/jsfa.9346 PMID: 30191570

Shuliang, Z., Brentuo, E., Horsey, E. M., & Osei, B. A. (2025). Farmers’ perceptions and government-farmer value co-creation: The roles of artificial intelligence adoption and government support. *International Journal of Innovation Studies*. https://doi.org/DOI: 10.1016/j.ijis.2025.10.004

Sreeram, R., Krishna, S. A., Kumar, A. S., Remya, S., & Cho, Y. Y. (2025). Soil moisture monitoring technologies in smart agriculture: A comprehensive review. *Farming System*, *4*(2), 100189. DOI: 10.1016/j.farsys.2025.100189

Uztürk, D., & Büyüközkan, G. (2024). Industry 4.0 technologies in Smart Agriculture: A review and a Technology Assessment Model proposition. *Technological Forecasting and Social Change*, *208*, 123640. DOI: 10.1016/j.techfore.2024.123640

Van Klompenburg, T., Kassahun, A., & Catal, C. (2020). Crop yield prediction using machine learning: A systematic literature review. *Computers and Electronics in Agriculture*, *177*, 105709. DOI: 10.1016/j.compag.2020.105709

Vijayakumar, S., Murugaiyan, V., Ilakkiya, S., Kumar, V., Sundaram, R. M., & Kumar, R. M. (2025). Opportunities, challenges, and interventions for agriculture 4.0 adoption. *Discover Food*, *5*(1), 265. Advance online publication. DOI: 10.1007/s44187-025-00576-3

Wolfert, S., Ge, L., Verdouw, C., & Bogaardt, M. (2017). Big Data in Smart Farming – A review. *Agricultural Systems*, *153*, 69–80. DOI: 10.1016/j.agsy.2017.01.023

Yang, Y., Chen, L., Diao, Z., Gao, P., Zhang, B., & Zhao, C. (2026). Recent advances in crop pest detection, forecasting and early warning: A review. *Artificial Intelligence in Agriculture*. https://doi.org/DOI: 10.1016/j.aiia.2026.03.009

Zhai, Z., Martínez, J. F., Beltran, V., & Martínez, N. L. (2020). Decision support systems for agriculture 4.0: Survey and challenges. *Computers and Electronics in Agriculture*, *170*, 105256. DOI: 10.1016/j.compag.2020.105256

Chapter 7
Artificial Intelligence for Integrated Environmental Resilience:
Agriculture, Water, and Biodiversity

Harshit Mishra
http://orcid.org/0009-0000-1951-7278
Acharya Narendra Deva University of Agriculture and Technology, India

Rashmi Mishra
http://orcid.org/0000-0002-8544-0149
University of Technology and Applied Sciences, Ibra, Oman

Fredrick Kayusi
http://orcid.org/0000-0003-1481-4016
Maasai Mara University, Narok, Kenya

Ioannis Adamopoulos
http://orcid.org/0000-0002-4942-7123
University of West Attica, Athens, Greece

ABSTRACT

Artificial intelligence is increasingly applied to environmental sustainability, yet research often treats agriculture, water management, and biodiversity conservation as separate domains. This chapter argues that AI can support integrated environmental resilience by enabling data-driven decision-making across these interconnected systems. It introduces a conceptual framework linking food production, water governance, and ecosystem protection through environmental informatics,

DOI: 10.4018/979-8-3373-9978-2.ch007

machine learning, and geospatial analytics. The chapter reviews key applications in precision agriculture, hydrological forecasting, intelligent irrigation, and AI-based biodiversity monitoring, while examining how these tools support climate adaptation, resource efficiency, and ecosystem management. It also evaluates limitations including data bias, infrastructure gaps, and governance challenges. The chapter concludes that responsible, cross-sector AI can strengthen coordinated climate action and sustainable environmental management.

INTRODUCTION

Sustainable development is no doubt one of the biggest challenges that the twenty-first century faces, and it has been pushing scientists, policy makers, and technologists to find transformative solutions for meeting growing human demands within the confines of the finite carrying capacity of Earth's environmental systems. The convergence of environmental challenges and computational developments has led to a new scientific era where artificial intelligence (AI) serves as a means of enhancing efficiency, as well as a methodological paradigm to explore, model, and manage complex nature-based (Nti *et al.,* 2022; Ruiz *et al.,* 2023) processes. The global food system has challenging task to feed an estimated 9.7 billion people by 2050, a pathway that requires around 70% increase in agricultural production- as indicated by the Food and Agriculture Organization (Sinha, 2024). In the meantime, the Intergovernmental Panel on Climate Change declares that the mean global surface temperature is estimated to have already risen approximately 1.1°C relative to pre-industrial levels, exerting cascading stress on agricultural soils, hydrospheres, and biodiversity-abundant ecologies. In this sense, AI-based ecosystem governance is no longer an entirely speculative concept but a diverse array of concrete tools that are producing measurable effects in a range of fields (Jones *et al.,* 2024).

The agriculture industry occupies about half of the world's habitable land and is responsible for around 19 percent of global greenhouse gas emissions when land-use change is factored in, and it is now at a turning point. Traditional agriculture practices, which remain prevalent in large parts of the developing world, are increasingly inadequate in the face of climate change, soil erosion, and dwindling supplies of resources (Raihan *et al.,* 2024). AI-based precision agriculture has been shown to have promising potential to mitigate these issues. However, ML-based models for crop yield prediction were able to improve up to around 10–20% in field trials while also decreasing the use of nitrogen fertilizer by up to 30% (Gryshova *et al.,* 2024; Zehouani *et al.,* 2024). The accuracy of deep learning frameworks in image-based pest and disease detection has already surpass 95% in controlled conditions, which can be used for early-warning systems that could mitigate crop losses amounting to 20–40% of global production (Singh & Jain, 2022; Shivaprakash *et al.*, 2022).

Another example is water resource management, with AI-based methods becoming more and more important. About 2.5% of the total water on Earth is freshwater, and less than 0.3% of it is available to humans on the surface in the form of lakes, rivers, and accessible underground aquifers (Abed *et al.*, 2025). Agriculture consumes about 70 percent of total global freshwater withdrawals, and poor agricultural irrigation practices are a common cause of losses of 30–50 percent through evaporation, runoff and percolation (Saikanth *et al.*, 2023). AI-driven predictive models are also increasingly being applied for irrigation scheduling, hydrological dynamics, and prediction of extreme events such as floods and droughts, which could aid in ensuring more efficient water resource utilization under climate uncertainty.

At the same time, the global biodiversity crisis is now at an all-time high. Recent evaluations indicate that approximately one million species are at risk of extinction owing to habitat destruction, climate change, invasive species, and wildlife trade (Alduweib *et al.*, 2025). "The tragedy of biodiversity loss is not just an ecological disgrace but also compromises ecosystem services that enable food production, water regulation, and climate stability," said report co-chair Robert Watson. AI-enabled monitoring technologies, such as satellite-based ecosystem monitoring and acoustic species identification solutions, are being leveraged to monitor the decline in biodiversity and guide conservation efforts at scale and temporal dimensions that are difficult for conventional monitoring systems to match (Mishra *et al.*, 2025; Shang & Xie, 2024).

These environmental pressures are seldom fulminated alone. Groundwater depletion compromises farm yields, land degradation worsens runoff and biodiversity decline, and deteriorating ecosystems impair the natural services that nourish both food and water supplies (Yousaf, 2024). Therefore, sustainability issues are more and more emerging at the convergence of agricultural practices, irrigation, and biodiversity protection rather than within any single discipline. Yet much of the current work on AI for environment sustainability is scattered across discipline areas including precision agriculture, hydrological modelling, and computational ecology etc. Bibliometric analyses reveal that the research outputs in the fields have been increasing rapidly, particularly in the application of machine learning in agriculture related publications increasing >400% from 2015 to 2022 (Shang & Xie, 2024; Wang *et al.*, 2025). Yet few studies focus on the potential of AI technologies to holistically influence the interdependent dynamics of food systems, water governance, and ecosystem protection.

This chapter accordingly develops the claim that artificial intelligence-AI1 in particular can support integrated environmental resilience by providing for data-enabled investigation and decision-making across interrelated environmental systems. Rather than seeing agriculture, water management, and biodiversity protection as disjointed fields of technology application, the chapter explores how AI-based analyses can inform integrated approaches to climate change adaptation, resource efficiency and ecosystem conservation. From this vantage point, AI is a type of environmental intelligence that processes multi-dimensional data sources such as remote sensing imagery, sensor networks, ecological monitoring systems, and crowdsourcing datasets. Using statistical models, predictive analytics and optimization algorithms, these systems can detect patterns, predict environmental hazards, and inform evidence-based governance decisions in many environmental areas (Kumari and Pandey, 2023).

New computational architectures are also reinforcing this integrative potential. Digital twin platforms, multi-objective optimization models and distributed edge-computing systems, among other technologies, can increasingly model environmental processes and real-time data streams to support multi-sectoral decision-making. Such tools can help policy experts, farmers, water managers and conservationists weigh the trade-offs in productivity, resource conservation and ecosystem protection. At the same time, their adoption poses significant issues for data governance, algorithmic transparency, institutional capacity and equitable access to digital environmental technology—especially in the parts of the world where environmental stress is greatest.

In this context, the chapter studies how AI for environmental sustainability can be conceived and realized as multi-layered services focusing on agriculture, water resources management, and biodiversity protection. It looks at the technical aspects of how AI systems work and at a higher level, the socio-technical dynamics that influence their function in actual environmental governance situations. On the analysis, they hold, it is clear that advancing technology alone cannot ensure sustainable output; how well AI can be deployed depends on interdisciplinarity, sound data infrastructures, governance frameworks that are inclusive, and the involvement of those local communities who are experiencing environmental change.

The rest of this chapter is organized as follows. Section 2 presents an overview of the foundations of artificial intelligence for environmental sustainability (e.g., key AI technologies, principles of environmental informatics) and the data ecosystems that enable ecological analyses. In Section 3, the topic of application of AI to sustainable agriculture is addressed, and in particular it is discussed how OI can be used to collect FP data on large scales and effective green spinning on small scales. Section 4 discusses AI based methodologies for integrated water resources management (IWRM), including hydrological prediction models, irrigation management, as well as water governance models. 5 The chapter assesses the potential of AI for biodiversity conservation, considering recent developments in ecological modelling, automated monitoring technologies, and conservation prioritization algorithms.

Section 6, finally, integrates these areas in a cross-sector analysis that explores the potential of unified environmental awareness, multi-objective optimization, and digital environmental infrastructures to foster coordinated sustainability policies. The chapter ends with a discussion on the AI implications for climate resilience and environmental governance, highlighting the need for responsible and just technologies development for enduring planetary stewardship.

THEORETICAL FOUNDATIONS OF AI FOR ENVIRONMENTAL SUSTAINABILITY

Artificial intelligence has been established as a powerful tool for modelling and controlling environmental complexity, including multi-scale interactions of such systems as agricultural production, hydrologic flow and ecosystem dynamics. Non-linear feedbacks, spatial heterogeneity and strong temporal variability are typical of environmental systems. Changes in one part of a system frequently cascade into others. For example, the demand for water and delivery of nutrients may be modified by intensification of agriculture, with consequent impacts on patterns of aquatic ecosystems and biodiversity. Conventional analytical approaches typically have difficulty addressing these cross-scale interactions, particularly if environmental data sets are large, heterogeneous, and generated in real time. AI provides a suite of computational methodologies that can learn from complex data sets, adjust to evolving conditions, and produce predictions or prescriptions that enhance integrated environmental decision-making. As a result, within sustainability science, AI is not only a technological instrument but a methodological lens to scrutinize related processes across the domains of food systems, water governance, and ecosystem management conservation. Machine learning algorithms, neural networks, and optimization methods enable scientists to find patterns in environmental data, model ecosystem responses, and inform adaptive management practices. This is particularly pertinent given that these processes challenge to mitigate climate functions of drought, land degradation, loss of soil organics, decline of biodiversity, and scarcity of resources.

The conceptual pillars of AI for environmental friendliness can be identified in three related ideas. The first is algorithmic learning, where models learn relationships from data themselves rather than relying only on explicit domain knowledge. This enables AI to uncover hidden patterns in environmental data, such as relationships among climate variables, crop yield, and hydrologic processes. The second is representational abstraction, which permits complex ecological and social-environmental processes to be represented by computational models. By way of mathematical modelling, AI can simulate the workings of an ecosystem, allowing scientists to model environmental conditions and assess interventions. The third is adaptive optimisation, whereby models are able to refine their predictions as further data are gathered. The ability to adapt enables AI systems to adapt to dynamic environmental factors, for example, altered rain regimes, active land use, or species distributions. These principles taken together allow the AI model to capture the cross-sector interactions and trade-offs between agricultural production, water security, and biodiversity protection. Instead of studying these domains one at a time, AI-driven techniques are enabling the modelling of environmental systems as multi-layered socio-ecological networks. In this way, artificial intelligence help integrated environmental assessment and robust adaptive resource management in (Ullah *et al.,* 2025).

Definitions and Taxonomy of Artificial Intelligence Techniques

Artificial intelligence (AI) can be broadly defined as a set of computational methods to tackle tasks that are traditionally regarded as requiring human intelligence, such as reasoning, learning, perception and decision-making. When considering environmental applications, rather than thinking of AI as a single technology, one should consider it as a myriad of approaches to methodologies that vary in their mode of learning, representation of knowledge, and logic of decision-making. These methods are the symbolic systems, machine learning algorithms, deep learning architectures, evolutionary optimization methods (EOMs) and hybrid intelligent systems. A detailed taxonomy of AI methods applicable to sustainability is provided in Table 1. As a result, within each class, environmental applications also have certain unique strengths and weakness when tackling problems such as crop prediction, water management, ecological monitoring and conservation planning.

Table 1. Classification of AI techniques used in environmental sustainability research

Category	**Core principle**	**Typical environmental uses**	**Strengths**	**Limitations**	**Representative accuracy / performance**
Symbolic AI	Rule-based reasoning	Policy modelling, expert decision support systems	Transparent logic; fully interpretable outputs	Limited adaptability to novel data patterns	~70–80% rule compliance in structured policy domains (Giarratano & Riley, 2005)
Machine Learning	Data-driven pattern discovery	Crop yield prediction, rainfall forecasting, land degradation mapping	Learns complex patterns from data; scalable	Requires large, labelled datasets; risk of overfitting	R^2 = 0.85–0.94 in crop yield prediction models (Van Klompenburg *et al.*, 2020)
Deep Learning	Multi-layer neural representation	Satellite image classification, pest detection, species recognition	High accuracy for complex, high-dimensional data	Computationally intensive; low interpretability	>95% classification accuracy in plant disease detection (Mohanty *et al.*, 2016)
Evolutionary Algorithms	Optimization via natural selection principles	Protected area design, irrigation scheduling, resource allocation	Handles nonlinear, multi-objective problems	Slow convergence; computationally expensive	15–30% improvement in irrigation efficiency (Goldberg, 1989; Naghibi *et al.*, 2016)
Hybrid Systems	Combined symbolic and sub-symbolic approaches	Ecosystem simulation, integrated watershed management	Flexible, robust, and context-adaptive	Complex design and validation requirements	Up to 20% improvement in forecast accuracy over single-model approaches (Sehgal *et al.*, 2021)

Source: Synthesized from Giarratano and Riley (2005), Van Klompenburg *et al.* (2020), Mohanty *et al.* (2016), Goldberg (1989), Naghibi *et al.* (2016), and Sehgal *et al.* (2021).

Machine Learning Paradigms

For now, ML is the dominant methodological tool in recent studies of AI and is also applied in the area of environmental sustainability. ML algorithms can directly learn from data and build predictive models, which can be used to support environmental decisions in agriculture, hydrology, and conservation, among other industries. Among others the environmental learning generally utilizes three major learning paradigms. Instead, supervised learning algorithms are given labeled datasets, meaning that each input data already has an associated output. This paradigm is widely applied in predicting crop yield, identifying crop disease, and forecasting flood risk based on environmental history (Mishra and Mishra, 2025). Supervised learning allows to develop predictive tools for agricultural planning and water risk management by learning from historical observations. Operating on unlabelled datasets, unsupervised learning seeks to find innate structures or patterns in environmental data. Unsupervised methods are frequently employed in ecological studies to cluster species distributions, identify ecological regions, or examine land degradation patterns. Such techniques allow researchers to identify connections that might be missed when running standard statistical analyses.

Reinforcement learning is a third paradigm, emphasizing the aspect of sequential decision making in dynamic environments. Here, algorithms learn optimal policies through repeated interaction with the environment, receiving reward or penalty. Reinforcement learning has also been used in water resources management, in which it can derive optimal reservoir release decisions at multiple reservoirs, irrigation scheduling, and water allocation schemes under changing climate and demand scenarios (Liu *et al.,* 2021). For each paradigm there are different epistemological means of producing environmental knowledge. Supervised learning is based on pre-existing labelled knowledge; unsupervised learning facilitates exploratory discovery of patterns and reinforcement learning simulates adaptive decision-making. Hence, the choice of suitable paradigm is determined by the type of environmental problem, the availability of data, and the level of desired interpretability of results.

Deep Learning Architectures

Deep learning is an advanced form of machine learning that utilizes multi-layered neural networks to learn hierarchial representations directly from data. These structures are well suited for high-dimensional data such as satellite images, sound, or long environmental time series. Convolutional neural networks (CNN) have been extensively applied in environmental image processing. They facilitate the automated land cover classification, deforestation detection, coral reef health monitoring and crop diseases identification from aerial or satellite images. RNN models, such as long short-term memory (LSTM) networks and gated recurrent units (GRUs), are hardware-friendly and suitable for sequential data processing applications, including flow forecasting (Singh *et al.,* 2022).

In more recent times, transformers, which were previously exclusively used for natural language processing (NLP), have been tailored towards modelling environmental time-series. Their capacity for long-range dependency capture renders them ideal for examining climate signals and ecosystem dynamics over extended temporal scales. The theoretical interest in deep learning stems from its ability to learn complex hierarchies of features from raw data. Deep learning models are explicitly known to learn feature hierarchies from data, in which higher level features are formed from lower-level features. This allows scientists to simulate ecological processes that otherwise would be difficult to observe or represent with traditional statistical methods (Ayoola *et al.,* 2024).

Hybrid Intelligent Systems

Hybrid intelligent systems integrate several AI techniques, taking advantage of the strengths of each individual methodology. Environmental systems tend to be subject to uncertainty, partial knowledge, and nonlinear processes which are not always sufficiently handled by a unique modelling approach. Hybrid methods may combine symbolic reasoning and machine learning, or employ neural networks with evolutionary algorithms. A specific illustration: a hybrid agricultural decision-support system -- by way of machine learning models, predicts soil moisture conditions, and through rule-based logic, checks whether its irrigation recommendations are sustainable and within regulatory guidelines. Likewise, evolutionary algorithms have been employed to optimize neural network architectures for climate and hydrological modelling. The rationale for hybrid systems is that environmental problems do not often fit neatly into one computational paradigm. By synthesizing complementary methods, hybrid class of models hold the potential to enhance robustness, interpretability, and flexibility in the context of complex environmental applications such as watershed management, ecosystem restoration planning, or integrated agricultural–hydrological systems (Balaska *et al.,* 2023).

Environmental Informatics and Ecological Modelling Principles

Environmental informatics is an emerging science that leverages information science and computational methods to address environmental issues. It acts as the conceptual interface between ecological theory and concrete AI solutions. It is underpinned by the perspective that environmental processes may be regarded as information systems with inputs, outputs, feedback mechanisms, and transformations. Ecological modelling has traditionally been built on mathematical formulae that describe processes such as population dynamics, nutrient cycling and hydrological flows (Efremova *et al.,* 2023). These models are theoretically sound, but they have found it difficult to accommodate the growing amount and diversity of environmental data being collected by satellites, sensors, monitoring programs.

AI techniques make contributions to ecosystem modelling with parameter estimation under data-driven modelling, uncertainty quantification, and prediction in real-time. Interoperability in environmental informatics implies the capacity to merge diverse datasets derived from satellites, field instruments, government monitoring systems and citizen science projects. To enable AI-driven environmental analysis, these datasets need to be standardized and harmonized. Also important is the principle of scalability, able to perform across spatial and temporal scales from local field-based measurements to global climate observations (AlZubi and Galyna, 2023). By integrating ecological theory with computer intelligence, environmental informatics provides the theoretical framework for studying complex environmental systems.

Data Ecosystems for Sustainability Science

Data quality, diversity and availability are at the core of AI's efficacy. A data ecosystem is a collection of data sources, storage and computing infrastructures, analytics tools and governance mechanisms that facilitate knowledge generation. In the science of sustainability, data ecosystems are naturally interdisciplinary as environmental problems have social, economic, and ecological facets.

Remote Sensing and Geospatial Data Streams

Remote sensing technology allows for the observation of the environment at regular intervals across large areas. Satellites, aerial drones, and radars produce geospatial datasets that describe land use, the state of vegetation, the composition of the atmosphere, and the availability of water. Such data sets are especially useful to AI in that they are of a high resolution, updated regularly and cover the world. Geospatial streaming data can be leveraged for predictive modelling of environmental change. For example, you can apply machine learning to time-series satellite images to identify early symptoms of desertification or glacier retreat. Geographic information systems that bring together several other environmental layers can also be supported by spatial data, enabling users to observe interactions between climate metrics, biodiversity proxies, and human endeavours (Isabele and Westerlund, 2022).

Sensor Networks and IoT Infrastructures

Sensor networks are formed by multiple devices, which monitor environmental parameters like temperature, humidity, soil moisture, water quality or air pollution. Once these sensors are networked through Internet of Things infrastructures, they are able to send real-time data to centralized or cloud platforms. These incoming streams can then be analysed by AI algorithms for identifying anomalies, predicting trends and automating reactions. AI-based decisions and systems Soil sensors connected to AI systems can provide farmers with best irrigation schedules, saving on irrigation water and increasing crop yield. In cities, on the other hand, air quality sensors can be used to feed adaptive traffic control schemes. The conceptual significance of sensor networks is that they have the potential to revolutionize environmental monitoring from periodic sampling to constant monitoring. Continuous data lead to better models and also to early warning systems for environmental hazards (Aijaz *et al.,* 2025).

Citizen Science and Participatory Datasets

Citizen science involves laypersons in the process of gathering data and making observations. Using mobile applications, they can log bird sightings, quantify rainfall, or notify of pollution incidents. Such participation-driven data sets dramatically enhance the spatial coverage and temporal resolution of environmental information. From an AI viewpoint, citizen science presents challenges and opportunities (Adanma and Ogunbiyi, 2024). While the large number of observations allows to better train the model, their quality can be highly variable, which means that strong data validation procedures are needed. Machine learning techniques like anomaly detection and probabilistic modelling are commonly employed to remove outliers. The assimilation of citizen science data also has epistemic value in that it opens up the processes of knowledge production and incorporates local ecological knowledge into scientific modelling.

Systems Thinking and Complexity Theory in Ecological Applications

Systems thinking is a way of thinking about and doing science that makes environmental phenomena viewed as connected parts of larger wholes. Rather than considering independent variables, it emphasizes the study of relations, feedback loops and emergent properties. Complexity theory also takes this perspective and investigates how relationships between components of a system give rise to phenomena that are unpredictable when considering the individual components alone. Ecological systems are good examples of complex, adaptive systems. Species interactions, nutrient cycles and climatic factors all generate nested, evolving networks of interdependence. These are processes that lend themselves to AI methodologies for modelling, as they may be non-linear, multi-dimensional interactions or radiations. Neural networks, for instance, can be used to approximate complex ecological functions without having to specify explicit equations. Agent-based models can represent individual organisms and examine the emergent collective dynamics. The systems thinking contribution to theory for AI applications is thus conceptual rather than computational. It leads researchers to develop models that incorporate ecological interdependence and that predictions consider indirect impacts and cascades. From such a vantage point, Classical models of AI are susceptible to generating accurate short-term predictions that aren't representative of long-term behaviour of the system (Mishra, 2025).

A Conceptual Framework: AI-Mediated Environmental Interdependencies

To give context to the function of artificial intelligence in sustainability science, environmental systems can be seen as networks of interactions between elements like agriculture, water resources, and biodiversity. These systems interact via feedbacks, e.g., land use change, nutrient flows, water withdrawals, and ecosystem services. Within this context, AI is positioned as an analytical catalyst that draws various datasets and decision-making processes within these related systems. Machine learning models are used to assess patterns of agricultural production, hydro-logical predictions are used for irrigating and reservoirs management, and biodiversity monitoring systems are employed to monitor ecosystem reactions to environmental change. By combining these analytical tools, AI provides scientists and policymakers with a way to "game" the trade-offs between food production, water security and ecosystem preservation conservation. Such integrative analysis facilitates the pursuit of sustainability strategies that simultaneously consider multiple, and possibly competing, environmental goals, rather than treating these goals as if they were isolated sectoral measures.

Ethical, Epistemological, and Governance Considerations

The use of AI in the environmental sustainability context also has its own unique ethical, epistemological and governance implications. Privacy and ethical issues related to fairness, accountability and potential societal outcomes of algorithmic decision-making. For instance, resource distribution systems that use AI may unintentionally discriminate against marginalized groups if the data that underpins them is a reflection of current inequalities. "Transparency" and "explainability" are thus crucial when AI-driven models impact policy decisions or livelihoods. Participants must have insight into the process of prediction and how uncertainty is addressed. In terms of their knowledge-generation, AI systems are in large part correlation-based, rather than causal-based. Therefore, their outputs need to be interpreted with caution along with expert knowledge and ecological theory (Shankar *et al.,* 2020).

Issues of governance, such as ownership of data, privacy and regulatory intervention, are also implicated. Environmental data may reveal details about land use, water, or biodiversity position. Well-defined governance mechanisms are also required to ensure responsible data use and institutional accountability. More recently, research has highlighted the trade-offs, both in terms of environmental and societal impacts, of AI adoption. Developing large-scale AI systems requires extensive computational infrastructure, which can be a source of energy use and carbon emissions. Meanwhile, the digital divide could further the divide between regions with solid technological capacity and those without, as they race to acquire solutions rather than build capacity. To tackle these issues, there is a need for cross-cutting governance models that marry technological innovation with ethical oversight, participatory decision-making processes, and just access to environmental data systems (Adewusi *et al.,* 2024).

AI APPLICATIONS IN SUSTAINABLE AGRICULTURE

In smart farming, AI is driving more and more the decisions at the farm level. AI platforms are no longer just an add-on to technology, but are becoming critical enablers to data-driven decision making in product planning, resource management, and climate risk management. The sustainable production of food requires a harmonious balance of productivity, environmental viability and social resiliency. AI serves these goals by processing vast quantities of agronomic, climatic, and biological data and converting them into usable information for farmers and agricultural planners. Machine learning and computer vision algorithms combined with decision support tools are enabling a new breed of precision farming technology.

Notably, these technologies are not confined to the large commercial farming enterprises. Mobile-based advisory tools, satellite-derived data services, and cheap sensors are beginning to deliver AI-informed recommendations to smallholder farmers and climate-vulnerable producers, who are often most exposed to environmental risks. When properly utilized, AI-enabled systems can alleviate uncertainty in farming decisions, enhance effectiveness, conserve natural resources, and build longer-term resilience to climate variability. The performance of these technologies, however, depends on the availability of data, local infrastructure, and the extent to which farmers are able to interpret and act upon digital advice. Here's how we break down some core AI applications for sustainable agriculture, looking also at practical implications for decision making and sustainability outcomes.

Precision Agriculture Frameworks

Precision agriculture is an approach to farm management that relies on information-based technologies to observe and react to spatial and temporal heterogeneities found within agricultural fields. AI enhances these systems by combining satellite images, sensor networks and field observations into analytical platforms that provide recommendations at the farm level. Farmers can use these tools to ensure that water, fertilizer and pesticides are only applied where and when needed. Such a focused management can mitigate environmental effects, reduce production costs, and enhance crop performance if implemented under the right conditions (Ali *et al.,* 2025). Precision agriculture is now shaping the decisions farmers make daily, not just how efficiently they use technology. Using the AI computed results in the form of field maps and predictions, farmers are able to know the best planting densities, irrigation regimens, and nutrient applications. For smallholding farmers with limited resources, simple mobile-based interfaces and advisory services can help transform complex AI findings into actionable guidance for work in the field. The system design of these methods in general follows through a data-processing pipeline with stages of raw data acquisition, preprocessing, model training and predictive validation, which is demonstrated in Figure 1 (Saha *et al.,* 2025). Precision systems can also enable the continuous monitoring and dynamic optimization of farm management. By ditching the old-time-fits-all approach of farming, precision agriculture empowered by AI advances productivity and environmental sustainability (Pimenow *et al.,* 2025).

Figure 1. Flowchart of data collection, preprocessing, model training, and validation steps for AI and machine learning-based crop yield prediction in precision agriculture

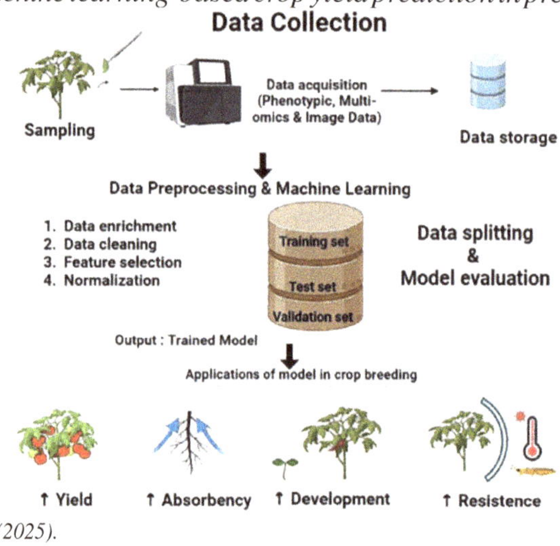

Source: Saha et al. (2025).

Predictive Analytics for Crop Yield Optimization

Predictive analytics is based on a combination of historical data and real-time environmental information to predict farming results. AI models predict yield considering weather, soil, crop variety, irrigation, and other factors. These predictions enable farmers to plan when to plant, what to plant and how to best use their resources. Machine learning methods are capable of capturing complex nonlinear interactions among environmental variables, which are often not detectable by traditional statistical methods. After implementation in a proper environment and based on reliable datasets, predictive analytics also enables the increase of yield stability and the decrease of wasteful input use. Nevertheless, the quality of the results of predictive models is highly dependent on the quality and representativeness of the training data. Models trained in one redecoration may not necessarily be robust when simulated in other climatic or edaphic regions. Therefore, local calibration and validation are still critical in order to make AI-yield predictions transferable among various farming systems (Ziesche *et al.,* 2023).

Soil Health Monitoring Algorithms

Soil health is a key aspect of sustainable agriculture due to its effect on nutrient cycling, soil water holding capacity, and long-term crop production. AI-based soil monitoring systems combine data received from sensors, satellite imagery, drones, and laboratory analysis to determine soil status in near real time. Algorithms can be used to identify nutrient deficiencies, salinity, organic matter and microbial activity patterns. These can inform site-specific recommendations for soil amendments and nutrient management. By detecting differences in soil quality across a field, AI systems could help farmers use fertilizers more efficiently and reduce the chemicals they apply. Enhanced soil management further aids in climate mitigation by enabling practices such as decreased fertilizer use, enhanced soil carbon sequestration, and regenerative agriculture practices like crop rotation and the addition of organic matter to soil.

Intelligent Pest and Disease Detection Systems

Invasive pests and diseases continue to pose a serious risk to agricultural production and food security globally. AI-based detection systems are becoming more common; to detect infestations at an early stage and farmers can act before damage is spread over a large area. These solutions leverage image recognition, environmental monitoring, and predictive analytics to identify the symptoms of diseases and pests' outbreaks. Early identification allows more precise responses, with less use of broad-spectrum chemical pesticides. This strategy not only reduces environmental impacts but also helps maintain honeybees and other beneficial organisms which play a role in the ecological balance of agricultural systems.

Image Based Diagnostic Models

Historically, image-based diagnostic systems rely on computer vision algorithms to process images of the leaves, stems or fruits of plants. By analysing visual patterns against extensive annotated datasets, AI models are able to identify symptoms such as discoloration, lesions or abnormal development. Convolutional neural networks (CNN) are well suited for this task because they learn hierarchical visual features distinguishing from healthy plants to infected plants. Farmers can take pictures with their mobile phones or drones and get a diagnosis immediately (Botero-Valencia *et al.,* 2025). Early diagnosis facilitates timely treatment, and can prevent disease spreading as well as mitigate the overuse of pesticides. Such approaches are particularly useful in isolated areas that lack access to expert agronomists.

Early Warning Forecasting Tools

Early warning systems combine climatic information, population dynamics of the pest, and past records of outbreaks to assess the probability of occurrence of infestations. Machine learning models process temperature, humidity, precipitation, and wind trends to predict cycles of pest emergence. Farmers and agriculture authorities are alerted when the risk thresholds are met or surpassed. These are forecasts for taking action, rather than responses to damage. Early action protects crops from infestation, decreases the use of chemicals and enhances the country's capacity to cope with biological threats. Through predictive warning mechanisms, they also play a positive role in stabilizing regional agriculture and protecting the environment (Padhiary *et al.,* 2025).

Climate Adaptive Agricultural Planning

Climate is variability is one of the major constraints to sustainable agricultural production. Uneven precipitation, changing temperatures and more extreme weather events are routinely interfering with traditional farming schedules, and are causing a decline in crop yields. AI-based technologies enable climate-adaptive agricultural planning through modelling of interactions under prevailing weather between weather patterns and crop growth and environmental risks. This type of analysis directly feeds into climate adaptation (to allow farmers to better predict environmental changes and adjust their agricultural practices accordingly). Similarly, better resource management and practices can also contribute to mitigation by reducing energy use, emissions of fertilizers, and water use associated with farming.

Seasonal Prediction Modelling

Seasonal prediction models use historical climate data Seasonal prediction models are based on historical climate data and real time meteorological inputs to predict the climate phase for a few months in advance. AI is able not only to identify droughts, floods, or heat waves – it can spot the early warning signs. Farmers can apply these predictions to choose crops and varieties adapted to anticipated conditions, to change planting dates, and to schedule irrigation. Seasonal predictions also help with food supply management and contingency planning for government authorities. Farmers can minimise risk and maximise yield stability by synchronizing farming practices to predicted climate conditions (Zhang *et al.,* 2021).

Risk Informed Decision Support Systems

AI-driven decision support systems can integrate various types of data and provide a succinct recommendation. These processes analyse the risk involved with the variability of weather, soil condition, potential pest, and market of commodities. By offering a scenario-based analysis they enable farmers to weigh potential options and identify the most sustainable course of action. Risk informed platforms, in particular, are extremely useful for smallholder farmers, as they offer an expert level of guidance without the need for extensive training. Bringing analytics of risk into routine decisions creates more robustly informed choices and more responsible use of resources (Zhang *et al.,* 2021).

Autonomous Agricultural Machinery and Robotics

Autonomous technology is a major development in the mechanization of farming. Robots enabled by AI can plant, weed and harvest, as well as monitor crops with little human involvement. These are sensor-based machines, navigation systems and learning algorithms are combined to further improve the adaptability of these robots in highly dynamic operations. Precision robotics diminishes labour intensity and guarantees uniform job quality on vast surfaces. For instance, detection systems on autonomous weed removal machinery can detect harmful plants and physically remove those instead of spraying chemicals. This strategy reduces herbicide application and safeguards soil and water conditions. In addition, robotic harvesters can operate non-stop and can pick up produce gently, thus decreasing post-harvest loss and enhancing the supply chain (Mishra and Mishra, 2024; Chang *et al.,* 2023).

Resource Efficiency and Carbon Footprint Reduction Strategies

Sustainable agriculture is the one that maintains its productivity and utility to society over time without having an adverse impact on environment. AI contributes to this goal by improving resource efficiency and reducing the emission of greenhouse gases. Smart irrigation systems process soil moisture information and weather predictions and irrigate only when it is needed. In the similar way, nutrient management schemes specify the number of fertilizers needed to avoid overuse with the risk of soil salinization and water pollution (Mumtaz *et al.,* 2022). AI is also facilitating carbon footprint analysis by calculating the emissions linked to agricultural activities. These data enable farmers to implement low emission options, for example, with minimal disturbance of the soil (reduced tillage), more efficient fertilizer usage, or by becoming energy-neutral. The Correspondence between Sustainability Outcomes and AI Approaches is indicated in Table 2.

Table 2. AI techniques and their sustainability contributions in agriculture

AI technique	Primary function	Sustainability benefit	Evidence	Reference
Machine learning prediction	Yield and climate forecasting	Reduces input uncertainty and resource waste	10–20% crop yield improvement; up to 30% reduction in nitrogen fertilizer use	Pantazi *et al.* (2016); FAO (2022)
Computer vision (CNNs)	Pest and disease detection from field imagery	Minimizes pesticide overuse and crop loss	>95% diagnostic accuracy; 20–40% reduction in projected crop losses	Mohanty *et al.* (2016); Phadikar *et al.* (2013)
Sensor data analytics (IoT)	Real-time soil moisture and water quality monitoring	Conserves soil health and freshwater resources	25–30% reduction in irrigation water use; 15% improvement in nutrient use efficiency	Wolfert et al. (2017); Liakos *et al.* (2018)
Robotics and automation	Precision planting, weeding, and harvesting operations	Lowers labour costs, herbicide inputs, and post-harvest losses	Up to 90% reduction in herbicide application in robotic weeding trials	Bac *et al.* (2014); Fountas *et al.* (2020)
Decision support systems (DSS)	Strategic farm and supply chain planning	Enhances resilience and operational efficiency	12–18% improvement in resource allocation efficiency reported across case studies	Shim et al. (2002); Kamilaris & Prenafeta-Boldú (2018)

Source: Synthesized from Pantazi *et al.* (2016), Mohanty *et al.* (2016), Wolfert *et al.* (2017), Liakos *et al.* (2018), Bac *et al.* (2014), Fountas *et al.* (2020), and Kamilaris and Prenafeta-Boldú (2018).

As we can observe in Table 2, every AI technique enhances sustainability through unique activity-specific contributions. Predictive models reduce overapplication of inputs, and monitoring projects make certain that environmental parameters stay within appropriate bounds, that is, within optimal environmental ranges. Robotics brings precision, and decision support platforms complicated data into actionable recommendations. Collectively, these innovations constitute a comprehensive system that encourages agricultural efficiency and at the same time safeguards natural harmony (Xing and Wang, 2024).

Implementation Barriers, Equity, and Smallholder Access

Although the potential benefits of AI for agricultural sustainability are quite promising, the benefits are unevenly distributed across farm systems. Use usually requires access to dependable digital infrastructure, good quality data and technical know-how, all of which are in short supply in many developing areas. Smallholder farmers, who underpin much of the food supply in the world, have to contend with challenges including poor internet connectivity, steep costs for digital tools, and barriers to training in farm management informed by data. Moreover, AI systems developed mostly using large-scale industrial farms data may not adequately represent the environmental and socio-economic conditions experienced by smaller farmers.

To reduce these divides, enabling-policy environments and inclusive technological design are needed. Public data platforms, open-source AI tools and mobile-based advisory services can contribute to reducing entry barriers for smallholder farmers. Farmer-governed data systems and participatory data collection efforts might also be better able to assist farmers in making sense of AI-based advice or in integrating AI-based advice into local farming contexts. Democratizing access to AI technologies thus has importance for sustainable and climate-resilient agriculture. If the efforts to address data gaps, infrastructure constraints and knowledge inequalities are not made, the promises of digital agriculture may remain benefits of high-tech farming systems and not much of help in environmental sustainability at large.

Institutional Adoption Pathways in Agriculture

Role of Ministries, Extension Services, and Cooperatives

The deployment of artificial intelligence in agriculture does not occur in a vacuum. It depends, rather substantially, on the institutional architecture that surrounds the farming community — the government ministries that set policy directions, the extension services that carry knowledge to the field level, and the cooperatives that aggregate smallholder interests into something actionable. In India and across much of South Asia and sub-Saharan Africa, these three institutional actors together determine whether AI reaches farmers or remains confined to pilot demonstrations and academic literature. Agricultural ministries occupy the most consequential position in this ecosystem. At the national level, ministries determine which technologies receive budgetary support, which regulatory frameworks get enacted, and which public-private partnerships are formalised. India's Ministry of Agriculture and Farmers' Welfare, for instance, has piloted AI-driven advisory platforms through schemes such as the Digital Agriculture Mission, which funds the development of crop monitoring tools, soil health dashboards, and automated advisory systems delivered via mobile interfaces. Without this kind of institutional backing, AI tools tend to remain the preserve of large commercial farms or well-funded research institutions. What is needed, and what the evidence from various national programmes suggests works, is a deliberate policy mandate that treats AI adoption as an extension of existing agricultural development priorities — not as a separate technology initiative sitting alongside them.

Agricultural extension services are the connective tissue between policy and practice. In most developing countries, extension workers visit farms, conduct demonstrations, and serve as trusted intermediaries between scientific knowledge and farm-level decisions. AI systems that bypass extension workers risk producing technically capable but institutionally orphaned tools. The more productive approach involves training extension agents to use and interpret AI-generated outputs — soil nutrient maps, pest risk forecasts, irrigation scheduling recommendations — and then translate these into locally appropriate advice. This integration is not without friction. Extension services in many countries remain understaffed, underfunded, and institutionally siloed from the agricultural research organisations that develop AI tools. Bridging this gap requires not only training but also organisational reform — specifically, creating feedback loops through which extension workers can report on tool performance, flag inaccuracies, and contribute locally gathered data back into model training pipelines. Cooperatives present a different but equally important pathway. Because smallholder farmers individually lack the purchasing power, data infrastructure, and technical capacity to adopt AI tools, collective institutions can serve as aggregating mechanisms. A farmers' cooperative with two thousand members, for example, can negotiate bulk access to a remote sensing advisory platform, pool agronomic data across member farms to generate statistically meaningful insights, and employ one or two technically trained staff to manage the AI interface on behalf of members. This model has been tested with some success in Maharashtra and Karnataka for cotton and sugarcane, where cooperative-level AI adoption proved more durable than individual farmer adoption. The cooperative model also addresses a trust deficit that often accompanies new technology: farmers who are sceptical of an AI recommendation issued by an unfamiliar company are more likely to accept the same recommendation when it is endorsed and explained by their own cooperative's field staff.

Integration with Existing Agricultural Extension Systems

Integrating AI tools into existing extension systems requires attention to both the technical and organisational dimensions of the problem. On the technical side, AI platforms must be designed to interface with the data formats, languages, and communication channels that extension services already use. In India, many extension workers interact with farmers through platforms such as Kisan Call Centres and state-level mobile advisory services. AI tools that feed intelligible, vernacular-language outputs directly into these existing channels have a far higher probability of adoption than standalone applications that require extension workers to learn a new interface, log into a separate system, and manually transfer information across platforms. Organisationally, integration demands that AI developers engage extension departments during the design phase rather than arriving at the implementation stage with a finished product. Participatory design — which involves extension workers, senior agronomists, and representative farmers in defining what information the system should produce, in what format, and at what temporal resolution — tends to generate tools that are both technically adequate and institutionally legitimate.

Cost, Affordability, and Access to AI Tools

Low-Cost and Open-Source vs. High-End Systems

The cost profile of AI tools in agriculture spans an enormous range, and the implications of this range for equity are difficult to overstate. At one end of the spectrum sit high-end precision agriculture platforms — commercial drone fleets equipped with multispectral sensors, subscription-based satellite imagery services, and proprietary crop modelling software — that cost lakhs of rupees annually to operate and are effectively accessible only to large landholdings or well-capitalised agribusinesses. At the other end sit open-source alternatives and low-cost mobile applications built on publicly available datasets, satellite imagery from Sentinel and Landsat missions, and machine learning frameworks distributed under open licences.

The practical performance gap between these two ends of the spectrum has narrowed considerably over the past decade, largely because of the growing availability of high-resolution freely accessible satellite data, the proliferation of pre-trained machine learning models that can be fine-tuned at low computational cost, and the widespread adoption of smartphones among rural populations. Tools such as Plantix for pest identification, SoilGrids for soil nutrient mapping, and various

NDVI-based crop monitoring applications now offer smallholder farmers a level of diagnostic capability that would have required expensive laboratory or aerial infrastructure a decade ago. The critical limitation of low-cost tools, however, is not their technical capability but their support infrastructure: open-source tools do not come with customer service, software updates tailored to local conditions, or agronomic interpretation support. This is where institutional actors — extension services, cooperatives, and non-governmental organisations — must step in to provide the human layer that commercial platforms include in their subscription fees. Table 3 summarises the general cost and accessibility characteristics of representative AI tool categories relevant to smallholder agriculture in India.

Table 3. Cost and accessibility profile of AI tool categories for Indian smallholder agriculture

AI Tool Category	Approximate Annual Cost ( )	Data Requirements	Technical Support	Typical Adopter
Commercial drone + analytics platform	3,00,000– 8,00,000	High (proprietary)	Included	Large farms, agribusinesses
Subscription-based satellite advisory	15,000– 60,000	Moderate	Limited	Medium farms, FPOs
Open-source mobile advisory apps	0– 2,000 (data costs)	Low (public datasets)	None	Smallholders with smartphones
SMS/IVRS-based AI advisories	0 (government-subsidised)	Minimal	Mediated via extension	Marginal and remote farmers

Source: Author's compilation
FPO = Farmer Producer Organisation; IVRS = Interactive Voice Response System.

As Table 3 illustrates, the tools best suited to the majority of Indian farmers — those operating less than two hectares of land — are concentrated at the lower end of the cost spectrum, where institutional support is thinnest. Addressing this mismatch is arguably more important than developing technically superior AI systems.

Financial Barriers for Smallholder Farmers

Even nominally low-cost AI tools impose real financial barriers when one accounts for the full cost of access: smartphone ownership, reliable mobile data connectivity, the time cost of learning a new interface, and the implicit cost of acting on an AI recommendation that turns out to be wrong. For a farmer operating on a household income of 80,000– 1,20,000 per year, a failed recommendation that results in crop loss or unnecessary input expenditure is not an abstract risk — it is a potentially catastrophic one. This asymmetry between the risks borne by farmers and the optimism embedded in technology adoption narratives is a persistent structural problem in agricultural AI discourse.

Financial instruments that reduce the cost of experimentation — input subsidies conditional on AI-guided practices, crop insurance products that incorporate AI-assessed risk, and government-sponsored access programmes that provide free or heavily subsidised subscriptions to vetted AI tools — have shown promise in reducing adoption barriers. Several state governments have begun bundling AI advisory services with existing subsidy schemes for seeds and fertilisers, effectively making AI access costless at the point of use for enrolled farmers.

Capacity-Building and Training Needs

Skills for Farmers and Extension Workers

Effective use of AI tools in agriculture requires a baseline level of digital and agronomic literacy that cannot be assumed across the farming population or, for that matter, across the extension workforce. The capacity-building challenge is therefore twofold: it involves building the digital literacy necessary to access and operate AI interfaces, and the agronomic literacy necessary to critically evaluate AI-generated recommendations rather than accepting them uncritically. For farmers, the most effective training approaches documented in the literature are those that are embedded in familiar social contexts — farmer field schools, cooperative training sessions, and peer-to-peer learning arrangements in which experienced users demonstrate tools to their neighbours. Abstract classroom training in AI concepts has consistently proven less effective than hands-on demonstration of specific, locally relevant use cases: using a smartphone application to identify a pest on a crop that is currently in the field, interpreting a soil health card generated by an AI-assisted testing device, or adjusting an irrigation schedule based on an AI-generated evapotranspiration forecast. The specificity and immediacy of these demonstrations make the technology legible in a way that general awareness programmes cannot.

For extension workers, training needs are more technically demanding. Extension agents who are expected to use, interpret, and communicate AI outputs need grounding in the data sources that underpin AI recommendations — satellite imagery, weather station data, soil sensor readings — as well as a working understanding of the conditions under which AI models perform poorly: unusual weather events outside the training data distribution, crop varieties not represented in model training sets, pest pressures that are locally prevalent but globally rare. An extension worker who understands these limitations is in a position to exercise appropriate scepticism, supplement AI recommendations with local knowledge, and explain to farmers why a recommendation may need adjustment in their specific circumstances. This kind of critical AI literacy, rather than uncritical deference to algorithmic outputs, is what sustainable adoption looks like in practice. Training programmes for extension workers should therefore include modules on data interpretation, model uncertainty, and the practical mechanics of reporting tool failures — because the quality of AI systems in agriculture will improve only if there are systematic mechanisms for capturing where and why they fail at the farm level.

AI IN INTEGRATED WATER RESOURCE MANAGEMENT

Integrated water resource management now calls for impersonal governance systems determining among the competing demands on scarce water reserves, including agriculture, domestic supply, ecosystem, and industry. However, under climate variability, population growth, and resource constraints, AI is becoming an important analytical tool to assist in the making of such governance decisions. Data-enabled decision making, which can revolutionize water management through AI across hydrologic, ecological, agricultural, and governance scales, has in enabled transformative analytics in the context of secured water resource management over the past two decades integrated w. Traditional methods of water management such as static modelling, manual monitoring, and reliance on historical averages, can limit the ability to adapt to changing climate conditions, land use, and population growth. Instead, AI systems have the potential to bring an array of data, such as satellite images, sensor networks, meteorological records, and hydrological measurements, to offer adaptive and predictive information. This allows us to use water more efficiently, to decrease our vulnerability to extreme events, to improve monitoring of water quality, and to assist with institutional planning. These capabilities led to more efficient prediction of system performance and could be useful for the prediction of other systems where similar optimization is desired. In addition to prediction and optimization, AI methods are also employed to support allocation prioritization, regulatory enforcement, and coordination among institutions in charge of basin governance. Importantly, AI is not a replacement for hydrological science, it expands existing models through pattern recognition, optimization and automated reasoning and as such is a valuable tool for integrated basin management (Silvestro *et al.,* 2022).

Computational Hydrology and Predictive Water Modelling

Computational hydrology is the application of modern numerical and data-focused techniques for simulating the transport of water through the land surface, river basins, underground aquifers, cloud-aerosol systems, and other components of the atmosphere. Artificial intelligence has a role to play in modelling the complex and nonlinear interactions between climate variables, soil properties, vegetation cover, and topography. Among machine learning models, in particular, deep neural networks and hybrid physics–AI models can achieve better performance than traditional deterministic models when adequate data are provided (Xiang *et al.,* 2021). These models allow for continuous updating as new observations become available, so predictions can stay accurate when the environment is changing. Predictive water modelling also takes advantage of the strength of AI to handle both spatial and temporal data sets at once. For example, the seasonal dependencies can be captured by recurrent neural networks, and the spatial patterns in the precipitation fields can be interpreted by convolutional networks. The two may be blended to achieve improved accuracy. Conclusive: "such integration is necessary for high-resolution forecasting that is required for watershed management, reservoir operation, and climate adaptation planning (Gul and Banday, 2024)." These forecasting capabilities also serve as the analytical basis for the derivation of policy decisions regarding reservoir release, drought allocation policies, and emergency flood response activities.

Rainfall–Runoff Simulations

Rainfall–runoff modelling is a central part of hydrologic analysis as it allows to determine the portion of precipitation that will appear as surface flow. AI-based methods improve this process by learning catchment-specific hydrological responses with little parameter calibration. Conventional models rely significantly on assumptions relating to infiltration, evapotranspiration and soil storage. AI techniques learn instead to infer these relationships from data – which can be especially valuable in places where there are few surface measurements. Decades of rainfall and discharge records can be processed by deep learning architectures to recognize latent hydrologic patterns. They also account for covariates such as changing land use and temperature that affect runoff. Consequently, rainfall-runoff modelling with AI is expected to be more reliable for extreme weather events, enhancing early warning systems and infrastructure planning and design (Garske *et al.,* 2021; Haleem *et al.,* 2023).

Flood and Drought Forecasting Algorithms

Hydrological extremes such as floods and droughts differ in terms of hydrological deficit and surplus, respectively, but both may be effectively simulated based on AI. Algorithms for flood prediction use precipitation intensity, upstream water levels, soil saturation, reservoir releases to determine the probability of inundation at multiple points downstream. These models frequently run in near real time, providing alerts several hours to days in advance of peak flow, and such lead time is essential for evacuation by river residents and dam operations for water release (El Jarroudi *et al.,* 2024). The forecasting system conventionally starts with diverse sourced data collection followed by preprocessing and deep learning model training — most significantly utilizing Long Short-Term Memory (LSTM) networks or Gated Recurrent Units — towards the best discharges at a reference downstream station, such a workflow is diagrammed in a structured way in Fig. 3, demonstrating this entire workflow for a comparative multi-model flood forecasting case study on a major river basin (Kim et al., 2024). Algorithms to forecast droughts are based on long term signals including rainfall deficits, groundwater depletion, vegetation stress, and temperature anomalies. Machine learning algorithms are able to forecast early warning signals several months ahead by capturing subtle patterns that possibly been ignored in typical statistical analyses. Because it integrates satellite-based vegetation indices with climate projections, it has risk detection capability for probable droughts that enable policymakers to organize water rationing and crop selection (Alotaibi and Nassif, 2024).

Figure 2. Flowchart for the data-driven LSTM-based flood forecasting model that includes data collection, preprocessing, training, hyperparameter optimization and discharge prediction output for a downstream reference station with multiple outputs from multiple upstream reservoirs

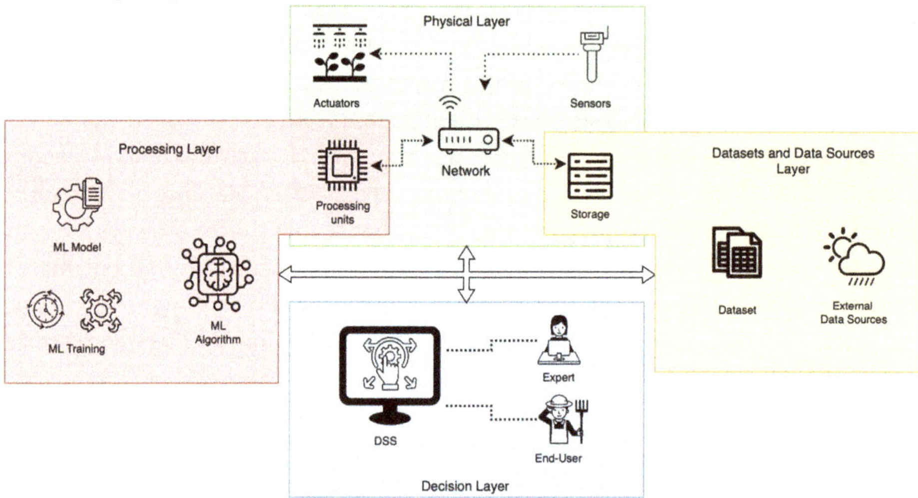

Source: Yi and Yi (2024).

Smart Irrigation and Allocation Optimization

Proper water distribution is critical to an agricultural and/or urban water system's sustainability. Using AI-based control methods, irrigation systems can be adaptive instead of running on predetermined schedules. Smart irrigation systems use information such as soil moisture, weather conditions and forecasts, stage of crop growth, and rates of evapotranspiration to estimate the required amount of water (Pasha Mohammed *et al.,* 2025). The physical sensor and actuator layer (P/SAL) deployed in the field, the IoT gateway with cloud processing layer (G/CPL), and the machine learning decision output layer generating real-time irrigation commands, are three hierarchically structured layers of the technical architecture considered herein for such systems, which is schematically illustrated in Figure 3 (Del-Coco et al., 2024). This multi-tiered system cuts down on wasted water, energy, and soil erosion related to overwatering. On a scale larger than the farm, AI can also guide basin-wide water allocation decisions by simulating competing demands between agricultural, domestic, industrial, and ecological use across space and time. Using the AI-based allocation models, it is possible to simulate multiple future scenarios and assess the trade-offs between reliability of water supply and environmental flow needs to enhance transparency in management of resources.

Figure 3. This functionality contributes toward equitable allocation while ensuring the environmental flow needs. physical sensor layer (bottom layer) IoT gateway cloud processing layer machine learning decision output for precision water management Schematic architecture of the smart irrigation platform

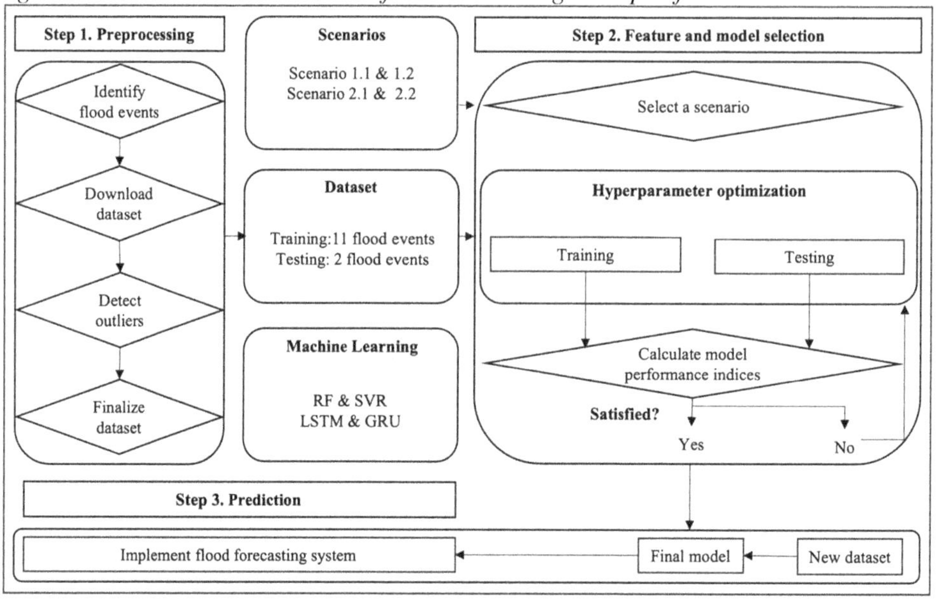

Source: Del-Coco et al. (2024).

Reinforcement Learning for Irrigation Scheduling

Reinforcement learning-based models are more suitable for irrigation scheduling as they allow systems to learn optimal decisions based on feedback through multiple iterations. In this process, an agent operates within a simulated or real farm environment, is awarded according to performance in terms of crop yield or water use efficiency, and learns over time. With the passage of time, it discovers a set of irrigation strategies that can increase production and reduce water use. Such models may be localized (specific soil, crop type, climate variability). Instead of rule-based controllers that are fixed and cannot be modified after design time, the RL policies are continuously updated as new environmental data stream in. This flexibility allows them to be particularly useful in areas with erratic rainfall patterns (Espinel *et al.*, 2024).

Demand Forecasting and Supply Balancing

Correct demand prediction is essential for efficient water distribution. AI methods including time-series analysis and probabilistic prediction can predict water demand for cities, industries and farms in the short-term and long-term. These projections take into account population growth, economic activities, seasonal consumption patterns, and climatic conditions. Supply balancing is the process of aligning predicted demand with some expectation of available water resources (Mishra and Dwivedi, 2025). Optimization routines may also suggest reservoir release schedules, groundwater extraction limits, and transfer volumes between storage facilities. Coupled with decision support systems, these instruments assist managers in circumventing shortages and the risk of aquifer over-exploitation.

Water Quality Monitoring through AI-Enabled Sensing

Historically water quality monitoring has relied upon periodic laboratory analyses, with a relatively slow turnaround and limited spatial coverage. AI-enabled sensing technologies address these challenges by integrating real-time sensors with automated data processing and analysis. Parameters such as turbidity, pH, dissolved oxygen, and contaminant concentrations are measured by sensors placed in rivers, reservoirs, and delivery systems. These data streams are analysed continuously by the AI models, which identify trends and anomalies. The addition of remote sensing technology enables even more options for monitoring. Processed satellite data by machine learning can detect algal blooms, sediment plumes and sources of pollution over vast geographic extents. This process allows for authorities to have a proactive approach in dealing with potential threats and keeping their water supplies safe.

Anomaly Detection in Water Systems

Anomaly detection algorithms look for patterns that are abnormal and differ from the normal behaviour of the system. In water systems an anomaly may signal a pipe leak, an illegal dump, sensor malfunction or an event of contamination. Historical operational data-trained machine learning models can learn what normal system signatures look like and automatically flag anomalies. Anomalies detection at the early stage mitigates economic loss and health risk for the general public. For instance, detecting a pipeline leak in hours instead of days will save a large quantity of water and repair expense. Identifying unusual chemical concentrations in drinking water systems enables officials to take action before contamination is widely circulated (Sharma and Shivandu, 2024).

Contaminant Prediction Models

Along with identifying current problems, AI can predict future contamination incidents. Predictive models use patterns of land use, rainfall forecasts, industrial activity, and hydrological flow paths to assess the probability of pollutants reaching water bodies. These predictions enable regulators to apply precautionary measures such as the limitation of industrial discharge during rainstorm events or by enhancing treatment processes ahead of time. Contaminant prediction models also contribute to protecting the environment by detecting areas within watersheds that have a higher potential risk. This information informs focused monitoring and remediation, enhancing both efficiency and environmental protection.

Decision Intelligence for Watershed Governance

Decision intelligence applies AI to noneconometric policy analysis and economic modelling and to stakeholder engagement and participation to facilitate integrated watershed management. Watersheds often span jurisdictional boundaries and are subject to multiple agencies that dictate water distribution, environmental conservation and infrastructure. Platforms enabled with AI include multiple information sources, such as hydrological data, legal constraints, infrastructure capacity and environmental indicators, and provide these through integrated analytical dashboards. These systems permit users to calculate competing demands on water such as between agriculture, urban utilities, and ecological flow needs. Scenario modelling can be used to simulate the effect of population and climate change, or the impact of a new piece of infrastructure, on basin water balances. In modelling the trade-offs between water supply reliability, ecological conservation, and economic costs, AI also allows institutions to coordinate decisions across borders and to prioritize allocation when resources are scarce. Evidence-based decision support of this variety fosters transparency and promotes fewer conflicting outcomes for stakeholders (Habila *et al.,* 2023).

Policy Analytics and Regulatory Compliance Monitoring

Policy analytics utilises AI to assess the efficacy of water policies and their compliance. Machine learning algorithms can also sift through years of administrative paperwork, inspection reports and sensor readings to assess if water users are staying within legal limits. Automated compliance monitoring streamlines the enforcement process and allows for more accuracy in enforcement. AI also helps inform policy by revealing the best performing regulatory actions for sustainable outcomes (Tiwari *et al.,* 2023). For example, predictive models can predict the effects of pricing or consumption restrictions on patterns of use. They can then adapt their policies in a forward-looking way, rather than waiting for resources to be over-exploited or the environment to be damaged.

As can be observed from Table 4, various AI techniques are applicable to various operational goals. Neural networks are best for complex hydrological forecasting while reinforcement learning is better for dynamic control problems (e.g. irrigation scheduling). Ensemble algorithms such as random forests provide a level of stability when the data is noisy or incomplete which is particularly useful in demand forecasting. The synergism between the two approaches suggests that IWRM should investigate using hybrid systems combining different AI approaches rather than a single algorithmic framework.

Table 4. Major AI techniques used in integrated water resource management and their functional roles

AI technique	**Primary function**	**Typical application area**	**Performance**	**Benefit**	**Reference**
Neural Networks (LSTM, CNN)	Nonlinear pattern recognition and time-series modelling	Rainfall–runoff modelling, streamflow forecasting	NSE = 0.87–0.96 in discharge prediction; RMSE reduced by 18–35% vs. traditional models	High predictive accuracy under non-stationary conditions	Kratzert *et al.* (2018); Shen (2018)
Reinforcement Learning	Sequential decision optimization	Irrigation scheduling, reservoir release management	~25% reduction in irrigation water consumption; 15% improvement in reservoir yield	Adaptive water uses with continuous policy improvement	Mnih et al. (2015); Ryu *et al.* (2020)
Support Vector Machines (SVM)	Classification and regression under small samples	Water quality assessment, drought classification	88–93% classification accuracy for water quality parameters	Reliable detection even with limited training data	Vapnik (1995); Najah *et al.* (2013)
Random Forests	Ensemble-based robust prediction	Demand forecasting, groundwater level prediction	R^2 = 0.89–0.95 in groundwater depth prediction; 10–20% lower RMSE than single-tree models	Stable performance with noisy, incomplete datasets	Breiman (2001); Naghibi *et al.* (2016)
Anomaly Detection Models (Autoencoders, Isolation Forest)	Real-time outlier identification in sensor data	Pipe leak detection, contamination event flagging	Leak detection within 2–6 hours vs. 24–72 hours manually; >91% recall rate	Rapid early warning, reduced water loss and health risk	Candelieri (2017); Ye *et al.* (2022)

Source: Synthesized from Kratzert *et al.* (2018), Shen (2018), Ryu *et al.* (2020), Najah *et al.* (2013), Breiman (2001), Naghibi *et al.* (2016), Candelieri (2017), and Ye *et al.* (2022).

Equity, Access, and Governance Capacity in Water AI

Although there is a great potential for improving water management with AI technologies, it is not the same for everyone, neither regionally nor institutionally. However, many water governance systems, especially those in developing regions, function with rudimentary technical infrastructure, minimal monitoring networks, and weak institutional capacity. Therefore, the implementation of AI-driven decision systems may exacerbate existing disparities if access to technology and data remain unequal. Rural communities, smallholder farmers and other marginalized water users may be excluded from accessing digital water management tools or engaging in data-driven governance processes. In addition, algorithmic decision systems may be based on datasets that are not sufficiently representative of local hydrology and community water needs. Meeting such challenges will take investment not only in digital infrastructure but also in participatory governance processes that engage local actors in data gathering, model interpretation, and policy enactment. Initiatives to build capacity in water agencies and community organizations are also critical to ensure that AI-enabled water management leads to fair and inclusive results, and does not exacerbate existing imbalance in water governance.

Institutional Adoption and Governance Frameworks

Role of Water Authorities and Basin Management Agencies

Water resource management in most countries operates through a layered institutional structure — national water ministries set broad policy frameworks, river basin authorities manage resources at the watershed scale, and local urban or rural water utilities handle distribution and treatment at the point of use. Introducing AI into this system is not simply a matter of deploying a better forecasting model; it requires that each layer of this institutional hierarchy understands what AI tools can and cannot do, has the data infrastructure to feed them, and possesses the decision-making authority to act on their outputs. Where these conditions are absent, AI systems tend to generate technically impressive outputs that accumulate in reports without influencing operational decisions.

River basin management agencies occupy the most strategically important position in this picture. They sit at the intersection of upstream catchment dynamics and downstream water demand, and they are typically responsible for allocating water among competing users — irrigators, urban utilities, industrial consumers, and ecological flows. AI tools that can model these competing demands simultaneously, forecast seasonal availability under different climate scenarios, and simulate the consequences of alternative allocation decisions are, in principle, highly valuable for basin agencies. In practice, however, many basin agencies in developing countries operate with fragmented data — gauge stations that have not been maintained, satellite imagery that has not been processed, and groundwater monitoring networks that provide only intermittent readings. Before AI can be meaningfully deployed in these contexts, significant investment in basic hydrological monitoring infrastructure is required. AI does not substitute for measurement; it processes and interprets measurement at scale.

In India, the National Water Mission and the Jal Shakti Ministry have initiated steps toward data integration across river basins through platforms such as the National Hydrology Project, which aims to consolidate river flow, groundwater, and rainfall data into a centralised repository accessible to basin managers. AI tools built on top of this data infrastructure have begun to support real-time flood forecasting and reservoir operation decisions in basins such as the Godavari and the Mahanadi. The institutional challenge going forward is not primarily technical; it is about data-sharing agreements between states, legal frameworks for acting on algorithmic recommendations in regulated water allocation decisions, and accountability mechanisms when AI-informed decisions lead to adverse outcomes.

Water utilities at the urban and peri-urban scale face a somewhat different institutional challenge. AI tools for leak detection, demand forecasting, and water quality monitoring are increasingly commercially available, but their adoption by public utilities is constrained by procurement processes designed for physical infrastructure rather than software services, by technical staff who lack the training to operate and maintain AI systems, and by political economies in which water tariffs are kept artificially low, leaving utilities with insufficient revenue to invest in digital infrastructure. Addressing these constraints requires regulatory reform alongside technology deployment — specifically, utility governance frameworks that create incentives for efficiency improvements and that allow a portion of the efficiency gains from AI adoption to be reinvested in further system upgrades.

Alignment with Watershed Management Plans

Watershed management plans, where they exist, provide the legal and technical baseline against which water resource decisions are made. AI systems deployed for water management should be designed to operate within these planning frameworks — generating outputs that are relevant to the specific objectives, targets, and constraints articulated in the plan — rather than as standalone analytical tools that produce insights disconnected from the regulatory context. In practice, this alignment requires that AI developers engage with watershed planning documents during the system design phase, identify the decision points at which AI outputs would be most useful, and design interfaces that deliver information in the formats and at the temporal resolutions that planners actually use.

There is also a more fundamental issue of plan currency. Many watershed management plans in India and other developing countries were prepared a decade or more ago, before significant changes in land use, groundwater abstraction rates, and climate patterns. AI systems trained on historical hydrological data that predates these changes may generate predictions that are systematically biased in ways that neither the AI developers nor the watershed planners immediately recognise. Regular recalibration of AI models against updated hydrological observations, combined with periodic revision of watershed management plans to incorporate AI-generated insights, is the kind of iterative institutional practice that makes AI adoption durable rather than episodic.

Cost and Infrastructure Constraints

Budget Limitations in Public Water Agencies

Public water agencies in developing countries operate under chronic budgetary pressure. Staff salaries, infrastructure maintenance, energy costs for pumping, and chemical costs for water treatment typically consume the majority of available budgets, leaving very limited discretionary funding for digital transformation initiatives. AI adoption in this context competes directly with basic operational expenditures, and it rarely wins that competition without external support — in the form of central government grants, multilateral development bank financing, or structured public-private partnerships.

The cost structure of AI tools for water management is also poorly matched to the financial planning cycles of public utilities. Most AI platforms are priced as annual software subscriptions, which require predictable recurring budget allocations. Public utilities in many states operate on annual budgets that are approved late, disbursed irregularly, and subject to revision in response to fiscal pressures. This mismatch between the subscription model of AI vendors and the budget realities of public utilities is a practical barrier that receives insufficient attention in the literature on AI for water management. Table 5 provides a comparative overview of the estimated costs associated with different AI tool categories for water resource management in the Indian context, alongside the typical institutional adopter and the primary budget source.

Table 5. Estimated cost ranges and institutional adopters for AI applications in water resource management in India

AI Application Area	Estimated Annual Cost ( )	Primary Budget Source	Typical Adopter
Real-time flood forecasting system	25,00,000– 80,00,000	Central/state government grant	River basin authority, CWC
Urban water demand forecasting	5,00,000– 20,00,000	Utility internal budget or Smart Cities funding	Municipal water utility
AI-enabled water quality sensors (network)	10,00,000– 40,00,000	AMRUT/Smart Cities Mission	Urban local bodies
Open-source groundwater monitoring tools	50,000– 2,00,000 (infrastructure only)	State groundwater authority budget	District groundwater offices
Irrigation scheduling AI (canal systems)	3,00,000– 12,00,000	Irrigation department budget	State irrigation departments

Source: Author's compilation
CWC = Central Water Commission; AMRUT = Atal Mission for Rejuvenation and Urban Transformation.

As Table 5 indicates, even the least expensive AI interventions — open-source groundwater monitoring tools — require dedicated infrastructure investment that exceeds the discretionary budgets of most district-level water offices. Centrally sponsored schemes and international climate finance mechanisms have therefore become the primary pathways through which AI adoption in water management gets funded. This dependence on project-based external funding creates its own problems: AI systems adopted under time-limited project grants frequently become non-operational once grant funding ends and no institutional budget line has been created to sustain them. Sustainability planning — identifying a recurring domestic budget source for AI operations before a grant-funded project concludes — should be treated as a non-negotiable element of any AI deployment in public water management.

A further infrastructure constraint worth noting is energy reliability. Many AI-enabled sensing and monitoring systems require continuous power supply to maintain data transmission and model operation. In rural and peri-urban areas of India and other developing countries, grid power supply remains intermittent, and the cost of battery backup or solar-powered alternatives adds meaningfully to the total cost of ownership of AI monitoring infrastructure. System designers working in these contexts need to account for energy infrastructure constraints from the outset, rather than discovering them at the commissioning stage.

Model Transferability and Regional Adaptability

Applicability Across Climatic and Hydrological Contexts

One of the more persistent challenges in applied hydrological AI is the transferability problem: a model trained on data from one river basin, climatic region, or hydrological regime frequently performs poorly when applied to a different context, even when the application appears superficially similar. This is not a deficiency unique to AI — traditional hydrological models face the same challenge — but it is particularly acute for machine learning approaches that derive their predictive structure from statistical patterns in historical data rather than from physical process equations. A long short-term memory network trained on monsoon-fed river systems in peninsular India, for instance, may have learned relationships between antecedent soil moisture, rainfall intensity, and peak discharge that do not hold in snowmelt-dominated Himalayan catchments, even if the general model architecture is identical.

The practical implication is that AI models for water management cannot be treated as off-the-shelf products that are valid across diverse contexts. Each deployment requires at minimum a validation exercise against locally observed hydrological data, and in many cases a partial retraining of the model using local data to adjust its internal parameters to the specific climatic and land surface characteristics of the target basin. This validation and retraining process requires both data — which may be scarce in data-poor river basins — and technical expertise — which may not be available within the implementing water agency. It also requires time: validation exercises that are rushed to meet project timelines tend to produce overconfident model performance assessments that do not hold up in operational conditions.

Arid and semi-arid regions present particular challenges for model transferability. Hydrological processes in dryland environments are characterised by high variability, episodic rainfall, and nonlinear runoff responses that are poorly captured by models trained on data from humid catchments. Conversely, models trained on arid-zone data tend to perform poorly in monsoon-dominated systems where sustained high-intensity rainfall drives very different runoff dynamics. This means that regional customisation is not optional but structurally necessary — and that the international transfer of AI models developed in data-rich temperate-zone contexts to data-scarce tropical or arid contexts requires careful validation rather than confident deployment.

Addressing the transferability problem at scale requires investment in regional hydrological data infrastructure — expanding gauge networks, maintaining existing monitoring stations, and systematically archiving observational data in accessible formats — alongside investment in the technical capacity of regional water agencies to conduct meaningful model validation. Physics-informed machine learning approaches, which embed known hydrological process equations as constraints within neural network architectures, have shown some promise in improving transferability by reducing the degree to which model performance depends on the statistical properties of the training dataset. These approaches are, however, more technically demanding to implement and require collaboration between hydrological scientists and machine learning specialists that is not yet routine in most water management agencies. Building this collaborative capacity — across disciplinary and institutional boundaries — is as important as the technical development of the models themselves.

AI-DRIVEN BIODIVERSITY CONSERVATION

In the era of big data, AI has become a powerful tool in biodiversity conservation, allowing researchers, environmental managers, and policy makers to make sense of ecological complexity at finer resolution and larger scale. Biodiversity systems are driven by nonlinear dynamics, and influenced by spatial, and temporal scales, that preclude the application of classical analysis tools. AI techniques, including ML, DNN models, probabilistic graphical models (PGM), and evolutionary algorithms (EAs), offer the computational capabilities to analyse large ecological datasets, identify patterns, and generate predictive knowledge (Aziz *et al.,* 2025). These functions enhance evidence-based conservation policy, diminish uncertainty in environmental decision-making, and lead to more effective adaptive management.

It is critical that farmland management and water management are linked to biodiversity conservation, which is not possible without dealing with agricultural production systems. Habitat of species and the resilience of ecosystem are affected by land-use change such as agricultural development, irrigation facilities, and watershed disaster. Hence, AI creates a possibility to combine biodiversity monitoring with decisions on managing agricultural landscapes and water resources, so that environmental managers can assess trade-offs among food production, water security and ecosystem protection. Now, the use of AI in biodiversity conservation has let go of conceptual explorations towards real-life implementation particularly in the context of monitoring of endangered species and detecting habitat degradation as well as guiding conservation planning.

Computational Ecology and Species Distribution Modelling

Computational ecology combines ecological theory with algorithms and computational structures to model the dynamics of ecosystems and the responses of species to environmental perturbations. Species distribution modelling is also among the most promising applications of AI in this field. Using environmental factors (temperature, precipitation, soil chemistry, vegetation cover, indicators of human disturbance) AI models can predict the likelihood of species occurrence over landscapes. Unlike classical statistical models, contemporary ML methods can naturally handle non-linear dependencies and high-order interactions among predictors and thus have potential to produce more accurate spatial predictions and to include uncertainty quantification, which is of paramount importance for conservation planning (Elbasi *et al.,* 2022).

In holistic environmental management, species distribution modelling also informs decisions in agricultural landscapes and watershed management by pinpointing regions in which land use practices potentially imperil focal habitats or ecological corridors. These types of information enable planners to balance protection of biodiversity with the expansion of agriculture and the development of water infrastructure.

A representative illustration of this approach is shown in Figure 4, in which a deep learning based species distribution modelling pipeline is presented that incorporates Global Biodiversity Information Facility occurrence records with Sentinel-2 satellite time-series imagery; the pipeline is constructed from raw occurrence and remote sensing data to convolutional neural network feature abstraction through to spatiotemporal habitat suitability prediction mapped at global scale for the Orchidaceae family (Estopinan *et al.,* 2022). This method serves as an example of the type of deep learning architecture that extends species distribution modelling by taking into account spatial and temporal environmental patterns simultaneously, producing output entirely within a function as greatly enhanced ecological realism than single variable modelling methods.

Figure 4. Visual abstract of a deep learning SDM (deep-SDM) method showing the pipeline from GBIF occurrence data and the Sentinel-2 satellite time-series image inputs through CNN feature extraction to prediction of spatiotemporal habitat suitability for the Orchidaceae family at the global level

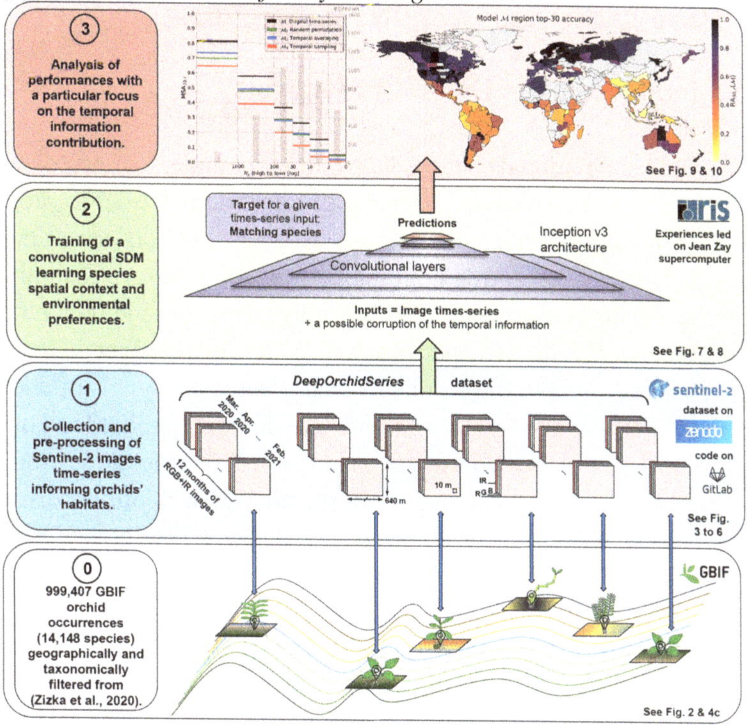

Source: Estopinan et al. (2022).

Habitat Suitability Prediction

Prediction of habitat suitability is to predict the region which can sustain a certain species taking species-related ecological conditions and environmental constraints into consideration. AI models trained on occurrence data and environmental layers can reveal sophisticated habitat affinities that may not be readily apparent. For instance, deep learning algorithms can process satellite images to identify fine-scale vegetation patterns that are predictive of particular wildlife populations. Such predictions are not only important for ecological assessment, but also have immediate policy relevance. Habitat suitability models enable governments and conservation organizations to identify priority landscapes were agricultural expansion, irrigation schemes, or infrastructure development may need to be tightly controlled to prevent ecosystem degradation. These predictions are particularly valuable for rare or cryptic species for which there are few field observations. Moreover, suitability models enable conservationists to delineate climate refugia, track movement of species corridors, and identify areas at risk of habitat fragmentation. In addition, AI facilitates the processing of large spatial data to detect habitat, making habitat assessment more efficient and less costly (Al-Raeei, 2024).

Migration and Population Dynamics Analysis

Migration and populational changes help build our understanding of ecosystem stability, and species' persistence. AI methods have the potential to process time series data from tracking devices, environmental sensors, and observational surveys to estimate movement paths and demographic processes. Recurrent neural networks and probabilistic state space models are especially adept at modelling temporal dependencies in population data (Miller *et al.*, 2025). Using these methods, scientists can predict changes in populations, identify early warning signals of decline, and assess the impact of environmental stressors like drought, habitat loss, or invasive species. Such information is increasingly used to develop water allocation, seasonal farm management, and land management practices that minimize disruption of ecosystems in the middle of critical migration corridors. Predictive insights from such analyses enable conservation action that is proactive rather than reactive.

Automated Biodiversity Monitoring Technologies

Monitoring biodiversity at the scale of entire landscapes has traditionally been labour-intensive and costly. Proposed is a set of AI-based monitoring tools that can be used for the continuous, scalable, and non-invasive monitoring of ecosystems. Combined with intelligent processing, automated data collection enables detection of ecological changes in near real time. They provide more frequent monitoring with less observer bias and logistical challenges. Significantly, broad surveillance can also provide environmental intelligence that may guide land-use policy, agricultural zoning, watershed management, and protected-area governance.

Acoustic and Visual Recognition Systems

Acoustic and visual recognition system are based on AI models that are trained to recognize species from audio clip or video frames. Acoustics sensors placed in forests, wetlands, or oceans can record the sounds of birds, amphibians, and mammals. Machine learning classifiers on audio spectrograms also differentiate species calls, predict abundance, and identify behavioural patterns. In a similar way, camera traps coupled with computer vision models can identify species, count individuals and determine behaviours such as feeding or nesting. Such automated recognition systems can open the door towards enabling large scale assessment of biodiversity with no or minimal human involvement. They also offer permanent digital records that can be reanalysed as algorithms advance. In conservation, these monitoring outputs are increasingly being employed by environmental ministries to inform the policies of habitat protection, to regulate the expansion of agriculture near areas of high biodiversity, and to assess the ecological effects of water infrastructure projects like dams and irrigation canals.

Satellite-Based Ecosystem Observation

Satellite imagery has been among the essentials for tracking the health of ecosystems. AI contributes to the analysis of satellite data through automated classification of land cover type and detection of deforestation and vegetation distress. Convolutional neural networks can operate on high resolution images to detect subtle ecological signatures like canopy gaps, wetland boundaries, or patterns of coral reef bleaching. The side effect is that this allows researchers to see long-term trends in the environment and to assess the impact of conservation policies. Such observations are particularly valuable in remote or politically sensitive areas where access to the field is denied (Chisom *et al.,* 2024). Spacecraft monitoring of the environment is beginning to be applied more frequently to assess the impact on the environment of changes in agricultural practice and water use and to provide governments with the evidence necessary to enforce environmental legislation and modify resource management policies.

Conservation Prioritization Algorithms

To Humanised Output Conservation funds are scant and decision makers need to prioritize actions that maximize the ecological returns. AI-based prioritization models consider ecological, economic, and social factors to identify the best investments in conservation. These approaches assess the relative trade-offs between competing goals, such as biodiversity, land use and community livelihoods. By running simulations of different options, they can teach us which ones are going to rake in the best conservation bang for the buck in rupees-balancing public funds. Designing a protected area is therefore a matter of choosing geographic borders that literally trap biodiversity without running afoul of human interests.

Optimization of Protected Area Design

Heuristic-based algorithms, such as genetic algorithms (GAs) and simulated annealing (SA), can assess a great number of spatial arrangements to find the optimal solution that captures most of the species and/or habitats. These algorithms introduce constraints, for example the cost of land, connectivity constraints, and ecological representation. These solutions support scientifically based reserve design and reduce reliance on subjective decision-making.

Multi-Criteria Ecological Valuation Models

These Multi Criteria ecological evaluation models typically integrate environmental indices along with economic and social factors to differentiate the priority of conservation areas. AI-based techniques can dynamically weight different criteria such as species richness, provision of ecosystem services, cultural importance, or feasibility of restoration. By combining different data types, these models offer a comprehensive assessment tool. Potential conservation regions can then be ranked by composite scores, allowing decision makers to direct their resources (Nti *et al.*, 2023). As can be seen in Table 6, diverse AI techniques have been used for several aspects of ecology for which they have complementary roles in overall conservation decision making.

Table 6. Major AI techniques and their applications in biodiversity conservation

AI technique	**Primary ecological function**	**Conservation outcome**	**Evidence**	**Reference**
Machine Learning Classifiers (RF, SVM, Gradient Boosting)	Species identification from occurrence and trait data	Rapid, large-scale biodiversity surveys	AUC = 0.88–0.97 in species distribution models across vertebrate taxa	Phillips *et al.* (2006); Elith *et al.* (2006)
Neural Networks (CNNs, ResNet)	Habitat suitability prediction from satellite imagery	Targeted conservation planning and corridor identification	>90% accuracy in land cover and habitat classification from high-resolution imagery	LeCun et al. (2015); Norouzzadeh *et al.* (2018)
Evolutionary Algorithms (GA, simulated annealing)	Spatial reserve design optimization	Cost-efficient protected area networks maximizing species coverage	20–35% improvement in species representation vs. ad hoc reserve selection	Moilanen *et al.* (2009); Margules & Pressey (2000)

continued on following page

Table 6. Continued

AI technique	Primary ecological function	Conservation outcome	Evidence	Reference
Time Series Models (LSTM, state-space models)	Population trend forecasting and demographic analysis	Early detection of species declines and risk assessment	Detection of population decline signals 2–5 years earlier than conventional monitoring	Fiske & Chandler (2011); Gimenez *et al.* (2021)
Computer Vision with deep learning	Automated wildlife identification from camera traps and drones	Reduction in field labour by 60–80%; elimination of observer bias	~96% species classification accuracy in camera-trap image datasets of >3 million images	Norouzzadeh *et al.* (2018); Tabak *et al.* (2019)

Source: Synthesized from Phillips *et al.* (2006), Elith *et al.* (2006), Norouzzadeh *et al.* (2018), Moilanen *et al.* (2009), Margules and Pressey (2000), Fiske and Chandler (2011), Gimenez *et al.* (2021), and Tabak *et al.* (2019).

Illegal Activity Detection and Prevention Systems

Biodiversity is threatened by unlawful activities including poaching, logging, and wildlife trafficking. AI-enabled surveillance linked to sensor feeds, drone and patrol data can be used to analyse suspicious activities. Predictive models help in delineating high risk areas for potential unlawful activities, where resources can be deployed more effectively by the concerned authorities. Pattern recognition algorithms contribute also to establish if some of the vehicles, material or people involved are linked to known unlawful activities. Moreover, anomalous detection techniques can highlight unusual changes in the environment that may be caused by illegal exploitation. These systems enhance environmental governance by allowing authorities to act faster, have better enforcement of conservation laws, and protect ecosystems that also support agricultural production and water supply protection.

AI for Ecosystem Restoration Planning

Ecosystem restoration seeks to repair damaged landscapes and restore ecological processes and functions. AI also supports restoration planning, in terms of identifying priority sites, predicting suitable species for reintroductions and expected restoration outcomes under alternative management options. Simulation models are used to assess the influence of soil conditions, climatic factors, and species interactions on restoration success. This enables them to develop strategies that promote maximum ecological resilience and long-term sustainability. The technology can also calculate financial estimates in rupees for different restoration plans, which can assist stakeholders in selecting cost effective options (Martos *et al.*, 2021). Restoration planning is increasingly intersecting with the management of sustainable agricultural lands and protection of watersheds, especially in areas where damaged farmland or deforested catchments need to be restored to re-establish ecosystem services (e.g., soil fertility, water regulation). Adaptive learning systems can further incorporate new data to continually update restoration strategies, thereby maintaining the relevance and effectiveness of management strategies in the face of changing environments.

From Monitoring to Policy Action: Closing the Governance Loop

Although AI-based monitoring systems have dramatically enhanced the ability to monitor the natural world, their greatest potential is in using environmental information to inform policy decisions. Adapting the monitoring output to the governance processes, conservation agencies are enabled to focus their protective measures on those areas most in need, to channel restoration effort investments appropriately and to further reinforce legal enforcement mechanisms. For instance, models of habitat suitability and biodiversity monitoring data may guide agricultural land-use planning, to assist governments in establishing conservation corridors that sustain ecological connectivity and disrupt food production systems the least. So too, satellite monitoring of deforestation and wetland loss can contribute to water management strategies that secure watershed health.

It is also crucial that local communities, Indigenous people, and grassroots conservation players help to contextualize AI-generated solutions and take them forward in environmental management. These individuals have place-based ecological knowledge that complements algorithmic analyses and ensures that conservation policies are both socially equitable and operationally effective. Through better integration of AI-powered ecological monitoring with decision-making bodies, conservation planning, and enforcement agencies, environmental governance could increasingly embrace an adaptive and holistic model of sustainable management.

Institutional Use and Policy Integration

Role of Conservation Agencies and NGOs

Biodiversity conservation operates through an institutional landscape that is, in many respects, more fragmented than either agriculture or water management. Responsibility for conservation outcomes is distributed across national wildlife authorities, state forest departments, international conservation organisations, local non-governmental organisations, community-based conservation groups, and an increasingly active private sector. AI tools must find their way into this complex, multi-actor system — and the pathways through which they do so are shaped as much by institutional culture, funding structures, and inter-agency relationships as by the technical merits of the tools themselves.

National conservation agencies — the Wildlife Institute of India, the Forest Survey of India, the National Biodiversity Authority, and their equivalents in other countries — are the institutions with the greatest formal authority over biodiversity governance. They set the standards for wildlife monitoring protocols, administer protected area networks, and produce the national assessments that feed into international reporting obligations under the Convention on Biological Diversity. When these agencies adopt AI tools — for species distribution modelling, habitat change detection, or wildlife crime monitoring — they create institutional precedents that cascade downward through state and local conservation administrations. The Wildlife Institute of India's work on tiger population estimation using camera trap image analysis driven by pattern recognition algorithms, for instance, has influenced how state forest departments across tiger range states think about monitoring technology and data management. Conversely, when national agencies remain sceptical of AI tools or lack the capacity to evaluate them, adoption at lower institutional levels tends to be patchy, project-driven, and unstained beyond the tenure of a particular motivated officer.

Non-governmental organisations occupy a different but equally consequential position. Large international conservation NGOs — the Wildlife Conservation Society, WWF, the Nature Conservancy, and others — have in many cases been earlier and more agile adopters of AI tools than government agencies, partly because they are not bound by government procurement processes, partly because they have access to international philanthropic funding that government agencies lack, and partly because they operate with technical staff who are more likely to have data science backgrounds. These organisations have deployed AI tools for acoustic monitoring of tropical forests, satellite-based deforestation alerts, and species identification from community-submitted photographs at a scale and speed that government agencies have found difficult to match. The challenge is that NGO-led AI adoption often occurs in parallel with, rather than integrated into, national conservation monitoring systems. Data collected by NGO-operated AI platforms may not flow into national biodiversity databases, and analytical outputs may not be formatted to meet the reporting requirements of national agencies. Closing this integration gap requires deliberate institutional effort — data-sharing agreements, standardised data formats, and joint working groups that bring government and NGO technical staff into regular contact.

Smaller, locally rooted NGOs and community conservation organisations present yet another dimension of the institutional picture. These organisations often work in the most biodiversity-rich and data-poor landscapes — remote forest areas, coastal wetlands, high-altitude grasslands — where large conservation agencies have limited operational presence. AI tools that are accessible to organisations with limited technical staff, low and irregular funding, and poor connectivity infrastructure can extend the reach of biodiversity monitoring into these underserved landscapes. Participatory species monitoring applications that allow community members to submit geo-tagged photographs for AI-assisted identification, low-cost passive acoustic monitors that can be deployed by local conservation groups without specialist technical knowledge, and satellite alert systems that notify local rangers of suspected encroachment or fire events are examples of AI tools designed with this institutional reality in mind. The effectiveness of these tools, however, depends on whether the data they generate is connected to decision-making processes — whether a deforestation alert actually triggers a ranger response, whether a community-submitted species record actually reaches a database where it influences conservation planning.

Integration with Biodiversity Action Plans

National Biodiversity Strategies and Action Plans, along with state and local biodiversity management plans, provide the formal planning framework within which conservation decisions are supposed to be made. AI tools that are designed and deployed without reference to these frameworks tend to generate outputs — species distribution maps, habitat quality assessments, invasive species alerts — that are technically valid but institutionally inert, because they do not connect to the specific targets, indicators, and governance mechanisms that the plans establish.

Effective integration requires that AI system designers engage with biodiversity action plans at two levels. The first is the level of indicators: national biodiversity strategies typically commit to monitoring specific indicators — forest cover change rates, population trends of threatened species, area under effective conservation management — and AI tools should be designed to generate outputs that are directly relevant to tracking these indicators, in formats that can feed into national reporting systems. India's commitments under the Kunming-Montreal Global Biodiversity Framework, which set ambitious targets for area-based conservation and ecosystem restoration to be achieved by 2030, provide a concrete set of planning objectives against which AI tool design can be calibrated.

The second level of integration is governance: biodiversity action plans establish decision-making processes — periodic reviews, adaptive management cycles, inter-ministerial coordination mechanisms — and AI tools should be designed to deliver outputs at the temporal and institutional points where decisions are actually made. A species distribution model that produces an annual update is only useful if there is an annual planning review at which that update can influence habitat protection or restoration decisions. Where these governance mechanisms are weak or non-functional, improving AI tools is unlikely to improve conservation outcomes; strengthening the governance mechanisms themselves is the more productive intervention.

Long-Term Monitoring and Maintenance Systems

Continuous Data Updating and System Sustainability

The biodiversity monitoring problem is fundamentally a long-term one. Species populations fluctuate over decades, habitat quality changes gradually in response to land use and climate pressures, and the effects of conservation interventions often take years to manifest in observable population or ecosystem indicators. AI systems designed for biodiversity monitoring must therefore be conceived and resourced as long-term infrastructure — not as projects with defined end dates — if they are to generate the longitudinal data streams that meaningful conservation assessment requires.

This is, in practice, one of the most challenging aspects of AI deployment in conservation. The funding landscape for conservation technology is heavily oriented toward project cycles of two to five years, after which funding must be renewed, often from a different source with different priorities. AI monitoring systems that are established under a time-limited project grant frequently encounter a sustainability crisis at the end of the grant period: the hardware requires maintenance, the software requires updates to remain compatible with changing operating systems and data formats, the models require retraining as new training data accumulates, and the human expertise required to manage the system must be retained or rebuilt. Each of these requirements has a recurring cost that must be met from somewhere, and in the absence of a dedicated institutional budget line, systems that were technically functional at the end of a project period quietly degrade into non-operation within a few years.

Addressing this sustainability challenge requires thinking about AI monitoring systems in the same way that physical scientific infrastructure — weather stations, hydrological gauges, seismic networks — is thought about: as capital assets that require operational budgets, maintenance schedules, and staffing plans over their full expected operational life. Conservation agencies and their funders need to move away from the project mentality — build it, demonstrate it, publish it, move on — toward an infrastructure mentality in which the long-term operational costs of AI monitoring systems are budgeted explicitly and covered by a stable institutional funding source.

The data dimension of system sustainability deserves particular attention. AI models for species distribution, habitat classification, and wildlife population estimation are only as good as the data on which they are trained and validated, and that data must be continuously updated to remain relevant as ecological conditions change. A habitat suitability model trained on remote sensing data from 2015 will progressively misrepresent actual habitat conditions as land use changes, invasive species spread, and climate-driven vegetation shifts alter the landscape over the following decade. Continuous data updating — integrating new satellite imagery, new field survey records, new citizen science observations, and new climate projections into model training datasets — is therefore not an optional enhancement but a functional requirement for AI systems intended to support ongoing conservation management.

The practical infrastructure for continuous data updating includes data ingestion pipelines that can automatically incorporate new satellite observations as they become available, data quality control processes that screen incoming field and citizen science records for positional accuracy and taxonomic reliability, and model retraining schedules that are triggered either at fixed intervals or when model performance metrics fall below defined thresholds. Building and maintaining this infrastructure requires a combination of data engineering expertise, ecological expertise to oversee data quality, and institutional commitment to treating AI monitoring as an ongoing operational function. Conservation organisations that have successfully built this infrastructure — including those operating large-scale acoustic monitoring networks in tropical forests and satellite-based habitat monitoring platforms in savannah ecosystems — consistently report that the institutional and organisational challenges of sustaining these systems over time are more demanding than the original technical challenges of building them.

There is also a model governance dimension to long-term sustainability that receives insufficient attention. As AI models are retrained on updated data, their predictions may shift in ways that complicate the interpretation of long-term trends. A change in a model's estimated species distribution between 2020 and 2025 might reflect a genuine ecological change, a change in the model's training data coverage, or a change in the model architecture resulting from a version update. Maintaining clear documentation of model versions, training datasets, and methodological changes is essential for distinguishing genuine ecological signals from analytical artefacts — and this documentation discipline is, in practice, difficult to maintain across institutional transitions, staff turnover, and the informal knowledge loss that inevitably occurs as organisations evolve over time.

CROSS-SECTORAL INTEGRATION AND SUSTAINABILITY INTELLIGENCE

Trans-sector collaboration is a core principle in modern environmental informatics, with AI-based systems integrating agricultural, hydrological, and ecological data streams to produce a combined sustainability intelligence. These have traditionally been disciplinary silos, with their own domain-specific tools for monitoring, modelling and governing. But environmental issues like climate change, soil erosion, water depletion, and species extinction lie beyond sectoral borders. Artificial intelligence contributes to convergence, by unifying diverse data types, by revealing hidden patterns, and by facilitating aligned decision processes across environmental domains (Kose *et al.*, 2022). Sustainability intelligence is the ability of automated agents to perpetually interpret environmental signals, forecast systemic reactions, and propose adaptive interventions within their operational scope of engagement. In contrast to traditional decision support systems designed for narrow programs of operation, the land use management, hydrological cycle and ecosystem dynamics models are integrated in a single AI platform and evaluated in real time for inputs uncertainties. This integrated approach is especially critical in areas where land use, water availability, and ecosystem health are interdependent. They do so by bridging data infrastructures and analytical workflows across domains, thereby AI systems diminish data silos, and they enable proactive environmental governance.

Interoperability of Agricultural, Hydrological, and Ecological Datasets

Interoperability is a measure of the technical and semantic capability of individual data to be used coherently as part of a common analysis platform. Typical agriculture datasets include crop yield data, soil nutrient data, irrigation data, weather data etc. Streamflow measurements, groundwater levels, precipitation intensity, and watershed characteristics are all included in hydrological data. Species distribution, habitat quality, vegetation indices and biodiversity are recorded by ecological datasets. Each type of dataset normally varies in file format, spatial resolution, temporal frequency, and metadata specification. Consequently, AI systems must perform data harmonization processes to interoperate. Standardization is key to enabling interoperability (Nautiyal *et al.*, 2024). Data schemas, ontologies, and controlled vocabularies allow them to make sense of data in the same way across sources. For example, to enable predictive modelling, soil moisture measurements from in situ sensors need to be temporally co-registered with precipitation observations and spatially with vegetation indices. In the absence of standardized metadata schemes, machine learning algorithms may generate spurious correlations or incomplete investigations (Sánchez *et al.,* 2022). Environmentally distributed data sets have been successfully integrated with cloud-based data lakes and federated data architectures. Such architectures enable data sets to be maintained in their original repositories and yet be made available for analysis through secure interfaces. It maintains the data ownership and privacy and at the same time enables cross-sector analytics. In addition, interoperability is beneficial for reproducibility as standardized datasets facilitate researchers in validating models at different geographic scales. Refer to Table 7 to demonstrate that the level of concern with interop requires addressing structural, semantic, temporal, spatial and access dimension(s) of the issue concurrently (Anjali *et al.,* 2026).

Table 7. Key dimensions of dataset interoperability for environmental AI systems

Dimension	Description	Significance for sustainability analytics	Practical challenge	Illustrative standard or tool	Reference
Structural compatibility	Alignment of file formats, schemas, and database architecture	Enables unified data ingestion pipelines; prevents parsing failures across heterogeneous sources	Fragmentation across CSV, NetCDF, GeoTIFF, and proprietary formats used by different agencies	INSPIRE Directive (EU); OGC standards (GML, WFS)	INSPIRE (2007); OGC (2022)
Semantic consistency	Standardized terminology, controlled vocabularies, and ontological alignment	Prevents cross-sector misinterpretation; enables meaningful aggregation of soil, climate, and ecological data	Divergent variable naming conventions (e.g., "soil moisture" vs. "volumetric water content") across agencies	ENVO environmental ontology; Darwin Core biodiversity standard	Buttigieg *et al.* (2016); Wieczorek *et al.* (2012)
Temporal synchronization	Matching timestamps, sampling frequencies, and temporal resolutions	Supports dynamic, multi-source system modelling; prevents lag-induced prediction errors	Satellite revisit cycles (5–16 days) misaligned with daily sensor readings	ISO 8601 time standard; NetCDF-CF conventions	Eaton *et al.* (2017); ISO (2019)

continued on following page

Table 7. Continued

Dimension	Description	Significance for sustainability analytics	Practical challenge	Illustrative standard or tool	Reference
Spatial alignment	Common coordinate reference systems, spatial scales, and projection standards	Facilitates geospatial overlay analysis and cross-regional model transferability	Inconsistent use of WGS84, UTM, and national reference systems across datasets	EPSG geodetic registry; WGS84 global datum	Longley *et al.* (2015); EPSG (2023)
Accessibility protocols	Secure APIs, authentication frameworks, and permission governance	Promotes collaborative research while protecting data sovereignty and privacy	Proprietary data portals, restricted government datasets, and licensing barriers	FAIR Data Principles; OGC API standards; Creative Commons licensing	Wilkinson *et al.* (2016); OGC (2022)

Source: Synthesized from INSPIRE (2007), Buttigieg *et al.* (2016), Wieczorek *et al.* (2012), Eaton *et al.* (2017), Longley *et al.* (2015), Wilkinson *et al.* (2016), and OGC (2022).

Multi-objective Optimization Across Environmental Sectors

Environmental management problems seldom have one goal. Agricultural planners need to consider crop production, water use, fertilizer use and environmental protection. Water managers must decide who will claim the dwindling water among irrigation, urban supply, and aquatic ecosystem. Ecology conservationists try to retain biodiversity and meet land use pressures. Multi-objective optimization algorithms offer a principled way of balancing those conflicting interests. Artificial intelligence uses evolutionary algorithms, reinforcement learning, and Pareto optimization methods to analyse trade-offs between objectives (Ngulube, 2025). Rather than providing a single best solution, these approaches return a set of feasible solutions demonstrating different trade-offs among objectives. Decision-makers can then choose among those results the options that best conform to policy priorities or societal values. For example, an optimization model could find irrigation practices that modestly reduce crop production while greatly conserve groundwater and protect wetland habitats. The benefits of the AI-based multi-objective schemes are that they can capture the complex interactions within the system. Conventional optimization techniques have been developed assuming linear systems while environmental systems are nonlinear with feedback loops. The nonlinearities these models learn from both historical and real-time data. As a result, they can also forecast how shifts in one industry affect criteria elsewhere, contributing to systematic and balanced policy design (Yadav and Singh, 2023).

Digital Twins for Environmental Systems Management

Digital twins are digital representations of physical systems that are updated continuously with sensor data and simulation algorithms. Digital twins can model farms, watersheds, forests, or even whole landscapes in environmental management. These models combine physical measurements, remote sensing data and predictive analytics to replicate real world scenarios with great detail. With digital counterparts running in lockstep, policymakers can test interventions in virtual world before implementing them in the real world. For farming, a digital twin of a field may simulate moisture in the soil, crop growth stages, nutrient cycling, pest emergence. In hydrology, runoff, infiltration, sediment transport, and reservoir storage can be modelled by a watershed twin. In ecology, a habitat twin can project how species populations may respond to environmental change. When these specialized twins are linked together, a full-scale environmental simulation platform results. Strategic implications for the digital twin are rooted in scenario analysis (Wani *et al.,* 2024). Under a range of climate scenarios, those with influence on policy are enabled to test the potential outcomes of irrigation use, land conversion and/or conservation practices. Since simulations take place in a virtual world, there's no ecological risk in experimentation. Additionally, by continuously updating predictions as new data become available, digital twins enable adaptive management, so that policy recommendations they generate remain relevant as conditions change.

Edge Computing and Real-Time Sustainability Analytics

Edge computing is the process of data being processed at the edge of the network, closer to the data source rather than in a data centre. In environmental sampling systems, sensors in fields, rivers or forests collect information continuously. Sending all raw data back to centralized data centres can be inefficient, expensive, and slow. Edge computing mitigates this by conducting initial processing on site rather than in the cloud, minimizing latency and bandwidth use age. Sustainability analytics in real time are based on quick interpretation of environmental signals. For example, soil sensors can sense moisture shortages and activate irrigation systems immediately to keep crops from being stressed. High-water-levels can be detected and floods warned by river gauging stations. Wildlife tracking tags can examine movement patterns to keep tabs on such threats as habitat breach (Kumari *et al.,* 2025). Since they process data locally, edge devices allow for immediate action, something that could be delayed if centralized systems were used. Another benefit of edge computing is its robustness. Distributed processing in order to guarantee system monitoring even when there is no network connectivity, it is essential to distribute monitoring process and not rely solely on one centralized server. This is critical, particularly in isolated rural or heavily wooded areas where the communication infrastructure might be weak (Rane, 2023). Additionally, edge computing improves the privacy of data because sensitive data can be handled within the local environment.

Socio-Technical Integration and Stakeholder Participation

Technological infrastructure by itself cannot guarantee sustainability; it is rendered effective or ineffective by social acceptance, institutional support, and participatory governance. Socio-technical integration highlights the alignment of technological innovation with human values, cultural practices, and policy instruments. Therefore, AI tools need to be co-designed with stakeholders from the outset of development. Farmers, water managers, conservationists, policymakers, and local communities all have their own expertise and priorities (Sanders *et al.,* 2021). Participatory design methodologies translate this knowledge into design principles that shape system design, making the outputs of AI interpretable, relevant, and actionable. For instance, rather than abstract probabilistic scores, farmers might want decision recommendations in practical terms such as when to irrigate. Likewise, conservation organizations may demand the same level of transparency when predictive models assess biodiversity threat. Trust plays a pivotal role in uptake. Stakeholders are more likely to trust AI systems when algorithms are explainable and the sources of data are transparent. Ethical implications around data ownership, consent and equity also need to be considered. Capacity-building programmes and training can help to facilitate further participation for stakeholders by providing users with the training they need to apply AI insights. In the end, socio-technical alignment turns AI from a strictly computational instrument to a collaborative one for sustainable development (Mishra and Mishra, 2023).

Scalability, Robustness, and Resilience of AI Systems

To be practically effective, environmental AI systems need to operate robustly at different spatial temporal scales under environmental and anthropogenic disturbances. Scalability is the property of a system to handle a growing amount of work or its potential to accommodate growth in data, geographic area, or the number of users. Robustness - Noise resistance, uncertainty and incompleteness of data. It accounts for the ability to withstand disruptions, e.g. sensor outages or severe environmental events), and recover Resilience describes the (and to recover from such disruptions). Scalable solutions are often based on modular architectures where data ingestion, processing, and visualization are loosely coupled but tightly orchestrated. This enables systems to grow incrementally without redesigning everything. Additionally, cloud-native and distributed computing paradigms add another layer of scalability when transferring computational resources on the fly as needed (Senoo *et al.*, 2024).

Robustness can be achieved by algorithmic approaches, such as filtering noise, detecting anomalies, and probabilistic modelling. Environmental data are subject to inevitable uncertainty as a result of measurement constraints and natural variation. Models of AI that take into account uncertainty estimates are able to provide confidence intervals in their predictions, which allows decisions to be made with a more refined risk evaluation. Resilience, finally, assures carrying on during interruptions. Redundancy, backup data links and adaptive algorithms enable systems to continue working when parts fail (D'Amore *et al.*, 2022). In applications to sustainability, resilience is paramount, not least because monitoring of the environment is typically conducted in severe if not unpredictable environments. When designed with resilience in mind, systems may continue to offer guidance during extraordinary events — when information is most needed.

Worked Integrated Example I: Dryland Farming, Groundwater, and Dryland Ecosystem Co-Management

In semi-arid areas, agriculture, groundwater, and soil-based ecosystems are tightly coupled. Extreme irrigation can lead to higher yields in the short term, but can speed up the drawdown of groundwater and destroy sensitive desert or grassland ecosystems. A tightly integrated AI system could help manage this challenge, integrating agricultural sensor networks, groundwater monitoring systems, and satellite derived vegetation indices. Machine learning models assess soil moisture, crop water needs and aquifer recharge to predict the best timing for irrigation. Concurrently, ecological surveillance models are monitoring vegetation and habitat status in adjoining dryland regions. By treating the trade-offs as a multi-objective optimization, the irrigation strategies that are recommended can sustain farm productivity, avoid overexploitation of groundwater, and maintain ecosystem resilience. These combined management approaches provide an example of how regional governments can integrate the management of agricultural water use and ecosystem conservation, demonstrating the potential for AI to reshape food security, water security, and biodiversity management concurrently.

Worked Integrated Example II: River Basin Management Under Climate Stress

River basins are commonly used for a variety of competing interests such as agriculture, urban water supply, flood control, and ecosystem conservation. These pressures are exacerbated by climate change, resulting in more frequent droughts and heavy precipitation events. In an AI-enabled river basin management system, hydrologic models play a vital role in analysing rainfall predictions, reservoir levels, and river discharge information for forecasting flood and drought. Agricultural data sets contain detailed information on irrigation demands and crop water needs within the basin. Ecological monitoring systems observe the status of riparian habitats and aquatic species. By combining the data within a single optimization framework, AI-enabled systems can manage reservoir releases, irrigation allotments, and natural flow needs. For instance, it may suggest temporary reductions of the irrigation withdrawals in drought conditions to ensure minimum river flows required to sustain aquatic ecosystems. Conversely, predictive models can recommend joint reservoir operations to protect downstream agriculture and wetland habitats during coincident drought conditions, and flood-forecasting models can be used to design coordinated reservoir operations that simultaneously protect downstream agriculture and wetland habitats during coincident drought conditions. Such integrated basin management provides an example of how AI can facilitate the coordinated governance of food, water, and ecosystem services.

Trade-Offs, Unintended Consequences, and the Limits of Integration

The toolkit of AI applications will undoubtedly grow with opportunities for cross-sectoral exploitation, which brings with it a need for careful consideration of the emerging risks and limitations related to AI in the context of environmental governance. A related worry concerns potential rebound effects, such that improvements in efficiency enabled through AI technologies may inadvertently lead to higher levels of resource consumption. For instance, more efficient irrigation systems may promote the expansion of irrigated agriculture, leading to an increase in total water withdrawals. The so-called cloak of digital infrastructure is also a problem. Broadly, machine learning-based systems rely on huge computational power that translates in high energy consumption, notably to train intricate models with extremely large volumes of data. When run on carbon-intensive energy sources, these computational processes could negate some of the environmental gains made through AI applications.

Data inequality is a huge problem as well. High-tech regions with well-established digital infrastructure along with dense sensor grids have much better opportunities to leverage AI-enabled sustainability analytics than rural regions that continue to be data deserts. Cross-sectoral data integration may also rely on good communication networks and institutional data sharing agreements, both of which may be difficult to come by in resource-poor environments. Finally, governance concerns remain non-negligible. To some extent integrated AI systems involve coordination among different institutions with varying responsibility for regulation and priorities of policies. In the absence of strong governance structures and well-articulated decision processes, AI-enabled environmental management may also be confronted with legitimacy challenges, or it may lead to unexpected policy clashes. The recognition of these constraints is key to making sure that the integration of AI in environmental governance as a whole serves sustainable and just environmental governance rather than perpetuating system inequalities or technological dependencies.

Institutional Coordination and Governance Integration

Cross-Ministerial and Multi-Agency Collaboration

The environmental challenges addressed in this book — agricultural sustainability, water resource management, and biodiversity conservation — do not respect ministerial boundaries. A decision made by an irrigation department about canal water allocation affects groundwater recharge rates that concern the groundwater authority, surface runoff patterns that matter to the flood management agency, and wetland hydrology that determines habitat quality for species under the mandate of the forest department. Yet in most countries, these agencies operate in institutional silos, with separate budgets, separate data systems, separate reporting lines, and limited routine communication. AI systems that could, in principle, model these cross-sectoral interdependencies and generate integrated management recommendations are of limited practical value if the institutional architecture through which decisions are made remains fragmented along sectoral lines.

Cross-ministerial coordination for AI-enabled environmental management is therefore not a technical problem — it is a political and organisational one. Getting multiple ministries to share data, agree on common analytical frameworks, and act on jointly generated recommendations requires sustained political commitment at senior levels, institutional mechanisms that create regular inter-agency contact below the political level, and legal or regulatory frameworks that mandate data sharing and joint planning. In India, inter-ministerial coordination on environmental issues occurs through mechanisms such as the National Mission for Sustainable Agriculture, the Jal Shakti Abhiyan for water conservation, and the National Action Plan on Climate Change, which providcs an overarching framework spanning multiple sectoral missions. These coordination mechanisms provide institutional hooks on which cross-sectoral AI initiatives can be anchored — but they have not, so far, been used systematically to drive integrated AI deployment across the agriculture-water-biodiversity nexus.

The experience of countries that have made more progress in this direction is instructive. The Netherlands' integrated water management system, which combines real-time hydrological monitoring, agricultural water use data, and ecological flow requirements in a single operational decision-support platform, was built over decades through a combination of legal mandates for data sharing, sustained investment in shared data infrastructure, and the cultivation of a professional culture in which water managers, agricultural regulators, and ecologists are trained together and routinely work in interdisciplinary teams. The technical AI components of the Dutch system are sophisticated, but they rest on an institutional foundation that took far longer to build than any individual algorithm. This sequencing — institution building first, technology deployment second — is a lesson that is frequently inverted in international development contexts, where technology projects are funded before the institutional conditions for their effective use are established.

At the sub-national level, district and watershed administrations often provide more tractable entry points for cross-sectoral AI integration than national ministries. A district collector in India, for instance, has administrative jurisdiction over agricultural extension, irrigation management, forest conservation, and rural water supply simultaneously — a span of authority that, at least in principle, allows cross-sectoral coordination without the inter-ministerial negotiation that national-level integration requires. District-level AI platforms that integrate crop monitoring, water availability forecasting, and forest cover change detection into a single decision-support interface have been piloted in several states under the Digital India Land Records Modernisation Programme and related initiatives, with mixed but instructive results. The technical integration of datasets is generally achievable within a project timeframe; the institutional integration of decision-making processes across the line departments that a district collector nominally supervises is considerably more difficult and takes longer.

Scaling Conditions and Success Factors

Data Pipelines, Funding Continuity, and Technical Support

Scaling AI systems from successful pilots to sustained operational deployment at regional or national scale is where the majority of environmental AI initiatives fail. The conditions that allow a well-designed pilot to function — dedicated project staff, active researcher involvement, flexible project funding, and the institutional attention that accompanies a new and visible initiative — do not automatically replicate when a system is handed over to a government agency operating under routine budget and staffing constraints. Understanding what scaling requires, beyond the technical replication of a working system, is therefore one of the most practically important questions in the field.

Data pipelines are the first and most fundamental scaling requirement. An AI system that works in a pilot context is typically fed data by a small team of technically skilled researchers who monitor data quality, handle exceptions, and manually correct problems as they arise. When the same system is scaled to operate across hundreds of monitoring stations, dozens of administrative units, and multiple data sources with varying quality and format standards, the manual data management approach that worked at pilot scale becomes completely unworkable. Automated data ingestion pipelines — with built-in quality control checks, exception handling protocols, and audit trails — are a prerequisite for scaling, and they are not trivial to build. They require software engineering expertise that is distinct from the machine learning expertise used to build the AI models themselves, and they require sustained maintenance as upstream data sources change their formats, update their APIs, or introduce new data quality problems.

Funding continuity is the second critical scaling condition. The transition from project funding to institutional budget is, as noted in earlier sections, a point of high failure risk for AI systems across all three environmental sectors considered in this book. At the scale of national deployment, the funding continuity problem is more acute because the operational costs are larger, the number of institutional stakeholders whose budget processes must align is greater, and the political economy of sustaining digital infrastructure funding across electoral cycles is more complex. Mechanisms that have shown some effectiveness in managing this transition include embedding AI operational costs within existing budget lines for monitoring and evaluation rather than creating new and vulnerable budget lines for technology, establishing multi-year funding agreements between central government and implementing agencies that span at least two budget cycles, and designing AI systems with explicit cost-reduction trajectories — anticipating that the per-unit cost of data processing and model operation will fall over time as open-source infrastructure matures and staff capacity builds.

Technical support infrastructure is the third condition. At pilot scale, technical problems are handled by the researchers or developers who built the system and who carry its full technical specification in their institutional memory. At operational scale, technical support must be provided by staff who were not involved in building the system, working from documentation that may be incomplete, and dealing with problems that arise in contexts the original developers did not anticipate. Building adequate technical support infrastructure for scaled AI systems requires investing in documentation that is genuinely useful rather than formally adequate, in training programmes that build real diagnostic and maintenance capability rather than surface familiarity, and in help desk or remote support arrangements that give field-level operators a reliable route to technical assistance when systems malfunction.

Stakeholder Trust and Adoption Incentives

Technical functionality is a necessary but not sufficient condition for AI adoption at scale. Across all three environmental sectors reviewed in this book, the evidence consistently points to stakeholder trust — or its absence — as a primary determinant of whether AI tools that are technically available are actually used to inform decisions. Trust in AI systems among farmers, water managers, and conservation practitioners is built through a combination of demonstrated accuracy over time, transparency about how recommendations are generated, responsiveness to feedback, and alignment between what the system recommends and what users already understand to be good practice from their own experience.

Demonstrated accuracy is the most important of these trust-building mechanisms, and it takes time. A flood forecasting system that issues its first accurate flood warning earns more institutional trust than a hundred technical validation reports, and a single high-profile false alarm can set back adoption by years. This means that AI systems in environmental management should be introduced with conservative confidence claims, with explicit communication about the conditions under which the system is likely to perform well or poorly, and with mechanisms for users to provide feedback on forecast accuracy that is then visibly used to improve the system. The visible use of feedback — showing users that their reports of system errors have led to model improvements — is particularly effective in building the ongoing trust that sustains adoption.

Adoption incentives matter alongside trust. In agricultural contexts, financial incentives — input subsidies conditional on use of AI advisory tools, insurance premium discounts for farmers who follow AI-guided practices, and market access benefits linked to AI-verified sustainable production — have shown effectiveness in driving initial adoption. In water management, regulatory requirements — mandating AI-based water accounting for large abstractors, or conditioning irrigation licence renewal on demonstrated use of demand forecasting tools — have been used in several water-stressed jurisdictions to accelerate adoption among users who would not have adopted voluntarily. In conservation, similar regulatory hooks exist through environmental impact assessment requirements and protected area management plan obligations. The appropriate mix of incentive and regulatory pressure varies by context, but the general principle — that intrinsic motivation to adopt AI tools is rarely sufficient at scale, and that external incentive structures are needed to accelerate and sustain adoption — is broadly applicable across environmental sectors.

Comparative Analysis of AI System Accessibility

Open-Source vs. Proprietary Platforms Across Sectors

The distinction between open-source and proprietary AI platforms is not merely a technical one — it carries substantial implications for equity, institutional autonomy, long-term cost, and the distribution of analytical capacity across the agriculture-water-biodiversity nexus. A comparative assessment across sectors reveals both common patterns and sector-specific differences that are worth examining in some detail.

Across all three sectors, the open-source ecosystem has matured considerably over the past decade. In agriculture, platforms built on open-source machine learning frameworks — TensorFlow, PyTorch, scikit-learn — combined with freely available satellite imagery from Copernicus, Landsat, and MODIS programmes have enabled the development of crop monitoring, soil health assessment, and pest detection tools that are technically comparable to commercial alternatives for many common use cases. In water management, open-source hydrological modelling frameworks such as SWAT, VIC, and OpenDrainageModel have been extended with machine learning components that substantially improve their predictive performance without requiring the licensing fees associated with commercial hydrological software. In biodiversity conservation, open-source tools for species distribution modelling — MaxEnt, Wallace, and the suite of R packages built around the sdm and biomod2 frameworks — are now standard in the peer-reviewed literature and are used by conservation practitioners in both high-income and low-income countries. Table 8 presents a structured comparison of open-source and proprietary AI platform characteristics across the three environmental sectors, covering key dimensions of accessibility, technical capability, support infrastructure, and long-term cost.

Table 8. Comparative characteristics of open-source and proprietary AI platforms across agriculture, water, and biodiversity sectors

Dimension	Open-Source Platforms	Proprietary Platforms	Sector Variation
Upfront cost	Minimal to zero	2,00,000–50,00,000+	Higher in water/agriculture; lower in biodiversity
Annual licensing	None	1,00,000– 20,00,000	Significant in agriculture and water sectors
Technical support	Community forums, documentation	Vendor helpdesk, SLAs	Proprietary advantage most pronounced in water sector
Customisability	Full source access	Limited to vendor APIs	Open-source advantage across all sectors
Data sovereignty	Complete	Variable, often cloud-dependent	Critical concern in biodiversity and water governance
Institutional dependence	None	High (vendor lock-in risk)	Most acute in agriculture (subscription models)
Update frequency	Variable (community-driven)	Regular (vendor-managed)	Proprietary more reliable for operational systems
Validated performance	Peer-reviewed literature	Vendor benchmarks, selective	Open-source more transparent across all sectors

Source: Author's compilation
SLA = Service Level Agreement.

Several patterns emerge from Table 8 that cut across all three sectors. Open-source platforms consistently offer advantages in cost, customisability, and data sovereignty — the last of these being particularly important for government agencies that have legal or policy obligations to maintain control over national environmental data. Proprietary platforms offer advantages in the reliability of technical support and the regularity of software updates, which matter most in operational contexts where system downtime has direct management consequences — a flood forecasting system that fails during the monsoon season, or an irrigation scheduling platform that goes offline during peak water demand, imposes real costs that a vendor service level agreement can partially mitigate.

The data sovereignty dimension deserves particular emphasis in the environmental management context. Many proprietary AI platforms for agriculture, water, and biodiversity operate on cloud infrastructure hosted outside the country in which the environmental data is collected. This means that nationally sensitive data — groundwater levels, forest cover change rates, endangered species location records — may be processed and stored on servers subject to foreign legal jurisdiction. For government agencies with statutory responsibilities for environmental data management, this arrangement raises genuine legal and policy concerns that are not always adequately addressed in vendor contracts. Open-source platforms deployed on domestically hosted infrastructure avoid this concern entirely, and several countries have begun mandating domestic data hosting as a condition of government AI procurement precisely for this reason.

The choice between open-source and proprietary platforms is rarely a single, system-wide decision. In practice, most sophisticated environmental AI deployments use a hybrid architecture — open-source components for core analytical functions where the open-source tools perform well and community support is robust, proprietary components for specialised functions where commercial tools offer a clear performance advantage, and custom-built components for the specific institutional interfaces and data integration functions that neither commercial nor open-source tools adequately address out of the box. Managing this hybrid architecture over time, keeping components compatible as each evolves independently, and maintaining the institutional knowledge of how the components fit together are significant operational challenges that receive insufficient attention in the technology adoption literature. Addressing them requires sustained investment in the technical capacity of environmental agencies — not just at the point of initial deployment, but as an ongoing institutional capability that evolves alongside the technology landscape.

CONCLUSION

Sustainable agriculture, community-based water resource management, and biodiversity conservation constitute a triad of interdependent components which determine the long-term viability of human and natural systems. Sustainable agriculture, community-based water resource management, and biodiversity conservation constitute a triad of interdependent components which determine the long-term viability of human and natural systems. The following analysis supports the view that in the coming years increasingly we may hear of (or from) application of AI techniques in these very fields. Results from the sections on agriculture, water, and biodiversity show that AI systems can enhance predictive ability, resource allocation, and large-scale environmental monitoring. In agriculture, precision analytics and machine learning models are used to predict yields, evaluate soil health and detect pests early. In the domain of water systems, AI-based hydrological predictions together with intelligent irrigation management have the potential to enhance water-use efficiency and adaptivity in the face of climate variability. Their use has been promoted in biodiversity conservation, where they facilitate large-scale species monitoring and help to prioritise conservation. When considered together, these findings suggest that the greatest benefit from AI may not be that which is gleaned from implementing sector-specific solutions but rather that which can be derived from weaving environmental data through food systems, water governance and ecosystem protection. The chapter also notes that environmental analysis across all sectors could help facilitate more aligned climate action. By leveraging remote sensing data and sensor networks, ecological observations, and predictive modelling, AI-based systems can offer decision makers integrated insights that inform climate adaptation strategies, enhance resource efficiency, and bolster ecosystem resilience. Such integrated analytic approaches can also identify trade-offs between agricultural production, water security, and biodiversity conservation and thus enable policy-makers and resource managers to develop more harmonious, science-based sustainable management options. Meanwhile, the chapter stresses that the transformative capacity of AI will be underpinned by responsible governance and inclusive deployment. Ongoing issues related to data quality, algorithmic bias, infrastructure reliance and uneven access to technology continue to be particularly prominent in the under-resourced regions of the world, where the stakes for the environment are generally highest. As such, future research and policy should focus on data justice, transparent algorithmic design, cross-sector governance arrangements, and environmentally sustainable digital infrastructure. Delivery on these agendas will be critical if AI systems are to be harnessed not only for technological efficiency but also for equitable climate action and long-term planetary stewardship.

REFERENCES

Abed, N., Murugan, R., Deldari, A., Sankarannair, S., & Ramesh, M. V. (2025). IoT and AI-driven solutions for human-wildlife conflict: Advancing sustainable agriculture and biodiversity conservation. *Smart Agricultural Technology*, *10*, 100829. DOI: 10.1016/j.atech.2025.100829

Adanma, U. M., & Ogunbiyi, E. O.Uwaga Monica AdanmaEmmanuel Olurotimi Ogunbiyi. (2024). Artificial intelligence in environmental conservation: Evaluating cyber risks and opportunities for sustainable practices. *Computer Science & IT Research Journal*, *5*(5), 1178–1209. DOI: 10.51594/csitrj.v5i5.1156

Adewusi, A. O., Asuzu, O. F., Olorunsogo, T., Iwuanyanwu, C., Adaga, E., & Daraojimba, D. O. (2024). AI in precision agriculture: A review of technologies for sustainable farming practices. *World Journal of Advanced Research and Reviews*, *21*(1), 2276–2285. DOI: 10.30574/wjarr.2024.21.1.0314

Aijaz, N., Lan, H., Raza, T., Yaqub, M., Iqbal, R., & Pathan, M. S. (2025). Artificial intelligence in agriculture: Advancing crop productivity and sustainability. *Journal of Agriculture and Food Research*, *20*, 101762. DOI: 10.1016/j.jafr.2025.101762

Al-Raeei, M. (2024). Artificial intelligence for climate resilience: Advancing sustainable goals in SDGs 11 and 13 and its relationship to pandemics. *Discover Sustainability*, *5*(1), 513. DOI: 10.1007/s43621-024-00775-5

Alduweib, E., Alromema, W., & Hezzam, E. (2025, April). Artificial intelligence for green agriculture and sustainable resource management: A comprehensive overview. In 2025 1st International Conference on Computational Intelligence Approaches and Applications (ICCIAA) (pp. 1-5). IEEE. https://doi.org/DOI: 10.1109/ICCIAA65327.2025.11013361

Ali, G., Mijwil, M. M., Adamopoulos, I., & Ayad, J. (2025). Leveraging the internet of things, remote sensing, and artificial intelligence for sustainable forest management. *Babylonian Journal of Internet of Things*, *2025*, 1–65. DOI: 10.58496/BJIoT/2025/001

Alotaibi, E., & Nassif, N. (2024). Artificial intelligence in environmental monitoring: In-depth analysis. *Discover Artificial Intelligence*, *4*(1), 84. DOI: 10.1007/s44163-024-00198-1

AlZubi, A. A., & Galyna, K. (2023). Artificial intelligence and internet of things for sustainable farming and smart agriculture. *IEEE Access : Practical Innovations, Open Solutions*, *11*, 78686–78692. DOI: 10.1109/ACCESS.2023.3298215

Anjali, P. K., Sreerekha, S., & Monu, N. C. (2026). Sustainable Resource Management with AI: Innovations for a Greener Future. AI for Our Planet: How Artificial Intelligence can Solve. *Global Challenges (Hoboken, NJ)*, •••, 151.

Ayoola, V. B., Idoko, I. P., Eromonsei, S. O., Afolabi, O., Apampa, A. R., & Oyebanji, O. S. (2024). The role of big data and AI in enhancing biodiversity conservation and resource management in the USA. *World Journal of Advanced Research and Reviews*, *23*(2), 1851–1873. DOI: 10.30574/wjarr.2024.23.2.2350

Aziz, D., Rafiq, S., Saini, P., Ahad, I., Gonal, B., Rehman, S. A., Rashid, S., Saini, P., Rohela, G. K., Aalum, K., Singh, G., Gnanesh, B. N., & Nabila Iliya, M. (2025). Remote sensing and artificial intelligence: Revolutionizing pest management in agriculture. *Frontiers in Sustainable Food Systems*, *9*, 1551460. DOI: 10.3389/fsufs.2025.1551460

Bac, C. W., van Henten, E. J., Hemming, J., & Edan, Y. (2014). Harvesting robots for high-value crops: State-of-the-art review and challenges ahead. *Journal of Field Robotics*, *31*(6), 888–911. DOI: 10.1002/rob.21525

Balaska, V., Adamidou, Z., Vryzas, Z., & Gasteratos, A. (2023). Sustainable crop protection via robotics and artificial intelligence solutions. *Machines (Basel)*, *11*(8), 774. DOI: 10.3390/machines11080774

Botero-Valencia, J., García-Pineda, V., Valencia-Arias, A., Valencia, J., Reyes-Vera, E., Mejia-Herrera, M., & Hernández-García, R. (2025). Machine learning in sustainable agriculture: Systematic review and research perspectives. *Agriculture*, *15*(4), 377. DOI: 10.3390/agriculture15040377

Breiman, L. (2001). Random forests. *Machine Learning*, *45*(1), 5–32. DOI: 10.1023/A:1010933404324

Buttigieg, P. L., Pafilis, E., Lewis, S. E., Schildhauer, M. P., Walls, R. L., & Mungall, C. J. (2016). The environment ontology in 2016: Bridging domains with increased scope, semantic density, and interoperation. *Journal of Biomedical Semantics*, *7*(1), 57. Advance online publication. DOI: 10.1186/s13326-016-0097-6 PMID: 27664130

Candelieri, A. (2017). Clustering and support vector regression for water demand forecasting and anomaly detection. *Water (Basel)*, *9*(3), 224. Advance online publication. DOI: 10.3390/w9030224

Chang, F. J., Chang, L. C., & Chen, J. F. (2023). Artificial intelligence techniques in hydrology and water resources management. *Water (Basel)*, *15*(10), 1846. DOI: 10.3390/w15101846

Chisom, O. N., Biu, P. W., Umoh, A. A., Obaedo, B. O., Adegbite, A. O., & Abatan, A. (2024). Reviewing the role of AI in environmental monitoring and conservation: A data-driven revolution for our planet. *World Journal of Advanced Research and Reviews*, *21*(1), 161–171. DOI: 10.30574/wjarr.2024.21.1.2720

D'amore, G., Di Vaio, A., Balsalobre-Lorente, D., & Boccia, F. (2022). Artificial intelligence in the water–energy–food model: A holistic approach towards sustainable development goals. *Sustainability (Basel)*, *14*(2), 867. DOI: 10.3390/su14020867

Del-Coco, M., Leo, M., & Carcagnì, P. (2024). Machine learning for smart irrigation in agriculture: How far along are we? *Information (Basel)*, *15*(6), 306. DOI: 10.3390/info15060306

Eaton, B., Gregory, J., Drach, B., Taylor, K., Hankin, S., Caron, J., Signell, R., Bentley, P., Rappa, G., Höck, H., Pamment, A., Juckes, M., & Raspaud, M. (2017). *NetCDF climate and forecast (CF) metadata conventions* (Version 1.7). CF Conventions Committee. https://cfconventions.org/cf-conventions/cf-conventions.html

Efremova, N., Foley, J. C., Unagaev, A., & Karimi, R. (2023). AI for sustainable agriculture and rangeland monitoring. In *The ethics of artificial intelligence for the sustainable development goals* (pp. 399–422). Springer International Publishing., DOI: 10.1007/978-3-031-21147-8_22

El Jarroudi, M., Kouadio, L., Delfosse, P., Bock, C. H., Mahlein, A. K., Fettweis, X., Mercatoris, B., Adams, F., Lenné, J. M., & Hamdioui, S. (2024). Leveraging edge artificial intelligence for sustainable agriculture. *Nature Sustainability*, *7*(7), 846–854. DOI: 10.1038/s41893-024-01352-4

Elbasi, E., Mostafa, N., AlArnaout, Z., Zreikat, A. I., Cina, E., Varghese, G., Shdefat, A., Topcu, A. E., Abdelbaki, W., Mathew, S., & Zaki, C. (2022). Artificial intelligence technology in the agricultural sector: A systematic literature review. *IEEE Access : Practical Innovations, Open Solutions*, *11*, 171–202. DOI: 10.1109/ACCESS.2022.3232485

Elith, J., Graham, C. H., Anderson, R. P., Dudík, M., Ferrier, S., Guisan, A., Hijmans, R. J., Huettmann, F., Leathwick, J. R., Lehmann, A., Li, J., Lohmann, L. G., Loiselle, B. A., Manion, G., Moritz, C., Nakamura, M., Nakazawa, Y., Overton, J. M., Peterson, A. T., & Zimmermann, N. E. (2006). Novel methods improve prediction of species' distributions from occurrence data. *Ecography*, *29*(2), 129–151. DOI: 10.1111/j.2006.0906-7590.04596.x

Espinel, R., Herrera-Franco, G., Rivadeneira García, J. L., & Escandón-Panchana, P. (2024). Artificial intelligence in agricultural mapping. *Agriculture*, *14*(7), 1071. DOI: 10.3390/agriculture14071071

Estopinan, J., Servajean, M., Bonnet, P., Munoz, F., & Joly, A. (2022). Deep species distribution modeling from sentinel-2 image time-series: A global scale analysis on the orchid family. *Frontiers in Plant Science, 13*, 839327. DOI: 10.3389/fpls.2022.839327 PMID: 35528931

Fiske, I., & Chandler, R. (2011). unmarked: An R package for fitting hierarchical models of wildlife occurrence and abundance. *Journal of Statistical Software*, *43*(10), 1–23. DOI: 10.18637/jss.v043.i10

. Food and Agriculture Organization of the United Nations. (2022). *The state of food and agriculture 2022: Leveraging automation in agriculture*. FAO. https://doi.org/ DOI: 10.4060/cb9479en

Fountas, S., Mylonas, N., Malounas, I., Rodias, E., Hellmann Santos, C., & Pekkeriet, E. (2020). Agricultural robotics for field operations. *Sensors (Basel)*, *20*(9), 2672. Advance online publication. DOI: 10.3390/s20092672 PMID: 32392872

Garske, B., Bau, A., & Ekardt, F. (2021). Digitalization and AI in European agriculture: A strategy for achieving climate and biodiversity targets? *Sustainability (Basel)*, *13*(9), 4652. DOI: 10.3390/su13094652

Geodetic Parameter Dataset, E. P. S. G. (2023). *EPSG registry*. International Association of Oil & Gas Producers. https://epsg.org

Giarratano, J. C., & Riley, G. D. (2005). *Expert systems: Principles and programming* (4th ed.). Thomson Course Technology.

Gimenez, O., Kery, M., & Schaub, M. (2021). Integrated population models: Theory and ecological applications with R and JAGS. *Methods in Ecology and Evolution*, *12*(3), 388–401. DOI: 10.1111/2041-210X.13524

Goldberg, D. E. (1989). *Genetic algorithms in search, optimization, and machine learning*. Addison-Wesley.

Gryshova, I., Balian, A., Antonik, I., Miniailo, V., Nehodenko, V., & Nyzhnychenko, Y. (2024). Artificial intelligence in climate smart in agricultural: Toward a sustainable farming future. *Access J*, *5*(1), 125–140. DOI: 10.46656/access.2024.5.1(8)

Gul, D., & Banday, R. U. Z. (2024). *Transforming crop management through advanced AI and machine learning: Insights into innovative strategies for sustainable agriculture*. AI, Computer Science and Robotics Technology., DOI: 10.5772/acrt.20240030

Habila, M. A., Ouladsmane, M., & Alothman, Z. A. (2023). Role of artificial intelligence in environmental sustainability. In *Visualization techniques for climate change with machine learning and artificial intelligence* (pp. 449–469). Elsevier., DOI: 10.1016/B978-0-323-99714-0.00009-1

Haleem, A., Javaid, M., Khan, I. H., & Mohan, S. (2023). Significant applications of artificial intelligence towards attaining sustainability. *Journal of Industrial Integration and Management, 8*(04), 489–520. DOI: 10.1142/S2424862223500331

INSPIRE. (2007). *Directive 2007/2/EC of the European Parliament and of the Council establishing an infrastructure for spatial information in the European Community (INSPIRE)*. Official Journal of the European Union. https://inspire.ec.europa.eu

ISO. (2019). *ISO 8601-1:2019: Date and time — Representations for information interchange — Part 1: Basic rules*. International Organization for Standardization. https://www.iso.org/standard/70907.html

Jones, J., Harris, E., Febriansah, Y., Adiwijaya, A., & Hikam, I. N. (2024). Ai for sustainable development: Applications in natural resource management, agriculture, and waste management. *International Transactions on Artificial Intelligence, 2*(2), 143–149. DOI: 10.33050/italic.v2i2.549

Kamilaris, A., & Prenafeta-Boldú, F. X. (2018). Deep learning in agriculture: A survey. *Computers and Electronics in Agriculture, 147*, 70–90. DOI: 10.1016/j.compag.2018.02.016

Kose, U., Prasath, V. B., Mondal, M., Podder, P., & Bharati, S. (Eds.). (2022). *Artificial intelligence and smart agriculture technology*. CRC Press., DOI: 10.1201/9781003299059

Kratzert, F., Klotz, D., Brenner, C., Schulz, K., & Herrnegger, M. (2018). Rainfall–runoff modelling using Long Short-Term Memory (LSTM) networks. *Hydrology and Earth System Sciences, 22*(11), 6005–6022. DOI: 10.5194/hess-22-6005-2018

Kumari, K., Mirzakhani Nafchi, A., Mirzaee, S., & Abdalla, A. (2025). AI-driven future farming: Achieving climate-smart and sustainable agriculture. *AgriEngineering, 7*(3), 89. DOI: 10.3390/agriengineering7030089

Kumari, N., & Pandey, S. (2023). Application of artificial intelligence in environmental sustainability and climate change. In *Visualization techniques for climate change with machine learning and artificial intelligence* (pp. 293–316). Elsevier., DOI: 10.1016/B978-0-323-99714-0.00018-2

LeCun, Y., Bengio, Y., & Hinton, G. (2015). Deep learning. *Nature, 521*(7553), 436–444. DOI: 10.1038/nature14539 PMID: 26017442

Liakos, K. G., Busato, P., Moshou, D., Pearson, S., & Bochtis, D. (2018). Machine learning in agriculture: A review. *Sensors (Basel), 18*(8), 2674. Advance online publication. DOI: 10.3390/s18082674 PMID: 30110960

Liu, T., Sun, Y., Wang, C., Zhang, Y., Qiu, Z., Gong, W., Lei, S., Tong, X., & Duan, X. (2021). Unmanned aerial vehicle and artificial intelligence revolutionizing efficient and precision sustainable forest management. *Journal of Cleaner Production, 311*, 127546. DOI: 10.1016/j.jclepro.2021.127546

Longley, P. A., Goodchild, M. F., Maguire, D. J., & Rhind, D. W. (2015). *Geographic information science and systems* (4th ed.). Wiley.

Margules, C. R., & Pressey, R. L. (2000). Systematic conservation planning. *Nature, 405*(6783), 243–253. DOI: 10.1038/35012251 PMID: 10821285

Martos, V., Ahmad, A., Cartujo, P., & Ordoñez, J. (2021). Ensuring agricultural sustainability through remote sensing in the era of agriculture 5.0. *Applied Sciences (Basel, Switzerland), 11*(13), 5911. DOI: 10.3390/app11135911

Miller, T., Durlik, I., Kostecka, E., Kozlovska, P., Łobodzińska, A., Sokołowska, S., & Nowy, A. (2025). Integrating artificial intelligence agents with the internet of things for enhanced environmental monitoring: Applications in water quality and climate data. *Electronics (Basel), 14*(4), 696. DOI: 10.3390/electronics14040696

Mishra, H. (2025). Artificial Intelligence Technologies in Agriculture. In AgriTech Revolution: Next-Gen Solutions and Challenges in Modern Agriculture (pp. 371-396). Singapore: Springer Nature Singapore. https://doi.org/DOI: 10.1007/978-981-95-1268-3_19

Mishra, H., & Dwivedi, S. Remote Sensing for Sustainable Agriculture: A Machine Learning Approach to Optimizing Farm Yield and Economic Returns. In Artificial Intelligence and Computer Vision for Ecological Informatics (pp. 116-150). CRC Press.

Mishra, H., & Mishra, D. (2023). Artificial intelligence and machine learning in agriculture: Transforming farming systems. *Res. Trends Agric. Sci, 1*, 1–16.

Mishra, H., & Mishra, D. (2024). AI for data-driven decision-making in smart agriculture: from field to farm management. In *Artificial intelligence techniques in smart agriculture* (pp. 173–193). Springer Nature Singapore., DOI: 10.1007/978-981-97-5878-4_11

Mishra, H., & Mishra, D. (2025). Robotics, drones, remote sensing, GIS, and IoT tools for agricultural operations and water management. In *Integrated Land and Water Resource Management for Sustainable Agriculture* (Vol. 2, pp. 21–49). Springer Nature Singapore., DOI: 10.1007/978-981-97-9800-1_2

Mishra, H., Mishra, R., Olayinka, O. T., Ghaib, A. A., Raza, M. Y., & Maroof, K. (2025). *The Role of Digital Agriculture in Mitigating Climate Change and Ensuring Food Security: An Overview. Climate Resilient and Sustainable Agriculture* (Vol. 2). Social and Transformative Strategies., DOI: 10.1007/978-3-032-04141-8_11

Mnih, V., Kavukcuoglu, K., Silver, D., Rusu, A. A., Veness, J., Bellemare, M. G., Graves, A., Riedmiller, M., Fidjeland, A. K., Ostrovski, G., Petersen, S., Beattie, C., Sadik, A., Antonoglou, I., King, H., Kumaran, D., Wierstra, D., Legg, S., & Hassabis, D. (2015). Human-level control through deep reinforcement learning. *Nature*, *518*(7540), 529–533. DOI: 10.1038/nature14236 PMID: 25719670

Mohanty, S. P., Hughes, D. P., & Salathé, M. (2016). Using deep learning for image-based plant disease detection. *Frontiers in Plant Science*, *7*, 1419. Advance online publication. DOI: 10.3389/fpls.2016.01419 PMID: 27713752

Moilanen, A., Wilson, K. A., & Possingham, H. P. (Eds.). (2009). *Spatial conservation prioritization: Quantitative methods and computational tools*. Oxford University Press. DOI: 10.1093/oso/9780199547760.001.0001

Mumtaz, N., Izhar, T., Pandey, G., & Labhasetwar, P. K. (2022). Utilizing artificial intelligence for environmental sustainability. In *Artificial intelligence for renewable energy systems* (pp. 259–279). Woodhead Publishing., DOI: 10.1016/B978-0-323-90396-7.00010-9

Naghibi, S. A., Pourghasemi, H. R., & Dixon, B. (2016). GIS-based groundwater potential mapping using boosted regression tree, classification and regression tree, and random forest machine learning models in Iran. *Environmental Monitoring and Assessment*, *188*(1), 44. Advance online publication. DOI: 10.1007/s10661-015-5049-6 PMID: 26687087

Najah, A., El-Shafie, A., Karim, O. A., & El-Shafie, A. H. (2013). Application of artificial neural networks for water quality prediction. *Neural Computing & Applications*, *22*(S1), 187–201. DOI: 10.1007/s00521-012-0940-3

Nautiyal, C. T., Nautiyal, P., Papnai, G., Mittal, H., Agrawal, K., & Nandini, R. (2024). Importance of smart agriculture and use of artificial intelligence in shaping the future of agriculture. *Journal of Scientific Research and Reports*, *30*(3), 129–138. DOI: 10.9734/jsrr/2024/v30i31864

Ngulube, P. (2025). Leveraging information and communication technologies for sustainable agriculture and environmental protection among smallholder farmers in tropical Africa. *Discover Environment*, *3*(1), 9. DOI: 10.1007/s44274-025-00190-1

Norouzzadeh, M. S., Nguyen, A., Kosmala, M., Swanson, A., Palmer, M. S., Packer, C., & Clune, J. (2018). Automatically identifying, counting, and describing wild animals in camera-trap images with deep learning. *Proceedings of the National Academy of Sciences of the United States of America*, *115*(25), E5716–E5725. DOI: 10.1073/pnas.1719367115 PMID: 29871948

Nti, E. K., Cobbina, S. J., Attafuah, E. E., Opoku, E., & Gyan, M. A. (2022). Environmental sustainability technologies in biodiversity, energy, transportation and water management using artificial intelligence: A systematic review. *Sustainable Futures : An Applied Journal of Technology, Environment and Society*, *4*, 100068. DOI: 10.1016/j.sftr.2022.100068

Nti, E. K., Cobbina, S. J., Attafuah, E. E., Senanu, L. D., Amenyeku, G., Gyan, M. A., Forson, D., & Safo, A. R. (2023). Water pollution control and revitalization using advanced technologies: Uncovering artificial intelligence options towards environmental health protection, sustainability and water security. *Heliyon*, *9*(7), e18170. Advance online publication. DOI: 10.1016/j.heliyon.2023.e18170 PMID: 37496916

Open Geospatial Consortium. (2022). *OGC API standards*. OGC. https://ogcapi.ogc.org

Padhiary, M., Hoque, A., Prasad, G., Kumar, K., & Sahu, B. (2025). Precision agriculture and AI-driven resource optimization for sustainable land and resource management. In *Smart water technology for sustainable management in modern cities* (pp. 197–232). IGI Global Scientific Publishing., DOI: 10.4018/979-8-3693-8074-1.ch009

Pantazi, X. E., Moshou, D., Alexandridis, T., Whetton, R. L., & Mouazen, A. M. (2016). Wheat yield prediction using machine learning and advanced sensing techniques. *Computers and Electronics in Agriculture*, *121*, 57–65. DOI: 10.1016/j.compag.2015.11.018

Pasha Mohammed, S., Deepika, J., Sritharan, N., Ravichandran, V., Prasanthrajan, M., & Kannan, P. (2025). A systematic literature review on artificial intelligence in transforming precision agriculture for sustainable farming: Current status and future directions. *Plant Science Today*, *12*(2), 1–13. DOI: 10.14719/pst.6175

Phadikar, S., Sil, J., & Das, A. K. (2013). Rice diseases classification using feature selection and rule generation techniques. *Computers and Electronics in Agriculture*, *90*, 76–85. DOI: 10.1016/j.compag.2012.11.001

Phillips, S. J., Anderson, R. P., & Schapire, R. E. (2006). Maximum entropy modeling of species geographic distributions. *Ecological Modelling*, *190*(3–4), 231–259. DOI: 10.1016/j.ecolmodel.2005.03.026

Pimenow, S., Pimenowa, O., Prus, P., & Niklas, A. (2025). The impact of artificial intelligence on the sustainability of regional ecosystems: Current challenges and future prospects. *Sustainability (Basel)*, *17*(11), 4795. DOI: 10.3390/su17114795

Raihan, A., Paul, A., Rahman, M. S., Islam, S., Paul, P., & Karmakar, S. (2024). Artificial intelligence (AI) for environmental sustainability: A concise review of technology innovations in energy, transportation, biodiversity, and water management. *Journal of Technology Innovations and Energy*, *3*(2), 64–73. DOI: 10.56556/jtie.v3i2.953

Rane, N. (2023). Roles and challenges of ChatGPT and similar generative artificial intelligence for achieving the sustainable development goals (SDGs). Available at *SSRN* 4603244. https://doi.org/DOI: 10.2139/ssrn.4603244

Ruiz, I., Pompeu, J., Ruano, A., Franco, P., Balbi, S., & Sanz, M. J. (2023). Combined artificial intelligence, sustainable land management, and stakeholder engagement for integrated landscape management in Mediterranean watersheds. *Environmental Science & Policy*, *145*, 217–227. DOI: 10.1016/j.envsci.2023.04.011

Ryu, J., Cha, D., & Oh, J. (2020). Development of irrigation water management system using deep reinforcement learning. *Applied Sciences (Basel, Switzerland)*, *10*(22), 8085. Advance online publication. DOI: 10.3390/app10228085

Saha, S., Kucher, O. D., Utkina, A. O., & Rebouh, N. Y. (2025). Precision agriculture for improving crop yield predictions: A literature review. *Frontiers in Agronomy*, *7*, 1566201. DOI: 10.3389/fagro.2025.1566201

Saikanth, K., Singh, B. V., Sachan, D. S., & Singh, B. (2023). Advancing sustainable agriculture: A comprehensive review for optimizing food production and environmental conservation. *International Journal of Plant and Soil Science*, *35*(16), 417–425. DOI: 10.9734/ijpss/2023/v35i163169

Sánchez, J. M., Rodríguez, J. P., & Espitia, H. E. (2022). Bibliometric analysis of publications discussing the use of the artificial intelligence technique agent-based models in sustainable agriculture. *Heliyon*, *8*(12), e12005. Advance online publication. DOI: 10.1016/j.heliyon.2022.e12005 PMID: 36478806

Sanders, C. E., Mayfield-Smith, K. A., & Lamm, A. J. (2021). Exploring Twitter discourse around the use of artificial intelligence to advance agricultural sustainability. *Sustainability (Basel)*, *13*(21), 12033. DOI: 10.3390/su132112033

Sehgal, D., Singh, R., & Dwivedi, S. (2021). Deep learning-based hybrid model for environmental forecasting: A review. *Journal of Environmental Informatics*, *37*(2), 89–104. DOI: 10.3808/jei.202100456

Senoo, E. E. K., Anggraini, L., Kumi, J. A., Karolina, L. B., Akansah, E., Sulyman, H. A., Mendonça, I., & Aritsugi, M. (2024). IoT solutions with artificial intelligence technologies for precision agriculture: Definitions, applications, challenges, and opportunities. *Electronics (Basel)*, *13*(10), 1894. DOI: 10.3390/electronics13101894

Shang, M., & Xie, J. (2024). Agricultural sustainable development: Soil, water resources, biodiversity, climate change, and technological innovation. *Advances in Resources Research*, *4*(2), 181–204.

. Shankar, P., Werner, N., Selinger, S., & Janssen, O. (2020, September). Artificial intelligence driven crop protection optimization for sustainable agriculture. In 2020 IEEE/ITU international conference on artificial intelligence for good (AI4G) (pp. 1-6). IEEE.

Sharma, K., & Shivandu, S. K. (2024). Integrating artificial intelligence and Internet of Things (IoT) for enhanced crop monitoring and management in precision agriculture. *Sensors International*, *5*, 100292. DOI: 10.1016/j.sintl.2024.100292

Shen, C. (2018). A transdisciplinary review of deep learning research and its relevance for water resources scientists. *Water Resources Research*, *54*(11), 8558–8593. DOI: 10.1029/2018WR022643

Shim, J. P., Warkentin, M., Courtney, J. F., Power, D. J., Sharda, R., & Carlsson, C. (2002). Past, present, and future of decision support technology. *Decision Support Systems*, *33*(2), 111–126. DOI: 10.1016/S0167-9236(01)00139-7

Shivaprakash, K. N., Swami, N., Mysorekar, S., Arora, R., Gangadharan, A., Vohra, K., Jadeyegowda, M., & Kiesecker, J. M. (2022). Potential for artificial intelligence (AI) and machine learning (ML) applications in biodiversity conservation, managing forests, and related services in India. *Sustainability (Basel)*, *14*(12), 7154. DOI: 10.3390/su14127154

Silvestro, D., Goria, S., Sterner, T., & Antonelli, A. (2022). Improving biodiversity protection through artificial intelligence. *Nature Sustainability*, *5*(5), 415–424. DOI: 10.1038/s41893-022-00851-6 PMID: 35614933

. Singh, A., Mehrotra, R., Rajput, V. D., Dmitriev, P., Singh, A. K., Kumar, P., ... & Singh, A. K. (2022). Geoinformatics, artificial intelligence, sensor technology, big data: emerging modern tools for sustainable agriculture. Sustainable agriculture systems and technologies, 295-313. https://doi.org/DOI: 10.1002/9781119808565.ch14

Singh, S., & Jain, P. (2022). Applications of artificial intelligence for the development of sustainable agriculture. In *Agro-biodiversity and Agri-ecosystem Management* (pp. 303–322). Springer Nature Singapore., DOI: 10.1007/978-981-19-0928-3_16

Sinha, G. (2024). Artificial intelligence in water management for sustainable farming: A review. *Journal of Scientific Research and Reports*, *30*(6), 511–525. DOI: 10.9734/jsrr/2024/v30i62068

Tabak, M. A., Norouzzadeh, M. S., Wolfson, D. W., Sweeney, S. J., VerCauteren, K. C., Snow, N. P., Halseth, J. M., Di Salvo, P. A., Lewis, J. S., White, M. D., Teton, B., Beasley, J. C., Schlichting, P. E., Boughton, R. K., Wight, B., Newkirk, E. S., Ivan, J. S., Odell, E. A., Brook, R. K., & Miller, R. S. (2019). Machine learning to classify animal species in camera trap images: Applications in ecology. *Methods in Ecology and Evolution*, *10*(4), 585–590. DOI: 10.1111/2041-210X.13120

. Tiwari, A. K., Mishra, H., Nishad, D. C., & Pandey, A. (2023). Sustainable water management in agriculture: irrigation techniques and water conservation. Dr. Ajay B. Jadhao, 53.

Ullah, F., Saqib, S., & Xiong, Y. C. (2025). Integrating artificial intelligence in biodiversity conservation: Bridging classical and modern approaches. *Biodiversity and Conservation*, *34*(1), 45–65. DOI: 10.1007/s10531-024-02977-9

Van Klompenburg, T., Kassahun, A., & Catal, C. (2020). Crop yield prediction using machine learning: A systematic literature review. *Computers and Electronics in Agriculture*, *177*, 105709. Advance online publication. DOI: 10.1016/j.compag.2020.105709

Vapnik, V. N. (1995). *The nature of statistical learning theory*. Springer., DOI: 10.1007/978-1-4757-2440-0

Wang, T., Zuo, Y., Manda, T., Hwarari, D., & Yang, L. (2025). Harnessing artificial intelligence, machine learning and deep learning for sustainable forestry management and conservation: Transformative potential and future perspectives. *Plants*, *14*(7), 998. DOI: 10.3390/plants14070998 PMID: 40219066

Wani, A. K., Rahayu, F., Ben Amor, I., Quadir, M., Murianingrum, M., Parnidi, P., Ayub, A., Supriyadi, S., Sakiroh, S., Saefudin, S., Kumar, A., & Latifah, E. (2024). Environmental resilience through artificial intelligence: Innovations in monitoring and management. *Environmental Science and Pollution Research International*, *31*(12), 18379–18395. DOI: 10.1007/s11356-024-32404-z PMID: 38358626

Wieczorek, J., Bloom, D., Guralnick, R., Blum, S., Döring, M., Giovanni, R., Robertson, T., & Vieglais, D. (2012). Darwin Core: An evolving community-developed biodiversity data standard. *PLoS One*, *7*(1), e29715. Advance online publication. DOI: 10.1371/journal.pone.0029715 PMID: 22238640

Wilkinson, M. D., Dumontier, M., Aalbersberg, I. J., Appleton, G., Axton, M., Baak, A., Blomberg, N., Boiten, J.-W., da Silva Santos, L. B., Bourne, P. E., Bouwman, J., Brookes, A. J., Clark, T., Crosas, M., Dillo, I., Dumon, O., Edmunds, S., Evelo, C. T., Finkers, R., & Mons, B. (2016). The FAIR Guiding Principles for scientific data management and stewardship. *Scientific Data*, *3*(1), 160018. Advance online publication. DOI: 10.1038/sdata.2016.18 PMID: 26978244

Wolfert, S., Ge, L., Verdouw, C., & Bogaardt, M. J. (2017). Big data in smart farming: A review. *Agricultural Systems*, *153*, 69–80. DOI: 10.1016/j.agsy.2017.01.023

Xiang, X., Li, Q., Khan, S., & Khalaf, O. I. (2021). Urban water resource management for sustainable environment planning using artificial intelligence techniques. *Environmental Impact Assessment Review*, *86*, 106515. DOI: 10.1016/j.eiar.2020.106515

Xing, Y., & Wang, X. (2024). Precision agriculture and water conservation strategies for sustainable crop production in arid regions. *Plants*, *13*(22), 3184. DOI: 10.3390/plants13223184 PMID: 39599396

Yadav, M., & Singh, G. (2023). Environmental sustainability with artificial intelligence. [IJMR]. *EPRA International Journal of Multidisciplinary Research*, *9*(5), 213–217.

Ye, G., Wang, Z., Yin, X., Tang, B., & Shen, H. (2022). A robust anomaly detection algorithm for water quality data based on autoencoder. *Journal of Hydrology (Amsterdam)*, *609*, 127686. Advance online publication. DOI: 10.1016/j.jhydrol.2022.127686

Yi, S., & Yi, J. (2024). Reservoir-based flood forecasting and warning: Deep learning versus machine learning. *Applied Water Science*, *14*(11), 237. DOI: 10.1007/s13201-024-02298-w

Yousaf, H. (2024). AI for environmental sustainability: Applications in resource management and conservation. *Journal of AI Range*, *1*(2), 13–26.

Zehouani, N., Ababou, M., & Faquir, S. (2024, May). Harnessing artificial intelligence for environmental sustainability: a comprehensive review of its applications in biodiversity, energy, transportation, and water management. In International Conference on Digital Technologies and Applications (pp. 410-419). Cham: Springer Nature Switzerland. DOI: 10.1007/978-3-031-68650-4_39

Zhang, P., Guo, Z., Ullah, S., Melagraki, G., Afantitis, A., & Lynch, I. (2021). Nanotechnology and artificial intelligence to enable sustainable and precision agriculture. *Nature Plants*, *7*(7), 864–876. DOI: 10.1038/s41477-021-00946-6 PMID: 34168318

Ziesche, S., Agarwal, S., Nagaraju, U., Prestes, E., & Singha, N. (2023). Role of artificial intelligence in advancing sustainable development goals in the agriculture sector. In *The ethics of artificial intelligence for the sustainable development goals* (pp. 379–397). Springer International Publishing., DOI: 10.1007/978-3-031-21147-8_21

Chapter 8

Artificial Intelligence in Climate Change Prediction and Sustainability Solutions

Sajad Ahmad Mir
http://orcid.org/0000-0001-6887-4925
University of Kashmir, India

Uzma Saleem Malik
http://orcid.org/0009-0005-0571-3493
Government College for Women, Jammu, India

Kajul Chauhan
http://orcid.org/0009-0003-3980-1577
Rayat Bahra University, Mohali, India

A. V. Senthil Kumar
http://orcid.org/0000-0002-8587-7017
Nehru Institute of Information Technology and Management, India

Aditya Bala
http://orcid.org/0009-0005-3054-6850
Rayat Bahra University, Mohali, India

Prabhleen Saini
http://orcid.org/0009-0004-6890-2358
Rayat Bahra University, Mohali, India

Sumanpreet Kaur
Desh Bhagat University, Punjab, India

Navneet Saini
http://orcid.org/0009-0002-5234-3200
Rayat Bahra University, Mohali, India

Latif Ahmad Mir
University of Kashmir, India

Chanika Chand
Rayat Bahra University, India

ABSTRACT

Climate change is a major global challenge requiring accurate prognostications and sustainably oriented decision-making frameworks. This chapter examines the

DOI: 10.4018/979-8-3373-9978-2.ch008

potential of Artificial Intelligence (AI) to enhance climate forecasting and support sustainable governance by integrating heterogeneous data streams such as satellite imagery, Internet-of-Things (IoT) sensor feeds, and historical climatological archives. Using a qualitative, secondary-data-based approach, it proposes an Integrated AI-Sustainability Framework with three hierarchical levels: Data Layer, AI Processing Layer, and Decision Layer, along with a Real-Time Adaptive Climate Prediction System. The chapter introduces an AI-Based Sustainability Impact Index to assess environmental, economic, and social outcomes. Emphasis on Explainable AI ensures transparency, accountability, and stakeholder trust while addressing data constraints, biases, and energy demands. Findings indicate AI improves predictive accuracy, resource optimization, and policy formulation for adaptive, transparent, and scalable climate solutions overall.

INTRODUCTION

Climate change is one of the major issues that characterize the modern-day era, affecting environment stability, economic growth, and people globally. Climate change can be solved by utilizing new models that facilitate predictions and reaction towards environmental changes (Edwards, 2010). There have been occurrences of extreme weather situations and an increase in temperature levels, among others, leading to the requirement for strategies that are able to mitigate climate change.

Traditionally, climate prediction and modeling have been based on scientific and physical approaches. However, since the techniques used are mostly based on structured data sets and certain assumptions, they cannot adequately process any kind of unstructured data set (Knutti, 2008). As a result, traditional climate prediction and modeling techniques are unable to cater for the dynamism that occurs in climate change processes.

In light of the problems faced by traditional prediction and modeling techniques in handling climate change, AI has been seen as the solution technology to aid in analyzing environmental data. This is because AI, especially machine learning and deep learning, is capable of studying vast volumes of multi-dimensional structured data and establishing new relationships or trends within them. This has enabled AI technology to provide more accurate predictions and forecasts of climate variables and extreme weather phenomena (Reichstein et al., 2019).

In addition to providing predictive power, artificial intelligence (AI) is important for driving decision-making in response to environmental changes because it allows for adaptive and immediate action against climate problems. For instance, AI systems can use data from satellite imagery, sensor networks, and various environmental databases to continuously monitor and analyze the current state of the environment to intervene in a timely manner in the case of floods, droughts, and abnormal temperatures (Rolnick et al., 2019). Moreover, the use of AI can help address the issues of sustainability through efficient energy consumption and allocation of resources as well as building climate resilient infrastructures in agriculture, water, and urban planning sectors (Vinuesa et al., 2020).

However, the literature on the topic is rather disjointed and fails to establish an integrative perspective on the subject matter. The vast majority of existing studies focus either on increasing the predictive capabilities of AI models or solving particular sustainability problems related to different fields, without discussing any links between AI-based predictions and relevant actions that need to be taken.

Moreover, aspects like explainability, accountability, and transparency tend to receive insufficient attention in current AI-assisted climate-related initiatives. There are issues with deploying highly sophisticated and intricate models when it comes to environmental policy-making. As stressed by Floridi et al., transparency and explainability play a crucial role in building trust between all relevant parties and supporting evidence-based policymaking (Floridi et al., 2018). In addition, it should be noted that the lack of adaptive governance frameworks, which would facilitate incorporating insights generated in real-time, is another major drawback.

A problem of considerable importance concerns the disparity in technological and institutional capacity among different countries. In some instances, there might be barriers that impede the adoption of an AI-enabled solution due to poor digital infrastructure, lack of data availability, and low institutional readiness in some countries.

To overcome some of these challenges, the paper proposes the creation of an integrated Artificial Intelligence (AI)-Sustainability Framework (IASF), a governance-based framework aiming at reconciling predictive analytics with decision-making that is sustainability-oriented. The IASF is comprised of three fundamental parts: data collection, data analysis through artificial intelligence, and decision making within a framework aimed at governing environmental issues. The inclusion of such aspects as explainable AI, sustainability measures, and feedback-based governance is expected to ensure transparency and sustainability through this framework. Besides enhancing climate prediction tools, this framework emphasizes that aspects such as governance, ethics, and institutions are equally important in ensuring sustainability. In this respect, this research contributes to the existing theoretical literature on the topic.

RESEARCH GAP

The critical analysis of existing literature reveals that the study of artificial intelligence in climate prediction and sustainability still suffers from a certain lack of coherence. Despite some success in terms of developing predictive models and using AI in sustainability issues, the scientific community does not provide any integrative approaches that could bring together these two fields into one comprehensive governance framework. In addition, no efforts have been made to fill in the need for explainable and adaptable artificial intelligence that can be used for decision-making processes.

PROBLEM STATEMENT

Absence of an integrated model that brings together climate forecast, sustainability goals, and governance structures is a major constraint in the current scientific environment. This makes it difficult for stakeholders to convert environmental information into practical plans for mitigating and adapting to climate change.

OBJECTIVES

The primary objective of this chapter is to develop an integrated AI-driven sustainability framework that enhances predictive capabilities while supporting data-driven and policy-oriented decision-making. Specifically, the study aims to:

(i) Examine the role of artificial intelligence in climate prediction and sustainability;
(ii) Analyze the application of AI techniques in environmental and governance contexts; and
(iii) Propose a comprehensive framework that integrates data systems, AI analytics, and decision-making processes to generate actionable insights for sustainable development.

REVIEW OF LITERATURE

Artificial Intelligence in Climate Prediction

It is clear that AI technology has become an important catalyst for the revolution in climate science and contributed greatly to improving the accuracy and effectiveness of climate projections. Generally, climate models have traditionally focused on computational modeling based on physical calculations, which provide a strong scientific basis. Nevertheless, computational models are usually costly and inefficient when dealing with non-linear processes (Edwards, 2010). This issue has arisen because of the strict mathematics used in the model and dependence on data structure.

In contrast to this approach, models based on machine learning (ML) and deep learning (DL) demonstrate flexibility by being able to process multi-source large amounts of heterogeneous data, recognize underlying patterns, establish non-linear relationships, and improve prediction accuracy. Numerous studies show the high accuracy of deep learning models, in particular, CNNs and RNNs in the predictions of temperatures, precipitation, and extreme weather events (LeCun et al., 2015). Moreover, LSTM models allow identifying temporal relationships in climate dynamics. For instance, they proved useful in predicting the El Nino-Southern Oscillation (Reichstein et al., 2019).

Inopportunely, the focus of current literature on the accuracy and predictive potential of AI algorithms makes researchers neglect some broader aspects of applying these technologies to solve problems related to climate change. AI models are seen predominantly as a technology that helps make more precise predictions and does not address issues of integrating this prediction into the existing system of governance (Rolnick et al., 2019).

AI in Governance and E-Governance Systems

The application of AI technology to governance systems has significantly revolutionized the field of public administration. This has been achieved through the automation of various governmental processes, efficient delivery of services, and the creation of evidence-based policies (Wirtz et al., 2019). Therefore, the use of AI and big data has paved the way for the emergence of predictive governance where governments deploy AI and big data to predict issues that may arise and develop proactive policy solutions.

Within the sphere of environmental governance, the application of AI in monitoring and governance of ecosystems requires critical analysis. The combination of data collected via satellite images, sensors, and environmental databases makes it possible to make constant assessments of environmental state and potential risks (Janssen & Kuk, 2016). Therefore, using AI provides an opportunity to address climate change-induced problems, namely, deforestation, air pollution, and water shortages.

Nonetheless, it should be stated that existing academic research provides a general understanding of key aspects of introducing AI into governance processes. On the one hand, concerns are raised about information security in light of possible data leakage, bias of algorithms, and lack of transparency. On the other hand, some organizations find it hard to implement AI due to lack of competence, resources, or organizational culture. Finally, such obstacles can be even more apparent in developing countries like India.

The analysis shows that the implementation of AI in the field of governance involves not only the utilization of advanced technological solutions but also organizational readiness. Consequently, the absence of this aspect demonstrates that the alignment of AI and governance requires further research and development.

Limitations of Traditional Climate Models

The conventional climate models are among the widely used models that occupy an important place within the field of climate studies. Such models have provided various scientific findings related to climate changes and their possible impacts on the environment (Knutti, 2008). They are based on deterministic techniques and utilize physical laws to predict climate changes. Nevertheless, despite the scientific value that can be obtained using the above models, there are several factors limiting their use in solving modern climate-related problems.

Firstly, the high costs associated with the use of such models reduce the possibility of conducting timely climate predictions. In particular, the lack of ability to analyze the received results in real time may result in delayed decision making, which will not allow coping with emergencies effectively. Secondly, traditional models lack capabilities to analyze various types of data, which are vital for climate research. Among these types of data are real-time sensor data, satellite information, and socio-economic data.

Furthermore, conventional models utilize some simplifying assumptions to simplify mathematical calculations, which limits the effectiveness of predicting certain local climate changes or extreme situations. Consequently, there is a necessity to develop new climate models that would allow obtaining more effective predictions. AI-based models can serve as potential tools in this process. However, according to the available literature, neither the traditional nor AI-based models can solve existing climate problems adequately without being supported by appropriate integrated models.

Ethical and Institutional Challenges in AI Deployment

There are significant challenges in the use of AI for predicting and managing the climate that should be considered. Data privacy, biases in algorithms, and opacity are recognized as important factors in such cases. They are especially relevant when dealing with applications in the public sector since their consequences affect people directly (Cath et al., 2018). It is even more difficult to ensure accountability in the context of opaque "black-box" systems. Thus, policymakers cannot understand what AI does or why certain results were achieved.

In addition, there are institutional barriers that prevent the application of innovative technologies. In many cases, lack of skills, poor technological infrastructure, and reluctance to changes play an important role. In developing countries, these become more apparent since there are socio-economic disparities, leading to inequality in terms of access to information, technology, and organizations. This calls for an important reflection on the need for ethical standards in the use of AI. In its absence, AI may exacerbate social inequalities and undermine people's confidence in authorities.

Despite the recognition of these challenges, many studies view ethical and governance problems as less important than technical progress in AI. It leads to incomplete knowledge about implementation of innovative solutions.

Research Gap and Need for an Integrated Framework

Several deficiencies in the current literature can be identified from a systematic review. First, there is a need for more attention towards AI-based climate forecasting systems and governance approaches, resulting in a technological-policy gap. Second, the literature lacks a consideration of ethics, institutional arrangements, and sustainability aspects, reducing the applicability of AI technology to the practice. Finally, there is a need for developing more holistic frameworks that incorporate data management, AI technology, and decision-making processes.

The identified literature gaps give rise to a necessity of developing an integrated, governance-based framework that will connect data collection, AI-based analysis, and decision making. The proposed model, the Integrated AI–Sustainability Framework (IASF), provides for such an approach, focusing on linking AI forecasting with sustainability governance in various aspects, including, but not limited to, explainability of AI, feedback and adaptation cycles, sustainability analysis, etc.

THEORETICAL AND CONCEPTUAL BACKGROUND

In this section, we define the theoretical basis of the Integrated AI–Sustainability Framework (IASF) that represents a confluence of AI, climate predictions, and sustainability governance. We apply a socio-technical-ecological perspective where we consider technology, such as artificial intelligence, as being neither separated nor independent from its social and environmental contexts. That is to say that our approach to technology stresses its coevolution and interdependencies with societal institutions and ecological phenomena, thus, providing a comprehensive basis for addressing various aspects of sustainability (Geels, 2004).

Artificial intelligence can also be considered a socio-technical system under the current model. Artificial intelligence is not only an independent device but also a system which is simultaneously affected and affects governance and the environment around it. Historically, artificial intelligence has been understood as a technological device used for various purposes. However, recent studies suggest that artificial intelligence impacts the process of governance by affecting decision-making, policy development, and organizational structures. Within climate predictions, artificial intelligence allows for the consolidation of different types of information and helps to make accurate forecasts about climate behavior (Huntingford et al., 2019).

In contrast, the increasing usage of artificial intelligence leads to questions about sustainability since it has significant effects on the environment. Due to high energy consumption, artificial intelligence contributes to the growing level of carbon dioxide emissions and higher resource consumption, thus bringing up questions about its sustainability. Some empirical studies have proven that the entire life cycle of AI models and algorithms should be considered when it comes to the alignment of its use with environmental targets (Henderson et al., 2020). Thus, it is crucial to understand that AI technologies can become both a solution and a problem for achieving sustainable development goals.

One of the main theoretical foundations of IAISF is explainable and transparent AI technologies, which can help improve the governance process' transparency and accountability. As climate change policies have far-reaching implications for society and the environment, the explainability of algorithms in such fields becomes essential. XAI allows decision-makers to determine how AI predicts the outcomes, thus making the decision-making process more transparent and reliable.

The proposed framework is based on the SDGs of the United Nations and focuses particularly on goals such as climate action, sustainable cities, and responsible consumption and production. Aligning technological advancements in terms of AI applications with the goals of social and environmental sustainability would ensure a more sustainable future for society. Considering the perspectives of environmental economics, organizational studies, and information systems will help develop an interdisciplinary approach in developing the proposed framework. Such an approach will facilitate the deeper comprehension of various aspects of sustainability and will contribute to creating efficient governance mechanisms (Bibri & Krogstie, 2017).

First of all, one needs to admit that the proposed framework will account for the relevance of institutional approaches related to responsible governance of artificial intelligence. In other words, responsible governance of artificial intelligence requires the introduction of sustainability accounting and reporting standards as well as reducing algorithmic biases. Moreover, the responsible use of artificial intelligence requires taking care of various ethical considerations, including the development of a solution to the issue of algorithmic fairness, accountability, and transparency (Floridi et al., 2018).

The last component of this dimension is the creation of a flexible framework to govern AI-related initiatives in the context of climate change since such initiatives cannot guarantee stable outcomes. The combination of artificial intelligence and feedback mechanisms helps in continuous learning and the adjustment of policy recommendations based on the analysis of new data. Thus, instead of adopting rigid, reactive strategies, the process of governance becomes more flexible and proactive.

From the point of view of the whole framework under discussion, this perspective will contribute to transforming artificial intelligence from a mere prediction tool to an agent of transformation of environmental governance through the incorporation of various aspects of sustainable development into it (Vinuesa et al., 2020).

ARTIFICIAL INTELLIGENCE IN BRIEF

Artificial intelligence (AI) refers to computing processes that can undertake tasks usually done by human beings with cognitive capabilities like learning, pattern identification, reasoning, and decision-making (Russell & Norvig, 2021). Unlike in previous decades when computer technology sought to replicate human mental processes, modern AI focuses on enhancing human decision-making through analysis of vast and sophisticated data sets that surpass human intellectual capabilities. Key sub disciplines of AI such as ML and DL help these computing processes identify patterns and extract information from structured and unstructured data to make accurate predictions (Goodfellow et al., 2016).

Currently, AI has evolved from a technical tool to a strategic factor in sustainability and governance. The ability of AI technology to merge the digital world with physical and biological systems has made it a key player in the development of sustainable practices and strategies. AI technology is highly involved in optimizing energy use, improving efficiency, and intelligent management of infrastructure within several industries that include energy, transport, and urban planning (Vinuesa et al., 2020). For example, AI-based optimization techniques have been extensively used in improving the efficiency of energy distribution networks and incorporating renewable energy into the power grid, which has helped reduce emissions and ensure sustainability (International Energy Agency, 2021).

Moreover, AI has played an important role in enhancing ESG frameworks. Through its ability to collect, analyze, and report data accurately, AI has improved the accuracy of sustainability assessments, helping stakeholders make sound decisions about sustainability initiatives (Eccles & Klimenko, 2019). Additionally, AI has provided opportunities for real-time monitoring and evaluation of sustainability indicators, thus fostering adaptive governance strategies.

Another significant aspect of AI revolves around the use of the technology for decision-making in uncertain conditions. The ability of AI tools to forecast regulatory changes, assess risks, and develop strategies in an ever-changing political environment is one of the key strengths that help companies remain competitive in today's market (Bughin et al., 2018). In terms of environmental governance, such a capability may prove quite useful when making decisions based on climate variability and other environmental issues.

The use of AI in relation to climate is associated with the improvement of both climate modeling and decision-making process. Processing large-scale environmental data obtained through satellite imagery, sensors, and geospatial data enables AI systems to increase the accuracy and resolution of climate forecasting thus making it easier to detect environmental trends and make policies informed by facts (Huntingford et al., 2019). However, at the same time, increased use of AI requires responsible action to ensure the positive impact on society.

AI may be viewed as both an innovative tool and an important element of integrated environmental management systems. Thanks to the potential to use predictive analysis and process big data in real time, AI has the potential to become a cornerstone in solving various sustainability issues.

CLIMATE CHANGE PREDICTION

Climate prediction is an important component of climate change mitigation and adaptation efforts. In view of the increased occurrence of extreme weather phenomena and changes in climate, there is a pressing need for prediction methods that are accurate and adaptable. Conventional climate models are widely considered scientific and based on the physical laws of nature. However, conventional climate models often face challenges in processing complex and high-dimensional data sets, thus failing to make real-time and place-specific forecasts. As a result, their usefulness and efficacy in changing environments are greatly compromised (Reichstein et al., 2019).

The application of artificial intelligence (AI) represents a valuable tool that helps integrate various data into the realm of climate studies. Indeed, satellite observations, sensor-based information, geospatial databases, and past climate data can be combined with the use of AI approaches (Rolnick et al., 2019). In this way, predictions concerning critical factors, including temperature and precipitation, can be made more accurate. Therefore, the contribution to risk assessment can be quite substantial.

Apart from predictive capabilities, the power of AI lies in its ability to help optimize energy consumption and increase energy efficiency. Smart buildings, intelligent energy networks, and green infrastructure serve as examples of how effective energy savings could be achieved with the help of AI (International Energy Agency, 2021). They demonstrate AI's ability to serve as an analytical tool and implement measures for climate change mitigation at the same time.

One of the key features of AI is its potential to reveal connections and anomalies in the environment. By analyzing multi-dimensional data sets, AI is able to detect the relationship between the processes in the atmosphere, oceans, and on land surfaces. It is especially relevant for modeling the feedback mechanisms and non-linear systems that cannot be easily detected using conventional methods (Huntingford et al., 2019). Thus, it improves the interpretability and scope of the models created.

Advanced approaches to AI technology, particularly machine learning algorithms and deep learning models, improve the prediction of extreme weather occurrences and climatic trends. Machine learning architectures, such as convolutional neural networks (CNNs), long short-term memory (LSTM) networks, and hybrid networks, exhibit remarkable skills in recognizing spatial and temporal correlations in climatic models. These algorithms exhibit high efficiency in dealing with uncertainties, resulting in more accurate predictions in the context of changing environments (Ham et al., 2019). The ability to handle large sets of data in real-time leads to better forecasting results. These results are important to help create an early warning system to prevent disasters. Moreover, the application of AI technology for real-time processing of information ensures the constant update of the climatic model in response to changing conditions. Adaptability is crucial not only for regular work but also for developing long-term climate policies (Rolnick et al., 2019). Through the integration of different types of datasets, AI technology helps develop more holistic climatic models.

However, there are several limitations related to applying AI technology in predicting climatic processes. First of all, issues concerning data credibility, model explainability, and uncertainty should be noted. The use of AI requires working with a lot of data, and this aspect increases the chance of creating biases. Second, due to the nature of AI models, it is difficult to understand the reason behind certain choices.

Integrating AI into traditional climate models still requires improvement since traditional models possess better theoretical foundations. However, through their flexibility, AI models ensure good performance and high precision. It is expected that the combination of two approaches will help enhance the results significantly.

To conclude, the use of AI technologies improves scenario planning, simulation, and disaster risk reduction measures. Besides, AI models contribute to implementing different initiatives and programs designed to improve environmental conditions by addressing climate change. In particular, it is worth mentioning the work of the

Intergovernmental Panel on Climate Change (IPCC), which emphasizes the significance of accurate scientific evidence for climate-related policy-making (IPCC, 2021).

SUSTAINABILITY DIMENSIONS

The sustainability concept is based on a combination of three interlinked aspects: environment, society, and economy. It should be mentioned that none of these concepts exists separately but rather interacts as parts of interconnected systems whose changes can influence each other's performance. The integration of the described dimensions through the analysis of complex systems and optimization processes becomes possible thanks to artificial intelligence (AI) which ensures better evidence-based decision making (Vinuesa et al., 2020).

In terms of environmental aspect, AI is crucial as it allows making climate models, monitoring the environment, and managing ecosystems. Using large volumes of data, AI makes it possible to track biodiversity, identify deforestation, and monitor such factors as air and water quality. In such a way, environmental problems become more evident and can be addressed before serious changes happen (Rolnick et al., 2019). What is more, AI allows optimizing resource usage in different spheres from energy consumption and waste reduction to agricultural practices and water management.

Just as important is the social aspect of sustainability, which centers on inclusiveness, equality, and improvements in living standards. In this regard, the role played by AI in this setting is through increased availability of vital services such as health care, education, and government services. For example, AI could be used to help improve the quality of diagnostics in health care, provide tailored education, and enhance the provision of services in public administration. AI therefore has the capability of increasing quality of life and reducing inequalities (Bughin et al., 2018). However, there are some limitations associated with the use of AI, including the digital divide, technological differences, and AI biases.

First of all, from an economic point of view, artificial intelligence serves as an accelerator of innovations and productivity as well as sustainability. The efficiency of operations is promoted through supply chain optimization, improvement of energy use efficiency, and promotion of business models based on data. Moreover, AI technologies help promote sustainable business practices as a result of improved resource management and reduced cost of operations (Wirtz et al., 2019). Finally, AI-based solutions assist in building smart industries characterized by increased efficiency of operations and environmental friendliness.

As far as circular economy is concerned, AI plays an important role in it and helps achieve sustainable production practices. With the help of AI-based solutions, predictive maintenance, efficient recycling procedures, and production cycles are possible, contributing to reducing the volume of waste and increasing the lifespan of products (Vinuesa et al., 2020). Additionally, with the help of AI, green finance becomes easier to promote due to improved risk management and the assessment of carbon footprints as well as investment choices aligned with environmental goals (Eccles & Klimenko, 2019).

Finally, it is important to note that a critical approach towards the contribution of AI towards sustainable business practices should be applied. Even though AI brings numerous benefits, it is worth noting that its operation poses certain risks and difficulties, including energy consumption and dependence on data. Therefore, the sustainability of AI technologies should be addressed when it comes to its design.

Furthermore, the application of artificial intelligence (AI) in relation to various aspects of sustainability would require adequate governance structures that would ensure an equitable trade-off between innovations on one hand and ethics and regulations on the other hand. Otherwise, there is a risk of uneven distribution of the benefits that arise from AI, which might further fuel the current problems.

Therefore, it can be concluded that the application of AI is quite effective when it comes to making decisions related to sustainability because of its ability to accommodate different situations. By making the environment, people, and economics more efficient and sustainable, AI helps create sustainable and resilient systems. Yet, in order for this full potential to be unlocked, proper governance is required.

AI TECHNIQUES FOR CLIMATE CHANGE PREDICTION

AI becomes extremely essential when predicting the climate since it allows for the analysis of various complicated and multi-dimensional sources of data which cannot be analysed by classical approaches based on physics and statistics. The climate phenomenon has some peculiarities including non-linearity and complicated interactions between various atmosphere, ocean, and land components. Classical modelling approaches do not allow for handling these issues, especially if they deal with massive data. On the other hand, AI technologies can provide for improvements in terms of both increased predictive power and decreased computational costs (Dueben & Bauer, 2018).

Machine learning algorithms (such as supervised, unsupervised, and reinforcement learning) constitute an important part of modern climate predictions. Supervised ML models can be used to forecast particular variables related to climate (e.g., temperature and precipitation) based on existing data. Unsupervised ML models are known for their ability to detect patterns and structures in the data sets thus allowing discovering features and trends which may go undetected by traditional means. Finally, reinforcement learning enables the decision making process under changing conditions and is especially useful when predicting in terms of natural resources management.

Deep learning (DL) algorithms have substantially advanced the modeling of climate change processes by effectively capturing both spatial and temporal dependencies within natural environments. Neural network approaches, particularly convolutional neural networks (CNNs), demonstrate high efficiency in managing geospatial data, including satellite imagery and remote sensing inputs. Moreover, recurrent neural network (RNN) architectures and long short-term memory (LSTM) networks are well suited for the analysis of time-series data, thereby facilitating the prediction of climate change processes and extreme events. These models enhance long-range forecasting by integrating intricate temporal dependencies and patterns that typify climate systems (Schmidhuber, 2015).

A central strength of artificial intelligence lies in its capacity to fuse diverse data sources within a unified analytical framework. Climate prediction necessitates the integration of multiple data streams, such as satellite observations, climate datasets, sensor measurements, and environmental monitoring records. AI models enable the combination of these data sources, supporting multi-dimensional climate change prediction endeavours (Reichstein et al., 2021). This integration improves the accuracy of projections related to climate change, temperature fluctuations, and extreme weather phenomena.

The increased access to high-quality, real-time information regarding the state of the environment has made the use of artificial intelligence (AI) techniques in climatic studies even more relevant. AI algorithms were developed specifically to work with continuous information streams, allowing for quick pattern detection and real-time analysis. It makes AI algorithms particularly useful for the creation of alert systems aimed at helping to combat natural catastrophes related to climate, such as floods, droughts, or heatwaves. Therefore, the introduction of AI techniques might greatly contribute to the improvement of disaster prevention and risk assessment (Shen, 2018).

At the same time, there are a number of challenges associated with the use of AI techniques in climate forecast that must be addressed. In particular, problems with data quality, uncertainty, and the need to produce explainable AI models become especially relevant at this point. The issue is that AI is often referred to as a black box, where predictions cannot be easily understood by decision-makers.

Finally, one should bear in mind that using AI for climate prediction implies making changes in the existing decision-making mechanisms. Even though AI models could generate quite accurate predictions, these predictions would not have an effect unless used in policy-making processes. Therefore, while developing AI-based predictive algorithms, researchers should take the current state of the governance process into account.

AI techniques provide great potentials to improve climate forecasts through advancements in data management, pattern identification, and predictive capacities. Nevertheless, the full realization of these potentials is only possible through the integration of these techniques with clear methodologies, governance structures, and inter-disciplinary processes that create a connection between technology and policy making.

Data Sources

Climate prediction using AI technology depends on various sources of data, such as satellite images, environmental monitoring sensors, and socio-economic data sets. Satellite data offers significant information about the state of the atmosphere and the earth's surface. Environmental monitoring sensors help with the immediate assessment of climate factors. Historical climate data is useful for identifying trends over time and training models, whereas socio-economic data enables an understanding of the effects of human activities on climate change (IPCC, 2022).

AI Models and Methods

There are many ways through which AI algorithms greatly aid the attainment of sustainability goals since these algorithms ensure optimal allocation of resources (Dignum, 2019). The common types of machine learning algorithms that are used to predict environmental and climate risks include ensemble algorithms and regression models.

Convolutional neural network (CNN) models and recurrent neural network models are efficient in modeling spatiotemporal climate changes and making predictions about them (Goodfellow et al., 2016). Such models enable better environmental modeling and help in the simulation of the climate. AI technologies have also been utilized extensively in remote sensing, enabling accurate land use mapping (Zhu et al., 2017).

Key Applications

AI technology spans a wide range of applications in the field of climatology that include weather predictions, natural disaster management, and environmental monitoring (Lary et al., 2016). AI technology is used to predict occurrences like floods and droughts that could happen, which will help in taking precautions in advance. Furthermore, there are many applications of AI technology when it comes to sustainable agriculture, managing natural water resources, and risks in climatic conditions because AI can help explain the intricate connections in the environment (Kamilaris & Prenafeta-Boldú, 2018).

RESEARCH METHODOLOGY

The methodology for creating this new IASF will be based on a conceptual research model that makes use of secondary data through climate databases, satellite imaging, Internet of Things (IoT) data, and institutional studies. The datasets will consist of several important variables in relation to climate data and include elements such as temperature, precipitation levels, air pollution, land use, and soil moisture.

Data processing includes cleaning, normalizing, and integrating data to create a coherent and complete dataset. This will be analysed using machine learning algorithms such as regression, decision tree, and random forest analyses, as well as deep learning approaches with CNN, LSTM networks, and their combinations. Reinforcement learning will be used to facilitate adaptive decision making.

In the IASF framework, there exist three levels of hierarchy that include the Data Layer (involving data integration and pre-processing), AI Processing Layer (involving modeling and prediction), and Decision Layer (involving policy and governance insights). An Adaptive Climate Prediction System in Real-Time mode is incorporated to enable constant updates in the models for improving predictions. In the evaluation process of the framework, an Artificial Intelligence-based Sustainability Impact Index including aspects like environment, economics, and society is used. Issues such as ethics in terms of transparency and accountability are included to make use of artificial intelligence sustainable.

AI-DRIVEN SUSTAINABILITY SOLUTIONS

Furthermore, artificial intelligence (AI) is one of the key elements that help make sustainability solutions effective. In the framework developed above, the effectiveness of AI solutions will be assessed on the basis of certain essential criteria. First, the emphasis will be placed on the predictive abilities, efficiency, and understand ability of solutions. In parallel, it is necessary to evaluate the adaptability and ethics involved in AI applications (Dignum, 2019).

From a systems perspective, AI supports sustainability efforts through dramatic improvements in environmental surveillance. The analysis of information obtained through sensor devices and satellite images makes it possible to trace changes in environmental indicators, such as air and water quality. As a result, it is possible to track environmental changes in real time and detect signs of environmental deterioration.

As a result, AI-powered monitoring contributes greatly to environmental sustainability as well (Lary et al., 2016). Furthermore, monitoring systems contribute to making more informed decisions based on the gathered data.

Finally, AI can be applied in order to enhance waste management and promote circular economy practices. With the help of AI solutions, better sorting of waste, optimization of waste processing, and recycling are made possible. As a result, there will be fewer resources wasted and less damage done to the environment. Thus, AI-powered waste management solutions will make it possible to promote sustainable development.

Smart Agriculture

In agriculture, artificial intelligence has helped introduce precision farming practices that ensure effective use of resources without causing harm to the environment. Through data analysis on factors such as soil quality, meteorology, health status of plants, and irrigation needs, the application of AI helps improve decision making when it comes to agriculture. The technology guarantees that there will be minimal use of water, fertilizers, and pesticides, making it sustainable without affecting efficiency (Kamilaris & Prenafeta-Boldú, 2018). In addition, artificial intelligence can be used to predict the output as well as detect any potential problems like pests and changes in climate. These factors improve the robustness of the practice and ensure food security in areas that are sensitive to climate change. It should be noted that, just like in other industries, the use of artificial intelligence in agriculture relies on availability of digital technology and data.

Energy Management

AI plays a transformative role in the energy systems through improving prediction of energy demand, optimization of energy distribution, and intelligent energy network control systems. The AI-powered systems provide high prediction accuracy in terms of energy demand predictions, which, in turn, contribute to resource efficiency and reduction of wasted energy. Such systems are especially important in relation to renewables like wind and solar energies, as they allow for regulation of energy flow and stabilization of energy supply (Siano, 2014).

In addition, AI contributes to the formation of smart grids and energy-efficient infrastructures, making a contribution to the implementation of the low-carbon development strategy. Through increasing energy efficiency in buildings, industrial production facilities, transportation systems, etc., the technology contributes to the mitigation of greenhouse gas emissions. At the same time, one should recognize the high amount of energy required by AI systems and technologies, hence the necessity to develop energy-efficient AI models.

Water Resource Management

Management of water resources is another important area where artificial intelligence plays a key role in promoting sustainability. This technology supports the monitoring and assessment of water resources by tracking their status, quality, and usage through advanced data analysis and remote sensing techniques (Zhu et al., 2017).

Moreover, use of AI technology will improve water irrigation management and the management of water infrastructure. Such efforts will promote resilience against climate change and water shortage. However, AI adoption requires that the technology should be integrated into the existing governance systems. Integrated Perspective and Governance Link

Whereas sector-specific implementations highlight AI's potential in addressing sustainability issues, such endeavours usually occur in a soloed manner compared to other sectors. One of the main drawbacks in the existing endeavours is the lack of integration across sectors and the governance systems, which restricts the potential for scaling up and implementing AI-powered sustainability solutions (Vinuesa et al., 2020). Hence, the AI solutions should be implemented within a bigger system that includes monitoring the environment, managing the natural resources, and policymaking.

The proposed IASF, in this case, attempts to tackle the identified drawback through presenting a uniform approach to incorporating technological innovations within governance frameworks. The inclusion of concepts like explainable AI, ethical accountability, and adaptive governance makes sure that the AI-powered solutions to sustainability problems are not only innovative technologically but also ethical and relevant to policies (Floridi et al., 2018).

CHALLENGES AND ETHICAL ISSUES

Although there are many possibilities and future applications, AI-powered climate forecasting and sustainability efforts face several challenges and limitations. The first of these challenges is the problem of inadequate data availability and selection biases during data collection, resulting in some parts of the globe being represented more accurately than others (Rolnick et al., 2019).

Secondly, there is a problem related to potential risks of algorithmic bias and lack of transparency. Given the complex nature of the climate and weather, it is vital that models are accurate and interpretable for proper policy-making purposes (Doshi-Velez & Kim, 2017).

In terms of ethics, AI may face several obstacles such as high electricity consumption and generation of additional carbon emissions. Besides, fairness, accountability, and inclusiveness should also be considered while deploying such systems (Floridi et al., 2018).

Finally, it is important to address all potential problems associated with using AI in order to create sustainable algorithms and climate models and ensure fair distribution of benefits.

CASE STUDIES AND PRACTICAL APPLICATIONS

The capacity of artificial intelligence in areas of climate change prediction, resource management, and sustainability is exemplified by its applications. Examples of large scale projects utilizing AI include Microsoft's AI for Earth aimed at providing climate-resilient decision making with the help of analyzing environmental data (Rolnick et al., 2019).

Among the most significant examples related to flood prediction, one should mention models incorporating satellite, hydrological data, and sensory inputs in order to improve predictions and create early-warning systems that will help with evacuation plans (Reichstein et al., 2019).

Regarding agriculture, AI technology promotes precision farming where data-driven systems optimize such processes as irrigation, crop pest control, and yield prediction (Kamilaris & Prenafeta-Boldú, 2018).

As far as water resource management is concerned, AI provides solutions for stream flow forecasting, smart irrigation, and infrastructure planning. Thus, there is a higher water efficiency in those areas.

However, despite the advantages, some issues arise such as those related to interpretability, transparency, fairness, and high energy consumption rates that need to be solved (Floridi et al., 2018). Additionally, one should note that all the mentioned cases show that artificial intelligence can produce efficient and scalable solutions for climate adaptation and mitigation if supported by appropriate data and policies.

PROPOSED MODEL: INTEGRATED AI–SUSTAINABILITY FRAMEWORK (IASF)

IASF is defined as a model which represents the combination of the concepts of artificial intelligence with sustainability through the integration of predictive analytics into actions taken on policy issues (Rolnick et al., 2019). Unlike many existing methods that focus either on predictive power or on the application areas for AI, the IASF creates a systematic means for turning predictions into actions in governance.

The three interconnected levels of the IASF include Data Layer, AI Processing Layer, and Decision Layer, which create a consistent cycle for turning data into effective environmental governance (Vinuesa et al., 2020). Rather than representing independent elements, the layers should be viewed as a system that complements and strengthens each other's efficiency.

Figure 1. Integrated AI–Sustainability Framework (IASF) for climate governance

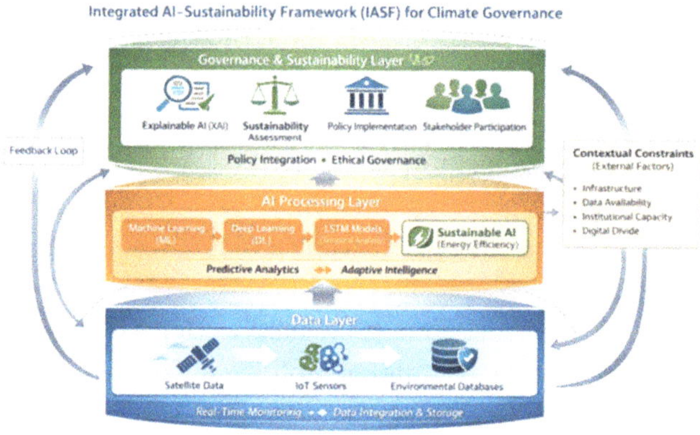

Source: Author's Own elaboration

Figure 1 illustrates that IASF represents a recursive and dynamic model in which environmental data is constantly gathered and analysed by means of AI technologies in order to derive relevant policies. Feedback is integrated into the model, allowing for continual learning and adaptation.

1. Data Layer: Foundation of Environmental Intelligence

The Data Layer represents the empirical backbone of the proposed framework and encompasses several data sources such as satellite images, IoT sensors, geo-spatial data, and cnvironmental databases (Reichstein et al., 2019). This allows gaining insight into the complexity of environmental systems and accounting for their variability at different scales and time periods.

The Data Layer relies on concepts that make use of scalable cloud infrastructures in order to ensure real-time environmental monitoring and allow for the effective management of massive amounts of data (Hashem et al., 2015). The regular update of data is expected to increase the accuracy of ecological processes and environmental changes modeling (Shen (2018)).

In addition to data acquisition, this layer focuses on the quality, consistency, and reliability of data, which are essential preconditions for successful analysis using AI technologies. Availability of various types of data increases the reliability of predictive models and results in higher sensitivity and specificity of insights obtained through analysis.

Nonetheless, the performance of this layer is conditioned by data availability and the existing infrastructure capacity, which can reduce its efficiency in developing countries. The absence of sufficient data and lack of adequate sensor networks can lead to certain difficulties in applying AI technology to environmental analysis.

2. AI Processing Layer: Predictive Analytics and Adaptive Intelligence

The AI Processing Layer represents an advanced analytics engine that transforms environmental data into valuable information via machine learning and deep learning techniques (LeCun et al., 2015). It enables the identification of complex patterns, non-linearity, and other important factors of climate change.

Distributed and edge computing systems have been adopted within the framework to process data in real-time and scale up the environmental modeling processes (Shi et al., 2016). Thus, the proposed solution allows processing large amounts of information in a time-efficient manner, as well as facilitating rapid decision making in the context of changing environment.

Temporal relationships in the climate systems can be identified using recurrent neural networks and, specifically, the long short-term memory algorithm that enhances the quality of forecasting (Hochreiter & Schmidhuber, 1997). Continuous learning can also be used to facilitate the updates in the AI model and, therefore, decrease the errors during predictions (Goodfellow et al., 2016).

Moreover, reinforcement learning algorithms will help enhance the adaptability of the AI models, allowing them to learn from the outcomes and decisions made within dynamic systems (Sutton & Barto, 2018). The mentioned capability seems relevant for the management of dynamic systems, including the energy grid, water resources, and climate risk mitigation measures.

At the same time, certain challenges will be associated with the use of the AI processing layer, namely data bias, model interpretability, and the computational costs (Goodfellow et al., 2016).

3. Decision Layer: Policy Integration and Sustainable Governance

The third layer, known as the Decision Layer, represents the governance dimension of the proposed framework. The main idea behind this layer lies in the transformation of insights produced by an artificial intelligence system into effective policy solutions.

In order to enhance transparency and trust, explainable AI or XAI technologies will be employed in the framework. Through the use of interpretability methods like SHAP and LIME, decision systems will be able to produce clear insight about how they behave and what outcomes can be expected from their actions, thus paving the way towards evidence-based governance and mitigation of the dangers associated with opaque decision systems (Molnar, 2020).

The implementation of dynamic feedback loops will be critical to the operation of this layer since it will foster the emergence of adaptive systems capable of adjusting and updating their policies based on real-time data from the environment (Janssen et al., 2017). As such, the limitations of static approaches in developing policies for an ever-changing environment can be addressed (Janssen & Kuk, 2016).

Another characteristic feature of the framework under discussion is its ability to support the evaluation of policies in terms of sustainability using data-based evaluation of environmental, economic, and social impact (Vinuesa et al., 2020). Moreover, the cost of implementing AI systems will include the environmental price of using them, thus encouraging more sustainable development.

Finally, ethical dimensions of AI deployment will be represented by embedding principles of accountability, fairness, and transparency in the framework and decision-making processes (Floridi et al., 2018).

It should also be noted that the implementation of AI-driven sustainability frameworks will be affected by such contextual factors as the digital divide, lack of necessary infrastructure, and limited institutional capacity in developing regions.

CONCEPTUAL SCOPE AND FUTURE VALIDATION

This theoretical framework outlined above is merely conceptual and acts as a basis for a comprehensive and integrated model comprising technology, environment, and governance elements. While it provides an effective framework and model, its usefulness lies on proving its empirical validity by applying it through real-life scenarios, simulations, and comparisons. Research work going forward should focus on validating the framework through a variety of environments and institutions.

RESEARCH CONTRIBUTION

The current research extends the existing body of knowledge through its innovative approach, which presents an integrated governance framework that ties artificial intelligence into sustainability decisions (Vinuesa et al., 2020). Even though there is a lot of evidence on the application of artificial intelligence technologies to predict sustainability issues and climate-related changes, the linkages have been very fragmented and do not present an overall framework connecting both areas.

While previous studies focused mainly on accurate predictions provided by artificial intelligence technologies, the presented approach offers a new angle as a direct connection of AI results into policy making. In doing so, the paper tackles one of the crucial flaws of the current literature, which is related to the inability of highly accurate prediction frameworks to turn into policies (Rolnick et al., 2019).

Another important contribution to knowledge in this area comes from the introduction of the three-layer structure for artificial intelligence-driven decision-making and policies (i.e., Data Layer, Artificial Intelligence Layer, and Decision Layer). Such a layered structure will facilitate real-time and adaptive decision-making in relation to the changing conditions. This way, the governance systems will be able to operate more effectively under constantly changing environmental circumstances (Reichstein et al., 2019).

Firstly, the use of XAI and sustainability assessment tools in the proposed model is a significant advancement of the current literature on the issue. Both approaches promote the notions of transparency, accountability, and interpretability. They are important characteristics for building trust in AI-based governance models.

Furthermore, the framework can be described as being multidisciplinary as it involves the aspects of artificial intelligence, environmental sustainability, and governance. Hence, a more complete understanding and evaluation of outcomes can be achieved because the model allows examining different impacts related to social, economic, and environmental factors.

Another advancement of the proposed framework is that it conceptualizes AI as a governance tool rather than a predictive approach. In other words, AI can facilitate governance through policy development and decision-making processes (Floridi et al., 2018). The approach brings new ideas to both areas of study and provides additional theoretical and practical benefits.

Finally, one more advantage is that the framework is conceptual and calls for future empirical verification. The proposed model can serve as a basis for further development of the topic under discussion as it may include various studies and experiments based on the introduced framework.

DISCUSSION

The findings of this study demonstrate an ever-growing importance of AI as a revolutionary tool in climate change forecasting and sustainable governance practices. Although previous literature on this topic primarily focuses on technological innovations and improvements that make AI better at predicting climate change patterns, this paper shows that this potential is fully realized only when AI is implemented in a broader governance structure. Thus, the proposed model is an innovative step toward more efficient climate science practices that could address some of the existing problems.

One of the key lessons offered by the study is that the use of AI should be considered from the perspective of governance rather than that of technology itself. Current AI practices in climate studies focus mainly on improving forecast accuracy, leaving no room for a more complex use of available data. However, the implementation of the proposed IASF ensures that AI tools will not only improve forecasts but become instrumental in climate governance.

As this research indicates, in order to implement AI into the field of climate science successfully, it is essential to ensure balance between accuracy and interpretability. While advanced AI models, especially those based on deep learning algorithms, offer excellent predictive performance, they still remain difficult to explain. As such, decision-making process involving the use of AI might be hindered. By including XAI in the discussed framework, the researcher provides an additional means of making the application of AI more successful, transparent, and efficient.

It should also be noted that this research demonstrates that when implementing AI into a given area, it is vital to consider sustainability implications. Despite the broad recognition that AI may act as a highly powerful tool to achieve sustainability goals, the vast amount of energy required to operate such tools needs to be further considered. Consequently, the inclusion of this parameter to the examined framework may contribute to the field.

Furthermore, another critical finding of the analysis is that the environmental and social context significantly impacts the effectiveness of using AI in climate science. The application of AI technology to climate governance is not only a technical challenge but also a socio-political and institutional one. For example, the digital divide, infrastructure issues, and differences in institutional preparedness could be major barriers to implementing the technology, especially in underdeveloped nations.

Lastly, the study highlights the significance of interdisciplinary for designing novel solutions to critical sustainability issues. It is evident due to the fundamental concept underlying the suggested model – the combination of AI, environmental science, and public administration offers an adequate solution to climate governance.

At the same time, while providing several contributions to the literature, the research has some limitations. First, the lack of implementation or simulation studies to test the efficiency of the presented framework might imply doubts about its credibility. Therefore, further analysis would be beneficial to evaluate the practicality of the framework discussed.

Nevertheless, the paper manages to provide a sound contribution to the area under consideration since it explicitly outlines the purpose of this model and suggests avenues for future research.

POLICY IMPLICATIONS AND GOVERNANCE PERSPECTIVE

In order to implement AI technologies in sustainable planning, the creation of a good governance framework will be necessary to assure the ethicality of use, transparency, and accessibility. As the AI's role is becoming larger and more important in relation to the environment, it becomes crucial to have such mechanisms for making the technology effective, sustainable, and responsible for its actions.

The use of an effective governance mechanism involves employing techniques of Explainable Artificial Intelligence (XAI). Indeed, being highly complex and advanced, AI technologies should allow for interpretability of the decisions made by a computer. In this way, those using such methods will be able to understand what is going on and base the decisions on that. Thus, the implementation of XAI increases the accountability and makes the AI process more understandable (Guidotti et al., 2018). Without XAI, it would be impossible to rely on black-box algorithms.

As regards regulation, the governments will need to develop comprehensive sets of rules that concern development, deployment, and supervision of AI. It is necessary to develop specific approaches to data governance, algorithm transparency, risks assessment, and accountability mechanisms. Regulatory oversight will be crucial in making sure that such AI does no harm to the community and does not perpetuate social injustices (Janssen & Kuk, 2016). Moreover, governments will be needed to facilitate the process of capacity-building and prepare organizations for utilizing AI tools.

Finally, international collaboration is yet another component of AI governance. Indeed, climate change and the problems associated with it are international problems, thus the use of international approaches and mechanisms will make sense. It will be possible to achieve greater success in developing AI and applying it to the problem if collaborative work is used in such spheres as data sharing and model validation (Vinuesa et al., 2020).

Major global organizations like the OECD and IEEE help shape ethical AI governance through the creation of standards, policy guidelines and collaborative work.

These organizations create frameworks for the responsible use of AI and establish best practices across all nations (Floridi et al., 2018). They are also instrumental in developing a set of shared principles for transparency, fairness and accountability. A human-centered approach to AI governance is also needed to enhance the social legitimacy and acceptability of AI interventions. This approach involves including diverse stakeholders in the decision-making process and is characterized by cross-disciplinary collaboration among technological, social and policy disciplines. The incorporation of local knowledge, social values and context will allow for the creation of governance frameworks that are relevant and effective in using AI (Janssen and Kuk, 2016).

Governance frameworks must equip themselves to meet major challenges associated with AI, such as data/class based inequalities, digital divides, and institutional limits. If AI systems are not created with multiple levels of safeguards, they could potentially enhance the gaps that already exist between those populations of developing countries that do not have access to digital infrastructure. For this reason, policy routes should include equitable data access, enabling equitable delivery of services regardless of location.

Overall, governance mechanisms are critical to ensure climate-based weather forecast AI applications will be not just high-tech but will also be ethical, transparent, and responsive in nature. The policymakers can maximize the benefits of AI while lowering the risks through the combination of regulatory frameworks, ethical regulations, and collaborative partnerships as ways to govern AI in a holistic manner. Ultimately, climate-based AI outcomes will succeed only if they are integrated into governance structures that build accountability, inclusivity, and sustainability into the implementation of applicable solutions.

LIMITATIONS OF THE STUDY

Conceptual orientation has been adopted in this research; therefore, there is no empirical validation of the IASF. The framework is developed theoretically and is integrative in that it provides a linkage between artificial intelligence and sustainability. This approach allows the formation of a well-structured foundation that will aid scholars and other researchers in developing and implementing the concept. However, this limitation is that the approach is restrictive since it cannot allow one to validate the framework effectively in practice.

Firstly, the fact that this study is based on the secondary data obtained from literature review means that the results obtained may be unable to reflect how context-specific applications of the framework could perform in specific environments. For instance, the applicability and efficiency of an AI tool will largely depend on data availability, technology, socio-political environment, and others. Therefore, the current study fails to empirically analyze those factors that can have a direct influence on the performance of the AI solution.

Secondly, the fast-evolving nature of AI technology means that any framework proposed in this area may eventually become obsolete because of continuous improvements being made by specialists. Machine learning algorithms continue to make advancements throughout the data ecosystem; therefore, there may come a time when it will be necessary to make enhancements updates changes to the framework to ensure it matches with future evolving technologies and sustainability objectives.

In addition, a lack of comparative analysis with many other existing models and or frameworks is another drawback that has to be rectified for this framework to improve its analytical capabilities, as it has not yet been compared against other solutions of the same type.

In order to get a better understanding of how well this new framework performs, and what type of governance effect it has, it is important that researchers perform an empirical investigation of the new framework by applying the new framework to real world case studies and comparing its performance via the use of simulation models to other frameworks. In doing so, researchers will gather additional knowledge that will assist in determining the viability and overall applicability of the new framework.

FUTURE RESEARCH DIRECTIONS

Future research efforts would benefit from developing and implementing thorough and standardised assessment metrics that identify attributes such as transparency and accountability of AI algorithms, among other things. Future studies should also analyse the sociocultural and ethical consequences of using AI systems for environmental decision making. By concentrating on both the technical aspects and the sociocultural ramifications of AI, researchers will be able to provide a more complete evaluation of the efficacy and dependability of AI tools for environmental administration purposes.

As stated above, one should pay more attention to the use of XAI in terms of promoting trust in AI and influencing policymakers' decisions and public perception. With AI tools being increasingly involved in climate-related decision-making, it is crucial to understand how various aspects of AI affect stakeholders' trust and willingness to adopt certain policies.

An important aspect of future research relates to improving the scalability, transferability, and flexibility of AI models. It is well-known that most of current AI models were developed in the context they operate. Thus, when it comes to applying them in other areas or regions, the accuracy of their predictions may suffer. To this end, further research should aim at devising AI algorithms that are sufficiently adaptable and scalable, thus providing better opportunities for application in diverse settings.

The fourth important trend pertains to the use of real-time data streams and decision-making in changing environments. Climate systems are highly uncertain and dynamic, requiring highly adaptable models. Therefore, the next step in developing AI tools for environmental decision-making should be aimed at devising models that could operate based on real-time data and adjust their predictions in response to changes.

Digital divide is an important issue that needs to be addressed. When adopting various AI systems, one has to pay special attention to the issues of unequal accessibility and distribution of technological advances. For example, the issue of infrastructure needs to be considered. In particular, developing regions may lack sufficient internet and power connectivity, hindering the adoption of new AI technologies. At this point, further research is required.

Uncertainty communication is an important part of effective decision-making process. When dealing with environmental data, one cannot overlook high degree of unpredictability and uncertainty involved. This is why future work should be concerned with developing techniques that would allow for effective communication of uncertainties inherent to climate and environmental data.

Finally, interdisciplinary collaboration can play a key role in making AI tools useful for environmental decision-making. In particular, insights from artificial intelligence, environmental science, public policy, and social sciences could be valuable in creating appropriate and efficient tools for sustainable climate and environmental governance.

CONCLUSION

This chapter discusses the way AI is utilized to forecast and promote sustainability in combatting climate change. From the text we know that advanced analytical tools help make accurate predictions; however, strong governance and management processes are necessary for implementing and executing AI effectively. There is an equilibrium of needing accurate prediction while also being transparent in order to create effective decision-making from those predictions. In this regard, explainable AI (XAI) allows for the building of trust and thus practicality.

This chapter also points out how environmentally sustainable AI development is important, especially from an energy and carbon footprint perspective. We also see that the governance of AI needs to be human-centered and incorporate technological progress as well as local knowledge and stakeholder involvement. This will ensure that AI will result in adaptive and inclusive solutions.

Additionally, successful governance of AI through the collaborative nature of institutions and across borders allows for continued innovation. Overall, this chapter suggests that AI will be a powerful tool to promote sustainable practices if it is built on a governance framework that is effective and AI is used responsibly.

REFERENCES

Bibri, S. E., & Krogstie, J. (2017). Smart sustainable cities of the future: An extensive interdisciplinary literature review. *Sustainable Cities and Society*, *31*, 183–212. DOI: 10.1016/j.scs.2017.02.016

Bughin, J., Seong, J., Manyika, J., Chui, M., & Joshi, R. (2018). *Notes from the AI frontier: Modeling the impact of AI on the world economy*. McKinsey Global Institute. https://www.mckinsey.com/featured-insights/artificial-intelligence/notes-from-the-ai-frontier-modeling-the-impact-of-ai-on-the-world-economy

Cath, C., Wachter, S., Mittelstadt, B., Taddeo, M., & Floridi, L. (2018). Artificial intelligence and the 'good society': The US, EU, and UK approach. *Philosophy & Technology*, *31*(4), 505–528. PMID: 28353045

Dignum, V. (2019). *Responsible artificial intelligence: How to develop and use AI in a responsible way*. Springer., DOI: 10.1007/978-3-030-30371-6

Dueben, P. D., & Bauer, P. (2018). Challenges and design choices for global weather and climate models based on machine learning. *Geoscientific Model Development*, *11*(10), 3999–4009. DOI: 10.5194/gmd-11-3999-2018

Eccles, R. G., & Klimenko, S. (2019). The investor revolution. *Harvard Business Review*, *97*(3), 106–116.

Edwards, P. N. (2010). *A vast machine: Computer models, climate data, and the politics of global warming*. The MIT Press. Doshi-Velez, F., & Kim, B. (2017). Towards a rigorous science of interpretable machine learning. *arXiv preprint arXiv: 1702.08608*. https://doi.org//arXiv.1702.08608DOI: 10.48550

Floridi, L., Cowls, J., Beltrametti, M., Chatila, R., Chazerand, P., Dignum, V., & Vayena, E. (2018). AI4People—an ethical framework for a good AI society: Opportunities, risks, principles, and recommendations. *Minds and Machines*, *28*(4), 689–707. DOI: 10.1007/s11023-018-9482-5 PMID: 30930541

Geels, F. W. (2004). From sectoral systems of innovation to socio-technical systems: Insights about dynamics and change from sociology and institutional theory. *Research Policy*, *33*(6–7), 897–920. DOI: 10.1016/j.respol.2004.01.015

Goodfellow, I., Bengio, Y., & Courville, A. (2016). *Deep learning*. The MIT Press.

Guidotti, R., Monreale, A., Ruggieri, S., Turini, F., Giannotti, F., & Pedreschi, D. (2018). A survey of methods for explaining black box models. *ACM Computing Surveys, 51*(5). *ACM Computing Surveys*, *93*(5), 1–42. DOI: 10.1145/3236009

Ham, Y. G., Kim, J. H., & Luo, J. J. (2019). Deep learning for ENSO forecasts. *Nature*, *573*(7775), 568–572. DOI: 10.1038/s41586-019-1559-7 PMID: 31534218

Hashem, I. A. T., Yaqoob, I., Anuar, N. B., Mokhtar, S., Gani, A., & Samee, U. (2015). The rise of "big data" on cloud computing: Review and open research issues. *Information Systems*, *47*, 98–115. DOI: 10.1016/j.is.2014.07.006

Henderson, P., Hu, J., Romoff, J., Brunskill, E., Jurafsky, D., & Pineau, J. (2020). Towards the systematic reporting of energy and carbon footprints of machine learning. *Journal of Machine Learning Research*, *21*(248), 1–43.

Hochreiter, S., & Schmidhuber, J. (1997). Long short-term memory. *Neural Computation*, *9*(8), 1735–1780. DOI: 10.1162/neco.1997.9.8.1735 PMID: 9377276

Huntingford, C., Jeffers, E. S., Bonsall, M. B., Christensen, H. M., Lees, T., & Yang, H. (2019). Machine learning and artificial intelligence to aid climate change research and preparedness. *Environmental Research Letters*, *14*(12), 124007. DOI: 10.1088/1748-9326/ab4e55

Intergovernmental Panel on Climate Change. (2021). *Climate change 2021: The physical science basis. Contribution of Working Group I to the Sixth Assessment Report of the Intergovernmental Panel on Climate Change*. Cambridge University Press., DOI: 10.1017/9781009157896

International Energy Agency. (2021). *Artificial intelligence and energy*. https://www.iea.org/reports/artificial-intelligence-and-energy

Janssen, M., & Kuk, G. (2016). Big and open linked data (BOLD) in government: A challenge to transparency and privacy? *Government Information Quarterly*, *33*(2), 363–368.

Janssen, M., Van Der Voort, H., & Wahyudi, A. (2017). Factors influencing big data decision-making quality. *Journal of Business Research*, *70*, 338–345. DOI: 10.1016/j.jbusres.2016.08.007

Kamilaris, A., & Francesc, X. Prenafeta-Boldú. (2018). Deep learning in agriculture: A survey. *Agricultural Systems*, *162*, 70–90. DOI: 10.1016/j.agsy.2018.02.016

Knutti, R. (2008). Should we believe model predictions of future climate change? *Philosophical Transactions of the Royal Society A: Mathematical, Physical and Engineering Sciences*, *366*(1885), 4647–4664. https://doi.org/DOI: 10.1098/rsta.2008.0169

Lary, D. J., Alavi, A. H., Gandomi, A. H., & Walker, A. L. (2016). Machine learning in geosciences and remote sensing. *Geoscience Frontiers*, *7*(1), 3–10. DOI: 10.1016/j.gsf.2015.07.003

LeCun, Y., Bengio, Y., & Hinton, G. (2015). Deep learning. *Nature*, *521*(7553), 436–444. DOI: 10.1038/nature14539 PMID: 26017442

Molnar, C. (2020). *Interpretable machine learning*. Lulu. Com.

Reichstein, M., Camps-Valls, G., Stevens, B., Jung, M., Denzler, J., Carvalhais, N., & Prabhat, . (2019). Deep learning and process understanding for data-driven Earth system science. *Nature*, *566*(7743), 195–204. DOI: 10.1038/s41586-019-0912-1 PMID: 30760912

Rolnick, D., Donti, P. L., Kaack, L. H., Kochanski, K., Lacoste, A., Sankaran, K., & Bengio, Y. (2019). Tackling climate change with machine learning. *ACM Computing Surveys*, *55*(2), 1–96. DOI: 10.1145/3485128

Russell, S., & Norvig, P. (2021). *Artificial intelligence: A modern approach* (4th ed.). Pearson.

Schmidhuber, J. (2015). Deep learning in neural networks: An overview. *Neural Networks : The Official Journal of the International Neural Network Society*, *61*, 85–117. DOI: 10.1016/j.neunet.2014.09.003 PMID: 25462637

Shen, C. (2018). A transdisciplinary review of deep learning research and its relevance for water resources scientists. *Water Resources Research*, *54*(11), 8558–8593. DOI: 10.1029/2018WR022643

Siano, P. (2014). Demand response and smart grids—A survey. *Renewable & Sustainable Energy Reviews*, *30*, 461–478. DOI: 10.1016/j.rser.2013.10.022

Sutton, R. S., & Barto, A. G. (2018). *Reinforcement learning: An introduction* (2nd ed.). MIT Press., https://mitpress.mit.edu/9780262039246/reinforcement-learning/

Vinuesa, R., Azizpour, H., Leite, I., Balaam, M., Dignum, V., Domisch, S., & Fuso Nerini, F. (2020). The role of artificial intelligence in achieving the Sustainable Development Goals. *Nature Communications*, *11*(1), 233. DOI: 10.1038/s41467-019-14108-y PMID: 31932590

Wirtz, B. W., Weyerer, J. C., & Geyer, C. (2019). Artificial intelligence and the public sector—Applications and challenges. *Public Management Review*, *21*(4), 596–615.

Zhu, X. X., Tuia, D., Mou, L., Xia, G. S., Zhang, L., Xu, F., & Fraundorfer, F. (2017). Deep learning in remote sensing: A comprehensive review and list of resources. *IEEE Geoscience and Remote Sensing Magazine*, *5*(4), 8–36. DOI: 10.1109/MGRS.2017.2762307

Chapter 9

Wind-Tree-Based Green Innovation Systems for Decarbonization Through Artificial Intelligence in Tanzanian Universities

Thompson Xavier Ananth
St. Joseph University in Tanzania, Dar es Salaam, Tanzania

Mohit Verma
http://orcid.org/0009-0005-3036-3894
O.P. Jindal Global University, India

Shashi Kant
http://orcid.org/0000-0003-4722-5736
St. Joseph University in Tanzania, Dar es Salaam, Tanzania

Duncan Mwakipesile
Independent Researcher, Tanzania

ABSTRACT

This paper explores how green innovation systems can be improved with the help of artificial intelligence (AI) to push towards decarbonisation, particularly when it comes to wind-tree-based systems of renewable energy. Based on the Technology-Organisation-Environment (TOE) paradigm, the study elaborates and estimates a structural equation model (SEM) in order to investigate the connections between AI, GIS, and decarbonisation results in Tanzanian universities. The results specify that

DOI: 10.4018/979-8-3373-9978-2.ch009

AI has a strong aptitude to enhance the performance of green innovation systems and is a mediator in the enhancement of the performance of decarbonisation. AI enhances efficient working and sustainable institutional practices by maximising energy consumption and promoting renewable energy sources. This paper offers empirical results in the case of a developing-country setting and illuminates the manner in which AI-based systems can facilitate climate action at the HEI level. Those findings are applicable as practical and policy-relevant information for sustainability in higher education.

INTRODUCTION

The growing exigence of decarbonizing the world has heightened the demand of viable and scalable solutions that fuse innovative technologies and sustainable energy systems. Here, the integration of the artificial intelligence (AI) and green innovation systems (GIS) can be viewed as a promising avenue to improve the environmental sustainability in the institutional context (Olawade et al., 2024). Such integrated solutions can be effectively applied in Tanzanian universities, which are environments of control and innovation, especially by adopting renewable energy technologies in line with the global sustainability frameworks, including the United Nations 2030 Agenda and the Paris Agreement (Olawade et al., 2024).

Wind-tree technology is one of the new renewable technologies that can be appropriately used in university campuses because it can be used in areas with low wind speed and low space. The wind-tree technology is small in size, modular and can be adapted to meet urban and semi-urban settings, unlike traditional wind turbines, which makes it especially applicable to the Tanzanian universities (Nugroho et al., 2025). By incorporating AI, these systems can be improved by real-time monitoring, predictive maintenance, and intelligent energy optimization, which can enhance efficiency and decentralized energy generation (Twabu, 2025). Meanwhile, AI facilitates greater institutional advantages by enhancing the efficiency of operations, optimizing resource allocation, and facilitating sustainability-driven decision-making (Khan et al., 2025).

In addition to technological uses, AI is also essential in enhancing sustainability capabilities in the higher education sector, as it allows the development of analytical skills, encourages innovation, and allows faster achievement of the Sustainable Development Goals through digital transformation and green transformation (Chun and Hwang, 2024; Saheb et al., 2022; Savec and Jedrinovic, 2024). Nevertheless, even in the light of these acknowledged advantages, the current body of literature mostly considers AI and sustainability as two distinct entities and provides little empirical support concerning how AI-enabled GIS can lead to direct decarbonization actions, especially in the context of developing countries.

This research fills this gap as it focuses on how AI can be used to enhance green innovation systems to decarbonize Tanzanian universities. In particular, it examines the way AI is used to improve renewable energy, like wind-tree systems, and assesses the connection between AI and GIS and the decarbonization results. The research will be valuable to the continued refinement of the idea of how AI-driven systems can be used to facilitate practical climate action in higher education institutions by going beyond the general discourse and introducing an empirical model. Moreover, the study focuses on a university context, providing context-specific findings on the implementation and scaling of AI-supported sustainability solutions in developing countries.

BACKGROUND

The use of artificial intelligence (AI) in higher education has been broadly acknowledged as able to enhance the efficiency, resource management, and sustainability outcomes (Opesemowo and Adekomaya, 2024; Filho et al., 2025). To a large extent, however, the extant literature is about general educational transformation, with little concern given to the contribution of AI to green innovation systems (GIS) and quantifiable decarbonization results. In colleges, AI-powered systems can optimize energy use, minimize waste, and assist in data-driven decision-making, thus establishing a platform to incorporate sustainability practices into the college operations (Filho et al., 2025; Twabu, 2025).

In this regard, GIS are institutionalized strategies that enable the implementation of environmentally friendly technologies and practices, including renewable energy systems such as wind-tree systems. To make them more efficient, AI improves these systems with predictive analytics, intelligent resource allocation, and real-time monitoring, which allow taking more efficient actions and running green innovations (Twabu, 2025). Simultaneously, successful implementation of AI requires institutional preparedness, stakeholder involvement, and lifelong learning, with the role of aligning technological capabilities to organizational and environmental realities being crucial (Opesemowo and Adekomaya, 2024).

The literature on Tanzanian higher education focuses on the transformative power of AI but also notes the issues of adoption, ethical implications, and institutional capabilities (Mambile and Mwogosi, 2024). The fact that several theoretical approach models, such as technology acceptance and technology-organization-environment models, are integrated, further indicates that the adoption of AI depends on individual, institutional, and external factors (Nugroho et al., 2025). These are valuable insights, but they are more conceptual and do not directly look at how AI capabilities can be converted into effective GIS practice and how the practice can be achieved in terms of decarbonization results.

Thus, there is an evident empirical gap in the knowledge of the mechanism relating to AI, GIS, and decarbonization, especially in the context of the developing countries, including Tanzania (Alkhawaja et al., 2025; Chan, 2023; Sebihi et al., 2025). To fill this gap, it is necessary to go beyond general discourses on AI in education to a systematic examination of how AI can help in the implementation of green innovation and improve environmental performance. This paper addresses this gap by discussing the connections between AI, GIS, and decarbonization in a university environment with an emphasis on the sustainability use cases of the context.

PROBLEMS IN IMPLEMENTATION

Sustained enforcement of the AI-based sustainability programs in the field of higher education is associated with considerable challenges, including the absence of a developed methodology to evaluate them thoroughly due to their dynamic and complex character. The existing assessment systems are usually insufficient to cover the environmental, economic, and social aspects that can be instrumental in developing a holistic view of the potential effects of the use of AI on sustainability. Moreover, even though AI has many opportunities to grow, digital infrastructure, connection problems, and lack of a regulatory framework and ethical concerns mostly hinder its full implementation (Sebihi et al., 2025). Besides, the difficulty in funding AI infrastructure and education alongside difficulty in applications through bureaucratic procurement procedures are also common barriers to AI adoption in institutions of higher education. All these are intensified by the fact that there are no holistic guide and best practice to college and university institutions to harness AI towards sustainable development in a manner that meets the SDGs (Opesemowo & Adekomaya, 2024).

On the one hand, there is a specific gap in the literature review on assessing the effectiveness of AI-driven sustainability initiatives in the educational sector, which implies that the existing methods do not have effective approaches to the comprehensive and multifaceted nature of AI technologies and their effects on sustainability. This highlights the need of coming up with new evaluation strategies that holistically address the diverse effects of AI in attaining sustainable development objectives in the realm of higher education. Such novel approaches must incorporate a solid accountability mechanism of evaluating the efficiency of AI-led sustainability initiatives and go past the promise of theory with the translation of theory into measurable business operations (Silva and Hamalainen, 2024). Such a gap serves to emphasize the necessity of new types of evaluation that would involve all three aspects, environmental, economic, and social, to consider AI role in promoting sustainability in the framework of the educational setting comprehensively.

The solution to these gaps would be to continue researching the long-term consequences of AI-enhances learning in terms of sustainability competencies, effectiveness of various AI-driven pedagogic strategies, and creating new methods of assessment that use AI powers. This study will attempt to fill this gap by proposing a holistic assessment system that could be used to evaluate AI application to Education to provide sustainable development in the higher education sector, thus helping in the further explanation of how AI can be successfully applied to support sustainability research and practices. This suggested approach would incorporate ethical frameworks of AI applications that safeguard privacy and reduce biases and cover a crucial under researched intersection in higher education.

PROGRAMMES AND INITIATIVES BY THE GOVERNMENT

Government initiatives are an important enabling factor to help achieve the successful deployment of AI-based green innovation systems (GIS) in higher education, especially in the creation of regulatory, financial, and institutional structures. The integration of AI in education should be backed by clear national policies to direct the adoption, provide ethical usage, and align institutional practices with sustainability goals (Opesemowo and Adekomaya, 2024). These frameworks are supposed to include the financial feasibility mechanism that assists in the development of infrastructures, capacity building, and research in AI-enabled decarbonization (Khan et al., 2025).

Besides this, there is a need to establish governance mechanisms that will resolve critical issues concerning data privacy, algorithmic bias, and responsible implementation of AI. Transparency and accountability may be reinforced with the use of effective regulatory policies that would allow institutions to feel confident about embracing AI technologies (Baskara, 2024; Khan et al., 2025). Such measures are especially significant to allow incorporating AI into GIS, which needs extensive data processing and optimization of the system.

Moreover, universities, industry, and government institutions should collaborate to enhance the adoption of sustainability initiatives supported by AI. These alliances aid in knowledge sharing, exchange of resources and creation of context specific solutions to decarbonization (Baskara, 2024). It is also possible to extend such collaborations to international networks to help align national strategies with global sustainability objectives (Mudenda, 2025).

On the whole, these state-driven programs offer the institutional framework that the successful implementation of AI into green innovation systems is possible, which helps to achieve better decarbonization results in higher education (Allam et al., 2025; Baskara, 2024; Nagaraj et al., 2023).

THEORETICAL BASE

The current study takes a multi-theoretical approach, although it narrows down its scope to the direct explanations of the connections between artificial intelligence (AI), green innovation systems (GIS), and decarbonization in universities.

The Diffusion of Innovation (DOI) Theory is used to describe the use of AI-based sustainability practices in institutions of higher learning. The knowledge, persuasion, decision, implementation, and confirmation stages assist in comprehending the implementation of AI in the green innovation systems in universities. DOI is especially applicable to the explanation of the adoption of AI-enabled GIS practices, such as wind-tree technologies, by various adopter groups in institutions (Abo-Khalil, 2024).

The Technology-Organization-Environment (TOE) Framework forms the main basis of this research as it helps to explain the impact of technological, organizational, and environmental factors on AI implementation. The technological, organizational, and environmental contexts are the abilities of AI tools, institutional preparedness and support by leadership, and regulatory and social pressures respectively. TOE specifically helps to establish the connection between the capabilities of AI and the implementation of GIS and their joint effect on the results of decarbonization (Frikha and Mrad, 2025).

Resource-Based View (RBV) adds to this perspective and focuses on the importance of internal resources like AI infrastructure, human expertise, and interdisciplinary collaboration. The resources can help universities successfully implement AI in GIS and attain better environmental performance. RBV thus demonstrates how AI can serve as a strategic resource that can improve the efficacy of green innovation systems and help decarbonize them (Nadeem et al., 2025).

The Ecological Modernization Theory offers an ecological perspective of sustainability theory, which postulates that technological innovation and environmental protection can be mutually supportive. Here, it describes that AI-empowered GIS, such as wind-tree systems, can enhance institutional performance and limit environmental impact at the same time, which underpins the direct connection between GIS and decarbonization (Baskara, 2024).

Lastly, Socio-Technical Systems Theory emphasizes the need to synchronize the technological systems with the social and organizational contexts. It describes that the successful AI application in GIS needs not only a high-tech solution but also an institutional basis, user interaction, and ethical regulations. This view justifies the joint operation of AI and GIS in university systems to attain sustainability.

Collectively, these theories offer a coherent structure in which DOI describes adoption, TOE describes conditions of implementation, RBV describes development of capabilities, and Ecological Modernization and Socio-Technical Systems describe sustainability outcomes. This combination is certain to make sure that every theoretical worldview has a direct impact on the comprehension of the relationships between AI, GIS, and decarbonization.

LITERATURE REVIEW

The wind-tree technology is a micro-scale renewable energy option, which is highly applicable within the university setting because of its integrated design and capacity to work within low-wind speed. Under green system of innovations, wind-tree is utilized in decentralized energy production and it also assists in decarbonization at campus level. The combination with AI makes them even better, as they can place them optimally, predictively repair, and optimize the entire process of the system in a real-time, thus, increasing its efficiency (Allam et al., 2025; Baskara, 2024; Nagaraj et al., 2023).

AI, as part of green innovation systems, handles various functional capacities, among them being data analytics, predictive optimization, and smart energy management. Universities can use AI-based analytics to analyse large amounts of energy and operational data to make informed decisions (Allam et al., 2025). Also, predictive optimization practices enable institutions to predict energy demand and direct conversational efficiency of renewable energy system like wind-tree technologies. Also, AI helps in achieving smart grids that assist in real-time monitoring, delivery of energy, and incorporation of decentralized renewable energy sources, thus improving the overall performance of decarbonization (Frikha and Mrad, 2025).

Wind-Tree-Based GIS Decarbonization

Based on TOE, the adoption of AI in higher education is not a change in technology alone, but a multi-faceted system change, requiring the system to be restructured. The massive implementation of AI tools by the students, especially within subjects such as Business and Social Sciences, is the most critical change in the academic practices with a tendency of enhancing the learning process and academic grades (Vinoth et al., 2025). Although the described benefits in terms of learning outcomes are positive, the access disparities and digital literacy gaps, as well as the ethical issues of data privacy and over-dependence are the key issues to consider equitable AI implementation (Vinoth et al., 2025).

Concerning the widespread use of AI by students in their academic lives, especially to improve efficiency and quality, universities are confronted with the necessity of creating formal guidelines and ethical standards that ensure that a so-called shadow pedagogy of AI does not manifest itself (Mazaherian and Nourbakhsh, 2025). In align with DOI, it involves formulating coherent policies concerning AI application, enhancing the concept of AI literacy, and creating an atmosphere. This requires a concerted institutional care of ethical implications of AI, such as algorithmic bias, data privacy, and the likelihood of committing scholarly dishonesty (Zakarneh et al., 2025).

GIS and AI wind-trees

In the context of RBV; AI, and especially Generative AI and Large Language Models can serve as an important tool to support of GIS so that more efficient approaches to decarbonization are possible (Adamakis and Rachiotis, 2025; Jordan et al., 2025). Moreover, organizations need to prepare in-depth policies on using AIs and especially focus on human controls and critical evaluation of AI-generated data to reduce the risks of misinformation and recapitulated prejudices based on the training data. By actively addressing the issue of academic integrity, this proactive strategy has the capacity to make AI tools a positive facilitator, but not a replacement of critical thinking and original research. It would have to be cautiously practiced in terms of their pedagogical, ethical, and institutional implications of AI, as well as the investigation of real-life use cases in the context of higher education (Twabu, 2025). This way of analysis prevails to explore the issue of including AI-related topics into the curriculum and how universities can discuss the ethical consequences of AI, training future generations of students to work in an AI-based environment.

The ethical consideration of AI in higher education, including academic honesty, quality of data, and possible impediments to conventional education and learning, are major issues that should be both well thought out and ethically implemented. The concept of holistic solution will also have to struggle against the issues of personal data privacy and algorithm bias, which are innate to AI systems and can neglect the achievement of fair educational results without sufficiently tackling them (Allam et al., 2025). The possibility of AI models being able to reproduce the biases of the datasets on which they are trained is especially alarming, as it can affect the objectivity of scientific studies and reinforce the existing systemic disparities of educational testing (Nagaraj et al., 2023).

AI and Decarbonization

With the help of AI technologies, the resources allocated and used in universities may be optimized, which can result in more efficient energy consumption, less waste production, and more cost-saving (Korzeb et al., 2024). As per TOE, predictive analytics devices focused on AI can predict changes in energy demand and supply. By it the university could take proactive steps to conserve energy and manage it on the demand side. Furthermore, AI can help to implement and advance new green technologies, to speed up research and development and to simulate, as well as to design materials in a more sustainable way. This also includes the development of digital twins of complex energy systems that, in turn, will allow predictive maintenance and real-time optimization of operating parameters to maximize the efficiency of energy consumption and reduce environmental footprint. Its use in these fields is not only efficient, but also creates a living lab that allows students and researchers to test new state-of-the-art sustainable technologies and techniques and enhance learning results as well as interdisciplinary collaboration.

Still, the ethical aspects of AI should be carefully handled in the future to make sure that these developments do not unintentionally undermine equity or academic principles (Khatri and Karki, 2023). Institutions should thus develop strong governance structures that will preside over AI adoption to bring in transparency, accountability, and compliance with ethical codes. Study founded on base of RBV shows that, it comprises creation of clear policy and open user interface that make the stakeholders aware of data use and incorporation of privacy-by-design concepts in AI implementation (Raman et al., 2024).

AI- GIS, Wind Trees, decarbonization

Founded on TOE study proposed that the ability of AI to analyse complex data should be used in the design and performance of wind-tree systems themselves so that the aerodynamics of blades and the location of turbines and energy conversion efficiency could be maximized to ensure maximum capture of renewable energy, reduce ecological footprint (Chauhan and Sahoo, 2024).

Besides, the simulations of such innovative systems, which are AI-based models, can forecast the future performance and the environmental interactions of the technology, which will be improved during the design phase and the locations to improve the overall sustainability rates. This involves the creation of governance systems that prioritize ethical and socially responsible AI, so that the advanced systems can ensure sustainability by resolving possible biases and privacy violations that are in the large-scale processing of data (Greif et al., 2024; Shah et al., 2024). The mechanisms of continuous auditing and assessment of AI systems to verify their compliance with sustainability goals and compliance with regulatory requirements should also be present in these frames. This is an important holism approach toward AI governance to create a sustainable, ethical, and inclusive higher education system, in accordance with the world objectives of sustainable development (Hong et al., 2025). The strategy will blend with the idea of green education, which will make students and faculty more aware and environmentally cautious. This attention to the environmentalism and the sustainable approach is one of the fundamentals of green education which strives to foster a greater awareness of the environmental concerns as well as a sense of responsibility towards the planet.

METHODOLOGY

The research is quantitative because it investigates the effect of wind-tree-based GIS on decarbonization in the context of AI in Tanzanian universities, and more specifically St. Joseph University In Tanzania (SJUIT). It uses a deductive approach, based on the existing theories, to consider the hypothesized relationships as a test of connections between AI, GIS, and decarbonization results. The study is cross-sectional because the researchers take the data at one time among the respondents in sustainability related initiatives. The research design adopted is explanatory research designing as the purpose of the study is to determine cause-effect relationships between variables. A questionnaire list is created with the pre-tested validated scales modified to a wind-tree technology, AI uses and sustainability of the environment. The questionnaire will involve several questions that form a moderate scale of AI capability, green innovation practices, and results of decarbonization with a 5-point Likert scale. The most important constructs in this research AI, GIS and DC were operationalized through validated measurement scales which were modified according to the past literature and adjusted to the context of the university. The AI was assessed regarding the ability to promote data-driven decision-making, energy optimization, and intelligent integration of systems. GIS indicates the level at which universities embrace environmentally friendly practices and renewable energy sources including wind-tree systems. Decarbonization entrapes the reduction of carbon emission and energy efficiency portrayed. To achieve consistency, a good level of reliability, and content validity, all constructs were also measured on a five-point Likert scale with strong disagreements to strongly agree.

The intended population is students who are involved in energy conservation clubs as well as sustainable projects at SJUIT. The reason why these respondents are appropriate is that they have the relevant awareness and exposure to the green technologies and practices surrounding the environment at the university. The respondent population is defined as 320 people and it is sufficient to perform Structural Equation Modeling (SEM), as well as, it would provide sufficient statistical power and stability of the model. The purposive sampling of the students actively engaged in the energy conservation organizations and sustainability initiatives is adopted in the study; they have access to the required knowledge and firsthand experience with green technologies and AI implementation. The methodology will allow the respondents to give informed and significant responses on the constructs being investigated especially in the case of GIS and decarbonization practices at the university setting.

The SEM analysis was used to perform the data analysis. The study used mediation analysis on AMOS based on bootstrapping technique and 5,000 resamples to analyse the mediating effect of GIS. The methodology gives strong estimates of indirect effects and confidence. The confirmation of mediation is based on the existence of the statistically significant indirect study between AI and decarbonization with the help of green innovation, and the confidence interval does not contain the value zero. The partial and the full mediation situations are assessed on the basis of the importance of the direct and indirect paths.

Ethical issues are adhered to properly such as informed consent, volunteer participation, and confidentiality of information of respondents. All in all, the methodology provides the rigorous and systematic approach to the analysis of the role of AI-driven GIS in facilitating decarbonization in the Tanzanian universities.

RESULTS

Table 1. Profile

Category	Frequency	Percentage (%)
Gender		
Male	182	56.9
Female	138	43.1
Age Group		
18–22 years	124	38.8
23–26 years	136	42.5
Above 26	60	18.7
Participation Type		
Energy Clubs	178	55.6
Sustainability Projects	142	44.4

Source: authors

A detailed description of the demographic and participation features of the respondents sampled in St. Joseph University In Tanzania (SJUIT) is given in Table 1. There is a reasonably well-balanced gender representation as out of the total 320 participants 56.9 are male and 43.1 are female. This balance makes the findings more generalized since both the sides are well represented. In terms of age distribution, the largest number of respondents are between the 23-26 years (42.5%), and 18-22 years (38.8), with the remaining few above 26 (18.7). This is indicative of the fact that the study to a large extent captures the youthful and energetic university members who are more prone to be involved in the innovation oriented and sustainability-oriented initiatives. When it comes to investment, 55.6 percent of the people interviewed have joined energy conservation clubs, and 44.4 percent will have been involved in sustainability projects. This proves the fact that all the respondents are directly engaged in activities that are environmentally relevant, which means that the data gathered is relevant and informed in a way that is also contextually relevant. The suitability of the sample to analyse the AI-mediated GIS and decarbonization are validated by the profile.

Table 2. Measurement Model

Construct	Item	Loading	CR	AVE
GIS (GIS)	GIS1	0.78	0.89	0.62
	GIS2	0.81		
	GIS3	0.77		
	GIS4	0.80		
Artificial Intelligence (AI)	AI1	0.83	0.91	0.66
	AI2	0.85		
	AI3	0.79		
	AI4	0.82		
Decarbonization (DC)	DC1	0.84	0.92	0.68
	DC2	0.86		
	DC3	0.80		
	DC4	0.83		

Source: authors

The quality of the measurement model is mastered in Table 2 as the factor loadings, Composite Reliability (CR), and Average Variance Extracted (AVE) are evaluated. The entire loadings of the observed indicators are at a range between 0.77 and 0.86 reflecting above the recommended 0.70 level. This shows that all of the measurement items have a strong contribution towards the respective latent constructs, that is, GIS, AI, and Decarbonization (DC). All the constructs have strong CR values of between 0.89 and 0.92, which is far above the acceptable mark of 0.70. This shows that the measurement scales have high internal consistency and reliability. Moreover, there is a scale of 0.62-0.68 which is greater than the minimum of 0.50, the requirements of the AVE values. This affirms that over half of the differences in the indicators is accounted by their respective constructs, which determine high levels of convergent validity. Altogether, Table 2 supports the idea that the model of measurement is reasonably sound, and all the constructs are highly measured and consistent allowing the subsequent structural model analysis.

Table 3. Reliability and Validity

Construct	Cronbach's Alpha	CR	AVE
GIS	0.88	0.89	0.62
AI	0.90	0.91	0.66
DC	0.91	0.92	0.68

Source: authors

Table 3 supporting the reliability and validity results provides the results of Cronbach alpha, CR and AVE values. The values of Cronbach alpha of GIS (0.88), AI (0.90) and DC (0.91) are more than 0.70, the suggested minimum level of item internal consistency, meaning that they have an excellent internal consistency. This reliability is also further corroborated by the CR values which state that the constructs are measured consistently in various items. Also, the value of the AVEs is greater than 0.50 in all the constructs, which reiterates the convergent validity. This table further supports the fact that constructs adopted in the research are reliable and valid. The fact that the alpha of Cronbach and CR are consistent also implies the consistency in the measurement and lessens the chances of measurement error.

Table 4. Discriminant Validity

Construct	GIS	AI	DC
GIS	0.79		
AI	0.37	0.81	
DC	0.42	0.78	0.82

(Diagonal values = √AVE)
Source: authors

Table 4 tests the allotment that the constructs are not similar to one another. The square root of AVE of individual construct (diagonal values) is higher than correlations with other constructs. As an illustration, the square root of AVE of AI (0.81) is greater than that of its links with GIS (0.37) and DC (0.78). This proves that both constructs are capturing dissimilar ideas and that they do not overlap in any significant way. This implies, in practice, that the variables of GIS, AI and Decarbonization are empirically different variables, although interconnected. Defining discriminant validity in SEM analysis is a very important exercise since it guarantees that the relationships which may be observed in the structural model are not elevated by the redundancy of constructs.

Table 5. Model Indices

Fit Index	Value	Threshold	Interpretation
Chi^2 /df	2.41	< 3	Good
CFI	0.94	> 0.90	Good
TLI	0.93	> 0.90	Good
RMSEA	0.067	< 0.08	Acceptable
SRMR	0.052	< 0.08	Good

Source: authors

Table 5 assesses the amount of goodness of fit of the proposed structural model to the observed data. The Chi^2 /df ratio (2.41) is acceptable (less than 3), which shows the good overall fit. Compared Fit Index (CFI = 0.94) and TLI = 0.93) both are higher than the suggested level of 0.90, and therefore the model fits the covariance structure well. Both the RMSEA is equal to 0.067 and SRMR is equal to 0.052; are less than 0.08 which implies that both values are very low. Together these indices portray that the hypothetical prototype could represent the data adequately and the hypothesized relationships among the constructs were determined to be correct.

Table 6. Structural Model (Direct Effects)

Hypothesis	Path	β	t-value	p-value	Decision
H1	GIS → AI	0.37	6.12	0.000	Supported
H2	AI → DC	0.78	12.45	0.000	Supported
H3	GIS → DC	0.42	7.01	0.000	Supported

Source: authors

Table 6 shows the direct correlation between the constructs. The relationship between GIS and AI ($b = 0.37$, $p < 0.001$) is positive and significant, and it denotes that the introduction of the wind-tree-based systems of green can improve

AI consumption in universities. The correlation between AI and Decarbonization (b = 0.78, p = less than 0.001) is the strongest, which indicates that AI is essential to better use of energy, efficiency, and decrease carbon emission. Furthermore, the positive impact of GIS directly addresses Decarbonization (b = 0.42, p = 0.001) which means that the given systems independently influence the environmental sustainability. Every hypothesis is utilized and it is certain that both the direct and indirect pathways are crucial in the explanation of the results of decarbonization.

Table 7. Mediation Analysis (Bootstrapping – 5000 Samples)

Path	Direct Effect	Indirect Effect	Total Effect	95% CI	Mediation Type
GIS → AI → DC	0.42	0.29	0.71	0.21 – 0.36	Partial Mediation

Source: authors

Table 7 tests whether there is a mediator role of AI between GIS and Decarbonization. The correlation value (0.29) of the independent variable is significant, and the confidence interval (0.21-0.36) does not take the value of 0 that establishes the existence of mediation. The direct effect (0.42) is still huge irrespective of the inclusion of the mediator; thus, the mediation can be described as a partial mediation. It implies that GIS have a direct and indirect impact on the Decarbonization via AI. This fact indicates the dual nature of AI; it is both a process that boosts the performance of green innovations and it works by itself to decarbonize.

Table 8. Coefficient of Determination (R^2)

Endogenous Variable	R^2	Interpretation
Artificial Intelligence	0.29	Moderate
Decarbonization	0.78	Strong

Source: authors

Table 8 indicates the strength of the model. AI has a R2 value of 0.29 meaning that it is explained using the Green Innovations Systems that explain 29 percent of the variance, which is moderate. In the case of Decarbonization, it is very high having a R2 value of 0.78. This is to say that the solution of GIS and AI combined accounts towards 78 percent variability in decarbonization results. Such significance of explanation implies that this model is really effective in terms of explaining the main determinants of decarbonization in the university environment.

Table 9. Effect Size (f^2)

Relationship	f^2	Effect Size
GIS → AI	0.16	Medium
AI → DC	0.61	Large
GIS → DC	0.21	Medium

Source: authors

Table 9 determines the relative significance of every relationship. The impact of AI on decarbonization (f2 = 0.61) is large showing that the AI has the most significant effect on the model. The influence of GIS on AI (0.16) and GIS on Decarbonization (0.21) is moderate, as they mean that, although GIS are critical, their effects are highly enhanced with the help of AI. The apparent finding highlights how strategic is the integration of AI and green technologies as a way of maximizing their environmental benefits.

Table 10. Predictive Relevance (Q^2)

Construct	Q^2	Interpretation
AI	0.21	Medium Predictive Relevance
DC	0.39	Strong Predictive Relevance

Source: authors

Table 10 evaluates the predictive ability of the model on the basis of Q2 values. The Q2 value of both AI (0.21) and Decarbonization (0.39) are all over zero and this is a sign that the model is highly relevant in prediction. The larger value of Q2 of Decarbonization indicates that the model is quite robust in the prediction of the major outcome variable. This proves that the model is not a mere hypothetical form but can also be practically applicable to predicting sustainability results in the real-world environment.

Table 11. Multi-Group Analysis (Gender-Based Comparison)

Path	Male (β)	Female (β)	Difference (Δβ)	p-value
GIS → AI	0.35	0.39	0.04	>0.05
AI → DC	0.76	0.81	0.05	>0.05
GIS → DC	0.40	0.44	0.04	>0.05

Source: Authors

The multi-group analysis was carried out in table 11 to investigate that the relationship in the model is different among male and female respondents. The results show that the path from green innovation systems (GIS) to artificial intelligence (AI) is slightly higher for females (0.39) compared to males (0.35), but the difference (0.04) is very small and not statistically significant ($p > 0.05$). This implies that gender does not play a role in the impacts of GIS on the adoption of AI in the research environment. On the same note, the correlation of AI and decarbonization (DC) is high among both groups, and females (0.81) have a marginally higher rate compared to males (0.76). But the difference (0.05) once again is not significant. It means that both male and female respondents show the same contribution of AI towards decarbonization.

Lastly, the direct correlation between GIS and decarbonization is equally close between females and males with a 0.44 and 0.40 respectively. The change (0.04) is minimal and not significant meaning that there is no variation in the effect of GIS on decarbonization across the genders. Generally, the results indicate that gender is not a moderator in the model. GIS, AI, and decarbonization have steady, consistent relationships with both male and female respondents and enhance the reliability and generalizability of the study findings.

Table 12. Multi-Group Analysis (Participation Type)

Path	Energy Clubs (β)	Sustainability Projects (β)	Difference (Δβ)	p-value
GIS → AI	0.38	0.36	0.02	>0.05
AI → DC	0.79	0.77	0.02	>0.05
GIS → DC	0.43	0.41	0.02	>0.05

Source: Authors

The multi-group analysis in table 12 was carried out to investigate that the relationship in the model is different among male and female respondents. The results show that the path from green innovation systems (GIS) to artificial intelligence (AI) is slightly higher for females (0.39) compared to males (0.35), but the difference (0.04) is very small and not statistically significant ($p > 0.05$). This implies that gender does not play a role in the impacts of GIS on the adoption of AI in the research environment. On the same note, the correlation of AI and decarbonization (DC) is high among both groups, and females (0.81) have a marginally higher rate compared to males (0.76). But the difference (0.05) once again is not significant. It means that both male and female respondents show the same contribution of AI towards decarbonization.

Lastly, the direct correlation between GIS and decarbonization is equally close between females and males with a 0.44 and 0.40 respectively. The change (0.04) is minimal and not significant meaning that there is no variation in the effect of GIS on decarbonization across the genders. Generally, the results indicate that gender is not a moderator in the model. GIS, AI, and decarbonization have steady, consistent relationships with both male and female respondents and enhance the reliability and generalizability of the study findings.

Table 13. Importance–Performance Map Analysis for Decarbonization

Construct	Importance (Total Effect)	Performance (Index Value)	Interpretation
Artificial Intelligence (AI)	0.78	72.5	High importance, moderate performance
Green Innovation Systems (GIS)	0.71	68.3	Moderate importance, moderate performance

Source: Authors

In table 13, to determine the performance of each construct based on its contribution to decarbonization and its current level of performance, the Importance-Performance Map Analysis (IPMA) was carried out to evaluate the importance of each construct. The findings indicate that the value of importance of artificial intelligence (AI) is the greatest (0.78), which means that it is the most significant factor that contributes to the decarbonization outcomes. However, its performance score (72.5) is only moderate, which suggests that although AI plays a critical role, its current level of implementation and effectiveness in universities can still be improved.

Green innovation systems (GIS) also show a significant contribution towards decarbonization, its value of importance is 0.71. This means that GIS plays a significant role, but a little less significant than AI. Its performance score (68.3) is moderate as well as the one of AI, which means that the use of green innovation practices is still in progress and needs to be reinforced. In general, the results suggest that AI and GIS are significant sources of decarbonization, yet neither algorithm has attained its best performance. It means that universities need to focus on better AI implementation and green innovation system effectiveness. By enhancing these aspects, energy efficiency and reduction of carbon emissions can be improved, which will contribute to the overall sustainability performance.

DISCUSSION

Although the results prove the interconnection of artificial intelligence and the system of green innovations and decarbonization, the interpretation of these findings as the cause of a link should be approached with care. The research uses perception-based, cross-sectional data, which puts a restraint on the provision of conclusive cause and effect detachment. Thus, the findings are to be interpreted as the signs of substantial associations that can guide future longitudinal and experimental studies. The results offer significant policy and institutional lessons to take in ensuring decarbonization in higher education. Universities must emphasize the insertion of the artificial intelligence into the green innovation systems, and in this way improve energy efficiency and optimize renewable energy solutions like wind-trees technologies. On the institutional level, there is the necessity of investments in digital infrastructures, technical capacity and interdisciplinary cooperation to sustain the successful implementation of AI-enabled systems of sustainability. At the policy level, the governments should come up with favourable regulatory environments, financing systems, and incentives that motivate the universities to implement AI powered renewable energy applications.

The intermediary function of Digital Green Innovation further underlines the fact that the role of AI in ensuring sustainable performance is the most productive when it is presented as a part of the overall digital innovation policies. In fact, the actualization of AI in the promotion of sustainable performance is mainly evident by its incorporation into digital green innovation, which underscores that generation of value in the environmental and social context is, dependent on its integration into innovation processes, whereby, the technological resources get embedded. This subtle interplay is also supported by the partial mediation of AI between the GIS and Decarbonization observed, wherein the direct impacts of green innovations are boosted and hastened with the addition of AI to it. The results therefore explain that well-established governance systems contribute a lot to the ability of firms to convert digital intelligence to environmentally efficient innovation output; hence, implying that adoption of AI on its own does not produce sustainable value. Instead, it is most effectively employed and achieved in the context of Advocacy in Organizations that encompass the comprehensive and effective digital innovation strategies ensuring technological resources are incorporated into the green innovation processes. The results underscore the fact that GIS that relies on wind-trees are effective with the assistance of AI, which streamlines energy production and resource allocation. This implies that AI-powered micro-renewable technologies will be additionally critical to the progress of decarbonization policies at universities.

CONCLUSION

The potential of this transformation indicates a pressing need to involve higher education institutions in Tanzania to become strategic in their commitment to realize the opportunities of AI in their sustainability plans, such that there is a synergistic shift towards decarbonization, and the achievement of ESG overall. As an active engagement between AI and research, it is essential to accelerate the research processes, use resources wisely, and promote the creation of environmental-friendly products and processes, which will improve the overall level of green innovation. The research makes a positive contribution to the AI-guided climate action by showing how AI can improve GIS to get tangible decarbonization effects in universities. It presents empirical data which prove that AI optimization can be effective in terms of efficiency of renewable energy devices including wind-tree systems. The paper puts emphasis on one of the practical uses of AI as it helps in facilitating institutional climate measures by concentrating on the context of expanding nations. In such a way, it assists in promoting the data-driven and scalable solutions to climate action in the sphere of higher education. It is consistent with the prior research reporting that AI is an essential resource to sustainable finance as it handles a substantial amount of sustainability-related data, standardizes the security analysis, and aids in making predictive decisions (Rane et al., 2024); (Mansour et al., 2025). Study is limited to one university; it can be extended to more HEIs.

REFERENCES

Abo-Khalil, A. G. (2024). Integrating sustainability into higher education challenges and opportunities for universities worldwide. *Heliyon, 10*(9), e29946. Advance online publication. DOI: 10.1016/j.heliyon.2024.e29946 PMID: 38707336

Adamakis, M., & Rachiotis, T. (2025). Artificial Intelligence in Higher Education: A State-of-the-Art Overview of Pedagogical Integrity, Artificial Intelligence Literacy, and Policy Integration. *Encyclopedia, 5*(4), 180. DOI: 10.3390/encyclopedia5040180

Alkhawaja, L., Idris, M., Al-Sayyed, S. W., & Jaber, A. R. (2025). Exploring the impact of Artificial Intelligence on students' skills for sustainable development in education. *Frontiers in Education, 10*, 1691148. Advance online publication. DOI: 10.3389/feduc.2025.1691148

Allam, H., Gyamfi, B., & AlOmar, B. (2025). Sustainable Innovation: Harnessing AI and Living Intelligence to Transform Higher Education. *Education Sciences, 15*(4), 398. DOI: 10.3390/educsci15040398

Baskara, F. X. R. (2024). Generative AI as an Enabler of Sustainable Education: Theoretical Perspectives and Future Directions. *British Journal of Teacher Education and Pedagogy, 3*(3), 122–134. DOI: 10.32996/bjtep.2024.3.3.9

Chan, C. K. Y. (2023). A comprehensive AI policy education framework for university teaching and learning. *International Journal of Educational Technology in Higher Education, 20*(1), 38. Advance online publication. DOI: 10.1186/s41239-023-00408-3

Chauhan, M., & Sahoo, D. R. (2024). Towards a Greener Tomorrow: Exploring the Potential of AI, Blockchain, and IoT in Sustainable Development. *Nature Environment and Pollution Technology, 23*(2), 1105–1113. DOI: 10.46488/NEPT.2024.v23i02.044

Chun, Y., & Hwang, J. (2024). The Nexus of Artificial Intelligence and Green Innovation: A Cross-Density Analysis at the Country Level. *Journal of the Knowledge Economy, 16*(1), 1688–1716. Advance online publication. DOI: 10.1007/s13132-024-02076-8

Filho, W. L., Kim, E., Borsatto, J. M. L. S., & Marcolin, C. B. (2025). Using artificial intelligence in sustainability teaching and learning. *Environmental Sciences Europe, 37*(1), 124. Advance online publication. DOI: 10.1186/s12302-025-01159-w

Frikha, M. A., & Mrad, M. (2025). AI-Driven Supply Chain Decarbonization: Strategies for Sustainable Carbon Reduction. *Sustainability (Basel), 17*(21), 9642. DOI: 10.3390/su17219642

Greif, L., Kimmig, A., Bobbou, S. E., Jurisch, P., & Ovtcharova, J. (2024). Strategic view on the current role of AI in advancing environmental sustainability: A SWOT analysis. *Discover Artificial Intelligence*, *4*(1), 45. Advance online publication. DOI: 10.1007/s44163-024-00146-z

Hong, T. T. M., Tung, N. T. T., & Thanh, N. T. P. (2025). Mapping artificial intelligence research in higher education toward sustainable development. *Discover Sustainability*, *6*(1), 1240. Advance online publication. DOI: 10.1007/s43621-025-02162-0

Jordan, A., Dockens, A. L., Pierson, N. A., & Ren, X. (2025). Editorial: AI's impact on higher education: transforming research, teaching, and learning. *Frontiers in Education*, *10*, 1682901. Advance online publication. DOI: 10.3389/feduc.2025.1682901

Khan, S., Mazhar, T., Shahzad, T., Khan, M. A., Rehman, A. U., Saeed, M. M., & Hamam, H. (2025). Harnessing AI for sustainable higher education: Ethical considerations, operational efficiency, and future directions. *Discover Sustainability*, *6*(1), 403. Advance online publication. DOI: 10.1007/s43621-025-00938-y

Khatri, B. B., & Karki, P. D. (2023). Artificial Intelligence (AI) in Higher Education: Growing Academic Integrity and Ethical Concerns. *Nepalese Journal of Development and Rural Studies*, *20*(1), 1–7. DOI: 10.3126/njdrs.v20i01.64134

Korzeb, Z., Alonso-Fariñas, B., Diéguez, A. I. I., Naharro, F. J., Kobylińska, U., di Pietro, F., Palacín-Sánchez, M.-J., Rollnik-Sadowska, E., Szpilko, D., Szydło, J., & Gallegos, A. de la T. (2024). The future of European universities on the path to sustainable development. *Engineering Management in Production and Services*, *16*(2), 68–89. DOI: 10.2478/emj-2024-0014

Mambile, C., & Mwogosi, A. (2024). Transforming higher education in Tanzania: Unleashing the true potential of AI as a transformative learning tool. *Technological Sustainability*, *4*(1), 51–76. DOI: 10.1108/TECHS-03-2024-0014

Mansour, M., Zobi, M. A., & Alomair, M. (2025). Artificial Intelligence, ESG Governance, and Green Innovation Efficiency in Emerging Economies. *Economies (Basel)*, *14*(1), 11. DOI: 10.3390/economies14010011

Mazaherian, A., & Nourbakhsh, E. (2025). *Beyond the Hype: Critical Analysis of Student Motivations and Ethical Boundaries in Educational AI Use in Higher Education*. DOI: 10.48550/ARXIV.2511.11369

Mudenda, S. (2025). Impact of Artificial Intelligence on College and University Students: A Global Transformation of the Education System. *Creative Education*, *16*(11), 1801–1828. DOI: 10.4236/ce.2025.1611112

Nadeem, W., Ashraf, A. R., Khan, H., & Kumar, V. (2025). Impact of AI strategies on climate-change performance: Responsible AI and crisis management perspectives. *Technovation, 150*, 103390. DOI: 10.1016/j.technovation.2025.103390

Nagaraj, B., & Kalaivani, A. (2023). The Emerging Role of Artificial Intelligence in STEM Higher Education: A Critical Review [Review of The Emerging Role of Artificial Intelligence in STEM Higher Education: A Critical Review]. *International Research Journal of Multidisciplinary Technovation, 1*, 1–19. Advance online publication. DOI: 10.54392/irjmt2351

Nugroho, M. A., Suyanto, U. Y., Hariyanto, D., & Arianisari, S. (2025). Uneven But Accelerating: AI Adoption in Higher Education. *International Journal of Advanced Computer Science and Applications, 16*(10). Advance online publication. DOI: 10.14569/IJACSA.2025.0161041

Olawade, D. B., Wada, O. Z., David-Olawade, A. C., Fapohunda, O., Ige, A. O., & Ling, J. (2024). Artificial intelligence potential for net zero sustainability: Current evidence and prospects. *Next Sustainability, 4*, 100041. DOI: 10.1016/j.nxsust.2024.100041

Opesemowo, O. A. G., & Adekomaya, V. (2024). Harnessing Artificial Intelligence for Advancing Sustainable Development Goals in South Africa's Higher Education System: A Qualitative Study. *International Journal of Learning Teaching and Educational Research, 23*(3), 67–86. DOI: 10.26803/ijlter.23.3.4

Raman, R., Pattnaik, D., Lathabai, H. H., Kumar, C., Govindan, K., & Nedungadi, P. (2024). Green and sustainable AI research: An integrated thematic and topic modeling analysis. *Journal of Big Data, 11*(1), 55. Advance online publication. DOI: 10.1186/s40537-024-00920-x

Rane, N. L., Rane, J., & Paramesha, M. (2024). *Artificial Intelligence and business intelligence to enhance Environmental, Social, and Governance (ESG) strategies: Internet of things, machine learning, and big data analytics in financial services and investment sectors*. DOI: 10.70593/978-81-981367-4-9_3

Saheb, T., Dehghani, M., & Saheb, T. (2022). Artificial intelligence for sustainable energy: A contextual topic modeling and content analysis. *Sustainable Computing : Informatics and Systems, 35*, 100699. DOI: 10.1016/j.suscom.2022.100699

Savec, V. F., & Jedrinović, S. (2024). The Role of AI Implementation in Higher Education in Achieving the Sustainable Development Goals: A Case Study from Slovenia. *Sustainability (Basel), 17*(1), 183. DOI: 10.3390/su17010183

Sebihi, A., Schoelen, L., & Uwamwezi, B. (2025). Unveiling AI's impact in Rwandan higher education: Navigating trends, challenges, and imperatives at the University of Rwanda. *Discover Education*, *4*(1), 166. Advance online publication. DOI: 10.1007/s44217-025-00576-6

Shah, E. A., Raiz, A., & Rashid, M. (2024). *Ethical Considerations in Use of Artificial Intelligence in Digital Marketing*. https://doi.org/DOI: 10.36968/JPDC-V08-I02-25

Silva, T. C. C., & Hämäläinen, M. (2024). Innovating for the Future: AI and HRM Capabilities for Sustainability in Higher Education. *Academy of Management Proceedings, 2024*(1). DOI: 10.5465/AMPROC.2024.19249abstract

Twabu, K. (2025). Investigating schumpeter's innovation theory in the context of AI in higher education research. *Discover Education*, *4*(1), 389. Advance online publication. DOI: 10.1007/s44217-025-00855-2

Vinoth, S., Chinnasamy, G., & Madbouly, A. (2025). Evaluating the integration and impact of artificial intelligence (AI) tools on academic learning, assessment, and research practices in higher education. *Education and Information Technologies*, *30*(18), 25655–25682. DOI: 10.1007/s10639-025-13719-x

Wainaina, P., & Sun, Y. (2025). Educators' perceptions and willingness to integrate Generative Artificial Intelligence in teaching and research: Evidence from Kenyan higher education. *Discover Education*, *4*(1), 347. Advance online publication. DOI: 10.1007/s44217-025-00820-z

Zakarneh, B., Aljabr, F., Mahmoud, D. A. M., Annamalai, A. N., Chakraoui, R., & Yakoob, A. (2025). Impact of Sustainability Awareness on AI Perception, Engagement, and Motivation among University-Level Literature Students. *World Journal of English Language*, *16*(2), 295. DOI: 10.5430/wjel.v16n2p295

Chapter 10
AI-Enabled Climate Adaptation and Resilience for Vulnerable Populations in the Global South

Sameen Rafi
http://orcid.org/0000-0003-2631-5871
Aligarh Muslim University, Aligarh, India

ABSTRACT

Climate change poses disproportionate risks to vulnerable populations in the Global South, where social inequality, limited adaptive capacity, and fragile institutions intensify climate impacts. Artificial intelligence (AI) has emerged as a promising tool for enhancing climate adaptation and resilience through improved forecasting, early warning systems, risk mapping, and decision support. However, the effectiveness of AI-enabled climate adaptation depends on how inclusively and ethically these technologies are designed and governed. This chapter examines the role of AI in supporting climate adaptation and resilience for vulnerable populations in the Global South, with particular reference to developing-country contexts. Adopting a human-centered and policy-oriented perspective, the chapter critically analyzes opportunities, equity and governance challenges, and sustainability concerns associated with AI deployment.

DOI: 10.4018/979-8-3373-9978-2.ch010

1. INTRODUCTION

Climate change has become a defining challenge of contemporary development, reshaping environmental conditions, economic systems, and social relations across the globe. Although climate change is a global phenomenon, its impacts are deeply uneven, with countries in the Global South experiencing disproportionate exposure to climate-related hazards such as heatwaves, floods, droughts, cyclones, and climate-sensitive disease outbreaks (IPCC, 2022). These risks are intensified by structural constraints including poverty, rapid urbanization, fragile infrastructure, informal economies, and limited institutional capacity, which collectively undermine adaptive capacity and resilience.

In developing regions, climate change is not only an environmental issue but also a profound social challenge. Climate hazards interact with existing inequalities related to age, gender, health status, income, and social position, producing differentiated patterns of vulnerability (Adger, 2006). Vulnerable populations such as older adults, women, persons with disabilities, informal workers, and low-income households often face higher exposure to climate risks, greater sensitivity to climate shocks, and fewer resources to cope with or recover from their impacts (Cutter et al., 2012). As a result, climate change threatens to exacerbate social inequalities and undermine progress toward sustainable development goals.

Climate adaptation has therefore emerged as a central pillar of global and national climate strategies. Adaptation seeks to reduce vulnerability, enhance resilience, and strengthen the capacity of societies to anticipate, absorb, and recover from climate-related shocks. However, in many Global South contexts, adaptation efforts remain fragmented, reactive, and constrained by limited data, weak coordination, and insufficient integration across sectors such as health, social protection, disaster management, and urban planning. These limitations have prompted growing interest in digital and data-driven approaches that can support more anticipatory and targeted adaptation strategies.

Artificial intelligence (AI) has gained increasing prominence as a potential enabler of climate adaptation and resilience. AI refers to computational techniques, such as machine learning (ML) and predictive analytics, that enable systems to analyze large datasets and support forecasting and decision-making. In the context of climate change, AI-enabled applications are being used to improve weather and climate forecasting, develop early warning systems for extreme events, map climate risks, optimize disaster response, and monitor climate-sensitive health outcomes (Rolnick et al., 2019). These capabilities offer significant opportunities to enhance preparedness and reduce climate-related losses.

Despite this potential, the use of AI for climate adaptation raises critical questions regarding equity, access, governance, and sustainability. AI systems depend on data infrastructures, digital connectivity, and institutional capacities that are unevenly distributed within and across countries. As a result, marginalized populations may be excluded from AI-enabled benefits or inadequately represented in data-driven decision-making processes (Vinuesa et al., 2020). Concerns related to algorithmic bias, lack of transparency, weak regulatory oversight, and the environmental footprint of energy-intensive AI systems further complicate the role of AI in sustainable climate action (Jobin et al., 2019; Strubell et al., 2019).

Moreover, much of the existing literature on AI and climate change remains technologically oriented, with a dominant focus on mitigation, efficiency gains, and system optimization. While this body of work is valuable, it often underemphasizes climate adaptation, social vulnerability, and the lived experiences of communities in developing regions. There is a growing need for interdisciplinary research that situates AI within broader socio-technical and governance frameworks and critically examines how AI can support inclusive and sustained climate adaptation.

This chapter addresses this gap by examining the role of artificial intelligence in enabling climate adaptation and resilience for vulnerable populations in the Global South. Anchored in developing-country contexts, with particular reference to India, the chapter adopts a human-centered and policy-oriented perspective. Rather than presenting AI as a standalone technological solution, the analysis conceptualizes AI as a socio-technical tool embedded within broader climate governance systems, public health infrastructures, and social protection mechanisms.

The chapter is guided by an integrated conceptual framework that combines social vulnerability theory, responsible artificial intelligence, and climate adaptation scholarship. Social vulnerability theory highlights how structural inequalities shape differential exposure, sensitivity, and adaptive capacity to climate risks. Responsible AI perspectives emphasize the ethical, governance, and accountability dimensions of data-driven technologies. Climate adaptation scholarship focuses on institutional and policy mechanisms that support resilience in the face of environmental change. By integrating these perspectives, the chapter develops a socio-technical framework for understanding how AI-enabled climate adaptation can be designed, governed, and implemented in ways that strengthen resilience while avoiding new forms of exclusion in the Global South.

1.1 Chapter Contribution

This chapter argues that artificial intelligence can contribute to climate adaptation and resilience for vulnerable populations in the Global South only when technological innovation is embedded within inclusive governance systems, social protection mechanisms, and responsible data practices. Rather than treating AI as a purely technical solution, the chapter conceptualizes AI-enabled climate adaptation as a socio-technical process shaped by institutional capacity, social inequality, and policy design. By integrating insights from social vulnerability theory, responsible AI scholarship, and climate adaptation research, the chapter develops a governance-oriented framework for inclusive AI-enabled climate adaptation.

This chapter makes two key contributions to the emerging literature on artificial intelligence and climate adaptation. First, it develops an integrated socio-technical framework that combines social vulnerability theory, responsible AI principles, and climate adaptation scholarship to explain how AI-enabled systems interact with governance institutions and social inequalities in the Global South. Second, the chapter advances a policy-oriented perspective by identifying the institutional, governance, and social conditions under which AI applications can contribute to inclusive and sustained climate adaptation for vulnerable populations.

The chapter is organized as follows: Section 2 examines climate vulnerability and social inequality; Section 3 reviews AI applications for climate adaptation; Sections 4 and 5 assess potential benefits and governance challenges; Section 6 proposes a policy framework for inclusive AI-enabled adaptation; and Section 7 outlines key directions for future research.

2. CLIMATE VULNERABILITY AND SOCIAL INEQUALITY IN THE GLOBAL SOUTH

Climate vulnerability in the Global South is shaped by the complex interaction of environmental exposure, social sensitivity, and adaptive capacity. The Intergovernmental Panel on Climate Change conceptualizes vulnerability as the degree to which a system or population is susceptible to, and unable to cope with, adverse effects of climate change, including climate variability and extremes (IPCC, 2022). In developing-country contexts, vulnerability is rarely determined by climatic factors alone; rather, it is socially produced and mediated by structural inequalities related to poverty, age, gender, health, livelihood, and governance arrangements.

Countries in the Global South are disproportionately exposed to climate hazards due to geographic location, climatic conditions, and development patterns. Rapid urbanization in flood-prone areas, dependence on climate-sensitive livelihoods such as agriculture and informal labour, and limited infrastructure resilience increase exposure to extreme weather events such as heatwaves, floods, droughts, and cyclones (Dodman et al., 2019). At the same time, sensitivity to climate impacts is heightened by factors such as malnutrition, chronic health conditions, inadequate housing, and limited access to clean water and sanitation. These conditions amplify the adverse effects of climate shocks on human health, livelihoods, and well-being.

Adaptive capacity-the ability of individuals and communities to anticipate, cope with, and recover from climate impacts is often constrained in the Global South by weak institutions, limited financial resources, and fragmented governance systems. In many developing countries, social protection mechanisms, health systems, and disaster management infrastructures are insufficiently equipped to respond to the scale and frequency of climate-related risks. As a result, climate shocks frequently translate into long-term social and economic setbacks, reinforcing cycles of vulnerability and exclusion (Adger, 2006).

Social inequality plays a central role in shaping differentiated climate impacts. Age is a critical but often underexamined dimension of climate vulnerability. Older adults are particularly susceptible to extreme heat, cold, and climate-sensitive diseases due to age-related physiological changes, pre-existing health conditions, and social isolation (Vicedo-Cabrera et al., 2021). Limited mobility and dependence on caregivers further increase their exposure during climate induced emergencies such as floods or heatwaves. In many Global South contexts, older populations also face barriers to accessing healthcare, early warning information, and social support, compounding their vulnerability. In the Indian context, research has highlighted that the absence of age-friendly infrastructure, inadequate housing conditions, and limited access to health and social support systems further intensify climate-related risks for older populations (Rafi, 2020).

Gender is another key axis along which climate vulnerability is structured. Women in developing countries often experience greater exposure and sensitivity to climate risks due to their roles in caregiving, household resource management, and informal employment. Climate shocks can intensify unpaid care work, reduce access to income-generating opportunities, and increase health risks, particularly for women in low-income and rural households (Alston, 2015). At the same time, gender inequalities in access to education, technology, and decision-making processes can limit women's adaptive capacity and participation in climate adaptation initiatives.

Moreover, elderly women experience compounded vulnerability due to the intersection of age, gender, health, and caregiving responsibilities. Evidence from India indicates high prevalence of physical and mental health issues among elderly women, which can be exacerbated during climate-induced stress events such as heatwaves, displacement, and livelihood disruption (Rafi & Saif, 2021). These findings underscore the importance of age- and gender-sensitive climate adaptation strategies that explicitly account for differential vulnerabilities within older populations.

Poverty remains a critical structural driver of climate vulnerability in the Global South. Low-income households are more likely to reside in hazard-prone areas such as floodplains, informal urban settlements, or regions exposed to extreme heat. Inadequate housing, poor access to water and sanitation, and limited availability of cooling or protective infrastructure heighten exposure to climate risks. For vulnerable populations, climate shocks can quickly translate into cascading crises, including health deterioration, income loss, food insecurity, and displacement. These impacts are particularly severe in contexts where social protection systems are fragmented or insufficient to buffer climate-related shocks.

Livelihood insecurity further exacerbates vulnerability in the Global South. A large proportion of the population relies on informal and climate-sensitive livelihoods, including agriculture, construction, street vending, and domestic work. Climate variability and extreme events can disrupt these livelihoods, leading to income loss, food insecurity, and displacement (Dodman et al., 2019). Informal workers typically lack access to insurance, social security, or formal safety nets, making recovery from climate shocks particularly challenging.

Climate-induced migration and displacement represent an increasingly important dimension of vulnerability. Extreme weather events and slow-onset climate stressors such as drought and sea-level rise can disrupt livelihoods and force temporary or permanent migration. While migration may function as an adaptive strategy for some groups, it often generates new vulnerabilities for older adults and those with limited mobility. Displacement can sever social networks, weaken informal support systems, and increase dependence on overstretched public services. In urban areas, climate-driven migration frequently contributes to the expansion of informal settlements, where residents face heightened exposure to flooding, heat stress, and environmental hazards (Dodman et al., 2019).

Health inequalities further amplify climate vulnerability across the Global South. Climate change exacerbates both communicable and non-communicable diseases, increases heat-related morbidity and mortality, and places additional strain on already under-resourced health systems (Haines & Ebi, 2019). Vulnerable populations often face barriers to accessing timely healthcare, including financial constraints, geographic distance, and inadequate service provision. Mental health impacts, such as stress, anxiety, and trauma associated with repeated exposure to climate hazards, displacement, and livelihood insecurity, add another layer of vulnerability that is frequently overlooked in adaptation planning.

India provides a particularly illustrative example of these intersecting vulnerabilities. The country faces rising temperatures, more frequent heatwaves, intensified monsoon variability, floods, droughts, and cyclones, all of which have significant social and economic consequences (Krishnan et al., 2020). These climate risks interact with deep-seated inequalities related to caste, gender, age, income, and urban–rural divides. Older adults, women, informal workers, and residents of informal settlements are among those most affected by climate-induced hazards, highlighting the need for targeted and inclusive adaptation strategies.

Understanding climate vulnerability as an intersectional and socially embedded phenomenon has important implications for climate adaptation policy. It highlights the need for strategies that prioritize populations most exposed and least able to adapt. In this context, artificial intelligence offers potential to improve climate risk assessment, early warning systems, and resource targeting. However, these benefits depend on whether AI systems are designed with attention to social inequality, data representation, and governance. Without such safeguards, AI applications may overlook or even reinforce existing vulnerabilities. The following section therefore examines current AI applications in climate adaptation, with particular attention to their relevance for vulnerable populations in the Global South.

3. AI APPLICATIONS FOR CLIMATE ADAPTATION AND RESILIENCE

Artificial intelligence has increasingly been positioned as a strategic tool for enhancing climate adaptation and resilience by enabling more anticipatory, data-driven, and targeted responses to climate risks. Unlike traditional climate information systems, which often rely on static models and delayed reporting, AI-enabled applications can process large and diverse datasets in near real time, allowing for improved forecasting, risk assessment, and decision support. In developing-country contexts, these capabilities offer potential pathways to overcome some of the informational and coordination challenges that constrain conventional adaptation efforts.

One of the most prominent applications of AI in climate adaptation is the development of early warning systems for extreme weather events. Machine learning algorithms can integrate meteorological data, satellite imagery, and historical climate records to improve the accuracy and timeliness of forecasts related to heatwaves, floods, cyclones, and droughts (Ren et al., 2021). Enhanced early warning systems can provide critical lead time for governments, health services, and communities to implement preventive measures such as evacuations, emergency healthcare preparedness, and resource mobilization. For vulnerable populations, timely warnings can significantly reduce mortality and injury, particularly in contexts where adaptive capacity is limited.

AI is also being used to support climate risk mapping and vulnerability assessment. By analyzing spatial, environmental, and socioeconomic data, AI-enabled tools can identify high-risk areas and populations, supporting more targeted adaptation planning and resource allocation (Rolnick et al., 2019). In the Global South, where data gaps and informal settlements often complicate risk assessment, AI-driven mapping can help visualize exposure and sensitivity patterns that might otherwise remain overlooked. When combined with demographic and health data, such tools have the potential to inform inclusive adaptation strategies that prioritize vulnerable groups.

Public health is another critical domain in which AI applications intersect with climate adaptation. Climate change influences the distribution and intensity of heat stress, air pollution, and vector-borne diseases, posing significant risks to population health. AI-enabled health surveillance systems can support early detection of climate-sensitive health threats by analyzing health records, environmental data, and mobility patterns (Haines & Ebi, 2019). For example, predictive models can be used to anticipate heat-related illness outbreaks or shifts in disease vectors, enabling health systems to prepare and respond more effectively. In developing-country contexts with constrained health infrastructure, such anticipatory approaches can enhance resilience and reduce the burden on already overstretched services.

Disaster response and recovery represent another area where AI is increasingly applied. AI-enabled systems can support rapid damage assessment using satellite imagery and remote sensing, enabling authorities to estimate the scale of impact and prioritize relief efforts more efficiently. Machine learning (ML) models can also optimize logistics and supply chain management during disaster response, improving the distribution of food, medical supplies, and emergency shelter (Vinuesa et al., 2020). For vulnerable populations, more efficient and targeted disaster response can reduce displacement duration, improve access to essential services, and support faster recovery.

Urban systems in the Global South present additional opportunities for AI-enabled climate adaptation. Rapid urbanization, often characterized by informal housing and inadequate infrastructure, increases exposure to climate risks such as flooding and extreme heat. AI applications can support urban climate adaptation by optimizing infrastructure planning, managing water and energy systems, and monitoring environmental conditions in real time. For example, AI-driven analytics can inform heat mitigation strategies in densely populated urban areas or support flood management through predictive modeling and sensor networks.

3.1 Limits of Technical Effectiveness Without Social and Institutional Integration

While artificial intelligence offers advanced capabilities for climate forecasting, risk mapping, and decision support, technical effectiveness alone does not guarantee meaningful reductions in vulnerability. In many Global South contexts, a persistent gap exists between the generation of climate risk information and the capacity to act upon it. Improved prediction accuracy does not automatically translate into protection for vulnerable populations if early warnings fail to reach those most at risk or if individuals lack the resources and institutional support necessary to respond. Research on climate adaptation has consistently shown that early warning systems are most effective when combined with preparedness planning, access to services, and social protection mechanisms (IPCC, 2022; Adger, 2006). Without these enabling conditions, AI-generated insights may have limited impact on reducing climate-related harm.

Institutional capacity and coordination play a decisive role in determining whether AI-enabled insights lead to effective adaptation outcomes. In many developing-country settings, climate governance is fragmented across multiple agencies responsible for environment, health, disaster management, urban planning, and social welfare. Weak coordination among these institutions can hinder the translation of AI-generated risk assessments into timely and coherent action (Dodman et al., 2019). Limited technical expertise, inadequate training, and resource constraints further restrict the ability of public institutions to interpret and operationalize AI outputs. As a result, even sophisticated AI systems may remain underutilized or disconnected from on-the-ground decision-making processes, reducing their contribution to resilience building (Vinuesa et al., 2020).

Social and political factors also shape the effectiveness of AI-enabled climate adaptation. Trust in technology, governance institutions, and information sources influences whether individuals and communities respond to AI-generated warnings and recommendations. In contexts characterized by social marginalization or weak state–citizen relations, automated systems may be viewed with skepticism, particularly when their functioning is opaque or externally driven. Language barriers, limited digital literacy, and unequal access to communication technologies further constrain the reach of AI-enabled alerts among vulnerable populations, including older adults, informal workers, and rural communities (Haines & Ebi, 2019). These barriers highlight that technological innovation alone cannot overcome deeply rooted social inequalities.

Another critical limitation concerns data representation and inclusivity. AI systems rely on large datasets that may inadequately capture informal settlements, marginalized groups, or remote regions where data collection is weak. When such gaps exist, AI-driven models may prioritize well-documented areas while overlooking populations with limited data visibility, leading to skewed risk assessments and uneven allocation of adaptation resources (Rolnick et al., 2019). This risk is particularly relevant in the Global South, where informality and data scarcity are widespread. Without deliberate efforts to incorporate socially disaggregated data and local knowledge, AI-enabled adaptation strategies may inadvertently reinforce existing inequalities rather than reduce them.

These limitations underscore the importance of embedding AI applications within broader socio-institutional systems rather than treating them as standalone technological solutions. Effective AI-enabled climate adaptation requires integration with climate governance frameworks, disaster management systems, public health infrastructure, and social protection mechanisms. It also necessitates participatory approaches that engage communities and local institutions in the design and deployment of AI tools (IPCC, 2022).

To illustrate how these applications operate in practice, the following subsection highlights several examples of AI-enabled climate adaptation initiatives emerging across the Global South.

3.2 Illustrative Examples of AI-Enabled Climate Adaptation in the Global South

Several emerging initiatives across the Global South demonstrate how artificial intelligence is being integrated into climate adaptation and resilience strategies, particularly in contexts characterized by high social vulnerability and institutional constraints.

In India, AI-enabled heatwave early warning systems have been developed to enhance the accuracy and timeliness of temperature forecasts and support coordinated public health responses. For example, predictive analytics and meteorological modelling have been incorporated into city-level heat action plans, enabling authorities to issue early warnings, prepare hospitals for heat-related illnesses, and conduct targeted outreach to high-risk populations such as older adults and individuals with chronic health conditions (Haines & Ebi, 2019; IPCC, 2022). These applications demonstrate the potential of AI to support anticipatory adaptation by reducing health risks associated with extreme heat events.

AI-based flood forecasting and cyclone prediction models represent another important area of application. By integrating satellite imagery, meteorological data, and historical climate records, AI systems can improve lead times for evacuation and disaster preparedness in flood-prone and coastal regions (Rolnick et al., 2019; Ren et al., 2021). Such systems have been piloted in several South Asian contexts where satellite imagery and hydrological data are used to generate near real-time flood forecasts for vulnerable river basin communities. In the Global South, where informal settlements and low-income communities are often located in high-risk areas, enhanced forecasting can support more timely evacuation planning and resource mobilization, thereby reducing displacement and loss of life (Dodman et al., 2019).

In the public health domain, AI-supported surveillance tools are increasingly explored to monitor climate-sensitive health risks, including heat stress, air pollution exposure, and vector-borne disease outbreaks. Predictive analytics applied to health and environmental data can assist authorities in anticipating climate-related health burdens and coordinating preventive interventions, particularly in resource-constrained health systems (Haines & Ebi, 2019; Vinuesa et al., 2020). For instance, AI models have been used to analyse temperature patterns, air pollution levels, and mosquito population dynamics to forecast potential outbreaks of climate-sensitive diseases. Collectively, these examples illustrate that the effectiveness of AI-enabled climate adaptation depends not only on technological capability but also on integration with public health systems, disaster management institutions, and inclusive governance frameworks.

4. POTENTIAL BENEFITS OF AI FOR VULNERABLE POPULATIONS

Artificial intelligence-enabled climate adaptation holds particular significance for vulnerable populations in the Global South, whose exposure to climate risks is intensified by social, economic, and institutional constraints. When designed and deployed with an explicit focus on inclusion and equity, AI applications can contribute to meaningful reductions in risk, improved access to services, and strengthened resilience among groups that are often underserved by conventional adaptation strategies.

One of the most immediate benefits of AI for vulnerable populations lies in improved access to timely and actionable climate risk information. Early warning systems powered by AI can provide advance notice of extreme weather events such as heatwaves, floods, and cyclones, allowing individuals and communities to take preventive action. For older adults, persons with disabilities, and those with limited mobility, timely warnings can be critical for evacuation planning, medical preparedness, and coordination with caregivers or support networks. In contexts where climate information is traditionally disseminated through fragmented or delayed channels, AI-enabled alerts if made accessible through appropriate formats and languages can significantly reduce mortality and injury risks.

AI-enabled health surveillance and telemedicine platforms can support early identification of climate-related health risks and psychological distress, enabling targeted interventions for vulnerable groups. Evidence from AI-supported elderly healthcare initiatives suggests that digital health tools, when ethically designed and supported by institutional frameworks, can enhance access to care and continuity of support for older adults in resource-constrained settings (Rafi, 2025).

AI-driven climate risk mapping also offers opportunities to improve the targeting of adaptation interventions and social protection measures. By identifying geographic areas and population groups that face heightened exposure and sensitivity, AI tools can support more equitable allocation of resources such as emergency shelters, cooling centers, healthcare services, and disaster relief assistance (Rolnick et al., 2019). For low-income households and informal settlements, which are often underrepresented in official data, inclusive risk mapping can help ensure that adaptation planning does not overlook those most in need.

Public health represents a critical domain in which AI-enabled climate adaptation can benefit vulnerable populations. Climate change exacerbates health risks associated with heat stress, air pollution, vector-borne diseases, and water-borne illnesses, disproportionately affecting older adults, children, and individuals with pre-existing health conditions (Haines & Ebi, 2019). AI-supported health surveillance systems can enable earlier detection of climate-sensitive health threats, allowing health authorities to mobilize preventive measures such as targeted outreach, medical preparedness, and community-level interventions. In resource-constrained health systems, anticipatory approaches supported by AI can enhance resilience by reducing peak demand and improving service coordination. Public health vulnerabilities associated with climate change extend beyond physical illness to include mental health and psychosocial stress, particularly among older adults and marginalized populations. Studies in the Indian context have documented risky health behaviours, stress, and declining mental well-being among elderly populations, especially under conditions of social disruption and limited support systems (Rafi & Saif, 2019; Rafi & Rafi, 2022).

Mental health and psychosocial well-being are increasingly recognized as important dimensions of climate vulnerability. Repeated exposure to climate-related disasters, displacement, livelihood loss, and uncertainty can contribute to stress, anxiety, and trauma, particularly among marginalized populations. While mental health impacts are often overlooked in climate adaptation planning, AI-enabled tools have the potential to support early identification of psychosocial stress patterns and inform targeted mental health interventions. When integrated into public health and social welfare systems, such tools can contribute to more holistic approaches to climate resilience that address both physical and mental well-being.

Livelihood security represents another domain where AI-enabled climate adaptation can benefit vulnerable populations. Informal workers and small-scale producers in the Global South often rely on climate-sensitive livelihoods and lack access to insurance or formal safety nets. AI applications supporting climate forecasting, risk assessment, and adaptive planning can help reduce livelihood disruptions by informing production decisions, work scheduling, and income diversification strategies. However, these systems must be designed carefully to avoid excluding populations with limited digital access while strengthening adaptive capacity among economically marginalized groups.

4.1 AI, Social Protection, and Welfare Targeting in Climate Adaptation

AI-enabled climate adaptation also holds significant potential to strengthen social protection and welfare delivery for vulnerable populations in the Global

South. Climate shocks frequently disrupt income, health, and housing security, increasing reliance on public assistance programs such as pensions, cash transfers, food distribution schemes, and emergency relief. However, conventional social protection systems often struggle to respond effectively to climate-induced needs due to fragmented data, administrative delays, and limited targeting capacity (Adger, 2006; IPCC, 2022). AI-supported analytics can help address these challenges by enabling more timely identification of households and individuals at heightened risk during climate-related events.

By integrating climate risk data with socioeconomic and demographic information, AI tools can support anticipatory social protection approaches that trigger assistance before shocks escalate into humanitarian crises. Predictive models can inform the temporary expansion of pension payments, food assistance, or healthcare support during heatwaves, floods, or droughts, particularly for older adults, persons with disabilities, and low-income households. Such anticipatory mechanisms align with broader climate adaptation goals by reducing negative coping strategies, including distress migration, asset depletion, and delayed healthcare seeking, which are commonly observed following climate shocks in developing-country contexts (IPCC, 2022; Dodman et al., 2019).

AI-enabled targeting also offers opportunities to improve the inclusiveness of welfare programs by identifying populations that are often underrepresented in official datasets, such as informal workers, residents of informal settlements, and socially marginalized groups. In contexts where informality and data gaps complicate conventional welfare delivery, AI-supported risk mapping and data integration can help reduce exclusion errors that limit the effectiveness of social protection interventions (Rolnick et al., 2019). When combined with local verification mechanisms and institutional oversight, such approaches can enhance equity in the allocation of climate adaptation resources.

For older populations, AI-supported social protection plays a particularly important role in climate adaptation. Climate-induced stressors such as extreme heat, displacement, and health emergencies disproportionately affect older adults who rely on fixed incomes or social pensions. Evidence from India highlights how age-related health vulnerabilities and limited access to care can intensify climate risks among older populations (Rafi & Saif, 2020; Rafi & Saif, 2021). AI-informed welfare delivery, when integrated with healthcare and disaster response systems, can help ensure continuity of support during climate shocks and reduce the risk of neglect or social isolation (Rafi, 2025).

Despite these opportunities, the application of AI in social protection for climate adaptation must be approached cautiously. Automated decision-making systems may reproduce existing social biases if underlying datasets reflect structural inequalities or if eligibility criteria are narrowly defined. Concerns related to transparency, accountability, and data governance are particularly salient when AI is used to determine access to essential welfare benefits (Vinuesa et al., 2020). Ensuring human oversight and rights-based governance is therefore essential. When embedded within inclusive social policy frameworks and aligned with public health and climate governance systems, AI-enabled social protection can contribute meaningfully to climate-resilient development by addressing both immediate climate risks and underlying social vulnerability.

At the community level, AI-enabled adaptation can support more coordinated and participatory responses to climate risks. By integrating data across sectors such as health, disaster management, urban planning, and social protection, AI systems can enhance coordination among institutions and improve the effectiveness of adaptation measures. For vulnerable communities, improved coordination can translate into faster response times, reduced duplication of efforts, and more consistent access to support during and after climate shocks.

However, it is important to emphasize that these benefits are not automatic. The extent to which AI-enabled climate adaptation improves outcomes for vulnerable populations depends on governance arrangements, institutional capacity, and the inclusiveness of design and implementation processes. AI systems that are developed without input from affected communities or that rely on biased or incomplete data may fail to capture the realities of vulnerability in the Global South. Moreover, benefits may accrue disproportionately to populations with better access to digital technologies, reinforcing existing inequalities if safeguards are not in place (Vinuesa et al., 2020).

In summary, AI-enabled climate adaptation offers a range of potential benefits for vulnerable populations, including improved access to risk information, enhanced public health preparedness, strengthened livelihood resilience, and more equitable resource allocation. When embedded within inclusive governance frameworks and aligned with social protection and public health systems, AI has the potential to support people-centered adaptation strategies that reduce vulnerability and enhance resilience in the Global South. The following section critically examines the equity, ethical, and governance challenges that must be addressed to ensure that these potential benefits are realized in a sustainable and inclusive manner.

5. EQUITY, ETHICAL, AND GOVERNANCE CHALLENGES OF AI-ENABLED CLIMATE ADAPTATION

While artificial intelligence offers promising tools for enhancing climate adaptation and resilience, its deployment in the Global South raises a range of equity, ethical, and governance challenges that must be carefully addressed to ensure sustainable and inclusive outcomes. These challenges stem not only from technological limitations but also from broader structural inequalities, institutional constraints, and power dynamics that shape how AI systems are designed, implemented, and governed.

A central concern relates to the digital divide that characterizes many developing-country contexts. Access to digital infrastructure, reliable electricity, internet connectivity, and digital literacy remains uneven across regions and social groups. Vulnerable populations including older adults, women, rural communities, and low-income households are often least likely to have access to digital technologies that underpin AI-enabled climate adaptation systems (Vinuesa et al., 2020). As a result, early warning systems, mobile-based alerts, or AI-supported decision tools may fail to reach those most at risk, limiting their effectiveness and potentially exacerbating existing inequalities.

Closely linked to issues of access are concerns regarding data availability and representation. AI systems depend on large volumes of data to generate accurate predictions and insights. In the Global South, data on informal settlements, marginalized populations, and non-market activities are frequently incomplete, outdated, or entirely absent. This lack of representative data can shape how risks are identified and prioritized, with important implications for equity in AI-enabled climate adaptation (Jobin et al., 2019). In the context of climate adaptation, such biases may result in inadequate protection for vulnerable groups or skewed prioritization of adaptation investments.

Transparency and accountability present additional ethical challenges. Many AI systems operate as "black boxes," producing outputs that are difficult for users, policymakers, or affected communities to interpret or contest. In climate adaptation and disaster management contexts, opaque decision-making can undermine trust, reduce public acceptance, and limit the ability of stakeholders to challenge or correct flawed outcomes. This is particularly problematic when AI-generated insights influence decisions related to evacuation, resource allocation, or access to social protection. Ensuring transparency, explainability, and accountability is therefore essential for responsible AI deployment in climate governance (Floridi et al., 2018).

Governance capacity represents another critical constraint in the Global South. Effective use of AI for climate adaptation requires institutional frameworks that can regulate data use, protect privacy, ensure ethical standards, and coordinate action across sectors such as environment, health, urban planning, and disaster management. In many developing countries, governance systems remain fragmented, with limited regulatory oversight and unclear lines of responsibility. This can lead to uncoordinated implementation of AI tools, duplication of efforts, or reliance on external technology providers without adequate safeguards or local capacity-building.

Environmental sustainability concerns further complicate the use of AI for climate action. AI systems, particularly those based on large-scale machine learning and data processing, can be energy-intensive and contribute to greenhouse gas emissions if powered by carbon-intensive energy sources (Strubell et al., 2019). In contexts where energy systems are already strained, the expansion of AI infrastructure may create trade-offs between technological innovation and environmental goals. Addressing the environmental footprint of AI is therefore an integral part of sustaining climate action rather than a peripheral concern.

Ethical considerations also extend to issues of participation and consent. AI-enabled climate adaptation initiatives are often designed and implemented by technical experts, government agencies, or private-sector actors, with limited involvement of affected communities. Excluding vulnerable populations from decision-making processes can result in solutions that are poorly aligned with local needs, cultural contexts, or lived experiences. Participatory approaches that involve communities in the design, deployment, and evaluation of AI systems are essential for ensuring legitimacy, relevance, and social acceptance. Ethical and governance challenges associated with AI deployment are particularly salient when addressing vulnerable populations such as older adults. Analyses of India's legal and institutional frameworks reveal gaps in protection, accountability, and enforcement mechanisms related to elderly welfare, which may be further strained by the introduction of AI-driven decision systems in public services (Rafi, 2019).

5.1 Algorithmic Bias, Data Inequality, and the Risk of Exclusion

Beyond general concerns regarding data availability, algorithmic bias represents a critical ethical challenge in AI-enabled climate adaptation. AI systems are shaped not only by the quantity of data available but also by the social assumptions embedded in data collection practices, model design, and variable selection. In the Global South, marginalized populations such as informal settlement residents, older adults, women, persons with disabilities, and informal workers are frequently underrepresented or misrepresented in official datasets. When AI models are trained on such incomplete or skewed data, they may systematically underestimate risks faced by these groups or exclude them from priority adaptation measures (Jobin et al., 2019; Vinuesa et al., 2020).

In climate adaptation contexts, algorithmic bias can have tangible consequences. Risk maps may fail to capture informal housing areas, early warning systems may prioritize regions with better sensor coverage, and welfare targeting mechanisms may exclude individuals lacking digital records. For vulnerable populations, these exclusion errors can result in delayed assistance, inadequate protection, or complete invisibility within adaptation planning processes. Addressing algorithmic bias therefore requires more than technical refinement; it necessitates deliberate efforts to incorporate socially disaggregated data, qualitative insights, and local knowledge into AI systems used for climate governance.

5.2 Energy Consumption and the Environmental Footprint of AI

A further ethical concern relates to the environmental sustainability of AI technologies themselves. While AI is often promoted as a tool for addressing climate risks, many AI systems rely on energy-intensive computing infrastructure, including large data centers and continuous data processing (Strubell et al., 2019). In regions where electricity grids are already under strain and heavily dependent on fossil fuels, the expansion of AI infrastructure may contribute to greenhouse gas emissions, creating a paradox in which technologies intended to support climate action generate additional environmental burdens.

For the Global South, these concerns are particularly salient. Limited access to renewable energy, unreliable power supply, and competing development priorities can constrain the sustainable deployment of AI systems. Future AI-enabled climate adaptation initiatives must therefore consider energy efficiency, low-resource computing models, and alignment with renewable energy strategies. Evaluating the full lifecycle environmental impacts of AI from data collection to deployment is essential to ensure that AI contributes to long-term climate resilience rather than undermining it (IPCC, 2022).

5.3 Accountability, Explainability, and Public Trust

Accountability remains a central governance challenge in AI-enabled climate adaptation. Decisions informed by AI systems such as evacuation orders, allocation of disaster relief, or prioritization of adaptation investments carry significant social consequences. Yet the opacity of many AI models limits the ability of policymakers, practitioners, and affected communities to understand how decisions are made or to contest erroneous outcomes. In contexts of disaster response or social protection, lack of explainability can erode trust and exacerbate perceptions of injustice (Floridi et al., 2018).

Ensuring accountability requires mechanisms that allow AI-generated recommendations to be scrutinized, questioned, and overridden when necessary. Human oversight, clear lines of responsibility, and transparent governance frameworks are essential, particularly when AI tools are integrated into public decision-making systems. For vulnerable populations with limited institutional voice, the ability to challenge AI-driven decisions is critical to preventing harm and maintaining legitimacy.

5.4 Participation, Co-Design, and Community Voice

Meaningful participation is a cornerstone of ethical AI deployment in climate adaptation. However, many AI-enabled initiatives remain expert-driven, with limited involvement of affected communities in system design or evaluation. Excluding local voices can result in solutions that fail to reflect lived realities, cultural practices, or practical constraints faced by vulnerable populations. Participatory and co-design approaches, which involve communities throughout the AI lifecycle, can enhance relevance, legitimacy, and effectiveness.

For older adults and marginalized groups, participation is especially important. Their experiences of climate risk, caregiving, health challenges, and mobility constraints may not be adequately captured through quantitative data alone. Integrating participatory methods with AI development can help ensure that technological solutions complement, rather than displace, existing community-based coping strategies and support systems (Adger, 2006).

5.5 Power Asymmetries and Technological Dependency

Power relations shape who designs, controls, and benefits from AI technologies used in climate adaptation. In many Global South contexts, AI systems are developed by external actors, including multinational corporations, international organizations, or Global North research institutions. This raises concerns about data ownership, accountability, and local capacity-building. Dependence on externally developed AI systems may limit local agency and reduce the flexibility of governments and communities to adapt technologies to context-specific needs (Vinuesa et al., 2020).

Addressing these asymmetries requires investment in local technical capacity, transparent data governance arrangements, and policies that prioritize domestic innovation ecosystems. Strengthening local ownership of AI tools is essential for ensuring that AI-enabled climate adaptation supports long-term resilience rather than reinforcing dependency.

Finally, there is a risk that AI-driven climate adaptation strategies may reinforce a form of techno-solutionism that prioritizes technological fixes over structural reforms. While AI can enhance information and decision-making, it cannot substitute for investments in basic infrastructure, healthcare, education, and social protection that are foundational to resilience. Overreliance on AI without addressing underlying drivers of vulnerability may lead to short-term gains without achieving long-term, transformative adaptation. From a social work and policy perspective, responsible AI implementation requires integration with existing care, legal, and welfare structures rather than parallel technological systems. Recent scholarship on AI in social work and psychology education emphasizes the need for ethical training, interdisciplinary collaboration, and human-centered governance models to ensure that AI enhances rather than undermines care and social justice objectives (Rafi, 2026). Recognizing these challenges does not diminish the potential of AI for climate adaptation; rather, it highlights the conditions under which AI can be deployed responsibly and equitably.

In summary, equity, ethical, and governance challenges pose significant barriers to the effective and responsible use of AI for climate adaptation in the Global South. Addressing these challenges requires moving beyond narrow technical considerations and adopting a holistic approach that integrates ethical principles, inclusive governance, and environmental sustainability. The following section builds on this analysis by proposing a policy-oriented framework for inclusive and responsible AI-enabled climate adaptation that centers vulnerable populations and supports sustained climate action.

6. TOWARD A POLICY FRAMEWORK FOR INCLUSIVE AI-ENABLED CLIMATE ADAPTATION

To ensure that artificial intelligence contributes meaningfully to sustained and equitable climate adaptation in the Global South, it is essential to move beyond ad hoc or technology-centric interventions and toward a coherent policy framework that integrates social inclusion, ethical governance, and environmental sustainability. Such a framework must recognize AI as a socio-technical tool embedded within broader climate governance systems rather than as a standalone solution (Adger, 2006; IPCC, 2022). This section proposes a policy-oriented framework for inclusive AI-enabled climate adaptation that centers vulnerable populations and supports long-term resilience.

The proposed policy framework is organized around five interconnected policy pillars that support inclusive AI-enabled climate adaptation in the Global South: (1) human-centered and inclusive design, (2) integration with social protection and public health systems, (3) ethical governance and accountability, (4) environmental sustainability of AI systems, and (5) institutional capacity building with multi-level governance. Together, these pillars provide a governance-oriented structure for embedding AI technologies within inclusive climate adaptation systems in the Global South.

6.1 Human-Centered and Inclusive Design

At the core of inclusive AI-enabled climate adaptation lies the principle of human-centered design. AI systems should be developed with explicit consideration of the needs, capacities, and lived experiences of vulnerable populations, including older adults, women, informal workers, and low-income households. Participatory approaches that involve communities in the design, testing, and evaluation of AI applications are essential for ensuring relevance, accessibility, and legitimacy (Adger, 2006; Vinuesa et al., 2020). Community engagement can help ensure that AI tools address locally relevant risks, use accessible communication formats, and align with cultural and social contexts.

Inclusive design also requires addressing barriers related to digital access and literacy. Policymakers should prioritize complementary investments in digital infrastructure, affordable connectivity, and capacity-building initiatives to ensure that AI-enabled adaptation tools are accessible to marginalized groups. Without such measures, AI systems risk disproportionately benefiting populations with existing digital advantages, thereby undermining equity objectives (Jobin et al., 2019).

6.2 Integration with Social Protection and Public Health Systems

AI-enabled climate adaptation is most effective when integrated with existing social protection, public health, and disaster management systems rather than operating as isolated technological platforms. Linking AI-driven early warning systems with health services can support the activation of heat-health action plans, while integration with social welfare programs can enable anticipatory cash transfers or food assistance during climate shocks (Haines & Ebi, 2019; IPCC, 2022).

Integrating AI with public health systems is particularly important in the Global South, where climate change increasingly affects disease patterns and health outcomes. Predictive analytics can support anticipatory health planning, but only if health institutions have the capacity, authority, and coordination mechanisms to act on AI-generated insights. Strengthening institutional linkages and data-sharing protocols across sectors is therefore a critical component of inclusive adaptation policy (Dodman et al., 2019).

6.3 Ethical Governance and Accountability

Robust governance frameworks are essential for ensuring the ethical and responsible use of AI in climate adaptation. Policymakers must establish clear standards for data governance, privacy protection, transparency, and accountability, including clear attribution of responsibility for AI-informed decisions and mechanisms for redress when harms occur (Floridi et al., 2018; Vinuesa et al., 2020).

Transparency and explainability should be prioritized, particularly when AI outputs influence decisions related to evacuation, disaster relief, or access to social support. Open and interpretable AI systems can enhance public trust, enable scrutiny, and support more democratic forms of climate governance. Regulatory frameworks should also ensure that private-sector involvement in AI-enabled climate adaptation aligns with public interest objectives and does not exacerbate inequality or technological dependency (Jobin et al., 2019).

6.4 Environmental Sustainability of AI Systems

Sustaining climate action with AI requires explicit attention to the environmental footprint of AI technologies themselves. AI systems, particularly those based on large-scale machine learning, can be energy-intensive and may contribute to greenhouse gas emissions if powered by carbon-intensive energy sources (Strubell et al., 2019). Policymakers should therefore encourage the development of energy-efficient AI models and promote the use of renewable energy to power data infrastructure.

Environmental sustainability criteria should be incorporated into public procurement processes for AI systems used in climate adaptation, ensuring alignment between technological innovation and climate mitigation goals. In resource-constrained contexts, prioritizing low-energy and context-appropriate AI solutions can help balance technological benefits with environmental considerations (IPCC, 2022).

6.5 Capacity Building and Institutional Strengthening

Within this framework, effective implementation of AI-enabled climate adaptation depends on coordinated action across multiple actors, including national governments, local authorities, public health institutions, civil society organizations, and technology providers. Governments in the Global South often face constraints related to technical expertise, financial resources, and coordination across agencies. Capacity-building initiatives including training programs, interdisciplinary collaboration, and partnerships with academic and civil society organizations are essential for enabling institutions to effectively interpret, use, and govern AI systems (Vinuesa et al., 2020).

Local governments and community-based organizations play a particularly important role in translating AI insights into action. Strengthening local capacity can enhance responsiveness, accountability, and alignment with community needs, thereby improving adaptation outcomes for vulnerable populations (Dodman et al., 2019).

6.6 Alignment with Global Climate and Development Agendas

Finally, inclusive AI-enabled climate adaptation should be aligned with broader global climate and development frameworks, including the Sustainable Development Goals and international climate agreements. Aligning AI initiatives with these agendas can help ensure policy coherence, mobilize resources, and promote shared learning across regions (IPCC, 2022). Such alignment reinforces the normative commitment to equity, justice, and sustainability that underpins long-term climate action in the Global South.

6.7 Local Governance, Implementation, and Accountability Mechanisms

Effective implementation of AI-enabled climate adaptation requires strong local governance mechanisms. Municipal governments, district administrations, and community-level institutions are often the first responders to climate risks and therefore play a central role in translating AI-generated insights into timely and context-sensitive action. Research on climate governance in the Global South emphasizes that localized institutions are better positioned to respond to differentiated vulnerabilities and contextual realities than centralized systems alone (Adger, 2006; IPCC, 2022).

Local institutions serve as critical intermediaries between technological systems and lived experiences. They are often responsible for identifying vulnerable households, coordinating health and welfare services, and ensuring that early warning information reaches populations with limited digital access, including older adults, women, and marginalized communities. Embedding AI-enabled tools within decentralized planning and decision-making processes can therefore enhance relevance, responsiveness, and social legitimacy, particularly in contexts where centralized AI systems may overlook informal settlements or socially excluded groups (Dodman et al., 2019).

Accountability mechanisms at the local level are equally important for ethical AI governance. Communities should have accessible channels to question, contest, and provide feedback on AI-informed decisions related to evacuation planning, disaster relief distribution, or welfare targeting. Such feedback mechanisms are essential for identifying errors, mitigating bias, and improving system performance over time. Transparency regarding how AI systems function and how decisions are made can further strengthen public trust and acceptance, especially when AI tools influence access to critical public services (Floridi et al., 2018; Vinuesa et al., 2020).

Partnerships between local governments, civil society organizations, academic institutions, and technology providers can further support inclusive implementation. Collaborative arrangements can facilitate capacity building, knowledge exchange, and participatory design processes that align AI-enabled climate adaptation with community priorities and ethical standards. From a social policy and social work perspective, such partnerships are particularly important for ensuring that AI systems complement existing care, welfare, and support structures rather than operating as parallel or exclusionary mechanisms (Rafi, 2019; Rafi, 2026).

By grounding AI-enabled climate adaptation within robust local governance and accountability frameworks, policymakers can enhance both the effectiveness and equity of climate action. Localized implementation that integrates technological innovation with institutional capacity, community participation, and ethical oversight offers a more sustainable pathway for leveraging AI to support resilient and inclusive climate adaptation in the Global South.

Taken together, these policy pillars constitute a governance-oriented framework for inclusive AI-enabled climate adaptation. The framework emphasizes that technological innovation must be accompanied by institutional coordination, ethical oversight, social protection integration, and local participation in order to generate meaningful reductions in climate vulnerability. By linking AI capabilities with governance capacity and social inclusion, this framework provides a practical foundation for policymakers seeking to harness AI for sustained and equitable climate adaptation in the Global South. At the same time, significant research gaps remain regarding the long-term effectiveness, governance, and inclusiveness of AI-enabled climate adaptation initiatives. The following section therefore outlines key directions for future research.

7. FUTURE RESEARCH DIRECTIONS AND CONCLUSION

7.1 Limitations

This chapter has several limitations that should be acknowledged. The analysis is primarily based on secondary literature, policy documents, and illustrative examples rather than original empirical data, which may limit the ability to assess the on-ground effectiveness of specific AI-enabled climate adaptation initiatives. While India is used as a contextual anchor, the diversity of social, institutional, and climatic conditions across the Global South means that findings may not be uniformly generalizable. In addition, the rapidly evolving nature of artificial intelligence technologies implies that applications and governance frameworks discussed here may change over time, underscoring the need for ongoing empirical evaluation.

7.2 Future Research Directions

While artificial intelligence offers significant opportunities to enhance climate adaptation and resilience in the Global South, substantial knowledge gaps remain regarding its long-term effectiveness, inclusiveness, and sustainability. Addressing these gaps is essential for advancing both scholarly understanding and policy practice in AI-enabled climate action.

First, future research should place greater emphasis on empirical evaluation of AI-enabled adaptation outcomes for vulnerable populations. Much of the existing literature focuses on technological capability or pilot initiatives, with limited assessment of how AI systems actually affect exposure, sensitivity, and adaptive capacity among marginalized groups. Longitudinal and mixed-methods studies are needed to examine whether AI-driven interventions lead to sustained reductions in vulnerability or whether benefits are unevenly distributed across social groups.

Second, there is a need for deeper investigation into the intersection of AI, climate adaptation, and social inequality. Future research should examine how age, gender, health status, disability, livelihood type, and socioeconomic position shape access to and outcomes of AI-enabled adaptation initiatives. Such analyses can help identify mechanisms through which AI may inadvertently reinforce inequality and inform the design of more inclusive systems.

Third, research on governance and institutional capacity is critical. Comparative studies across Global South contexts can provide insights into how different governance arrangements, regulatory frameworks, and institutional capacities influence the effectiveness and ethical deployment of AI for climate adaptation. Particular attention should be paid to the role of local governments, community-based organizations, and public–private partnerships in translating AI-generated insights into actionable and equitable outcomes.

Fourth, future scholarship should address the environmental sustainability of AI technologies within climate adaptation strategies. While AI is often framed as a tool for addressing climate risks, its own energy and resource requirements raise important questions about long-term sustainability. Research is needed to explore low-energy AI models, context-appropriate technological solutions, and the integration of AI systems with renewable energy infrastructure in developing-country settings.

Finally, interdisciplinary research that bridges technical, social, and policy perspectives remains essential. Collaboration between computer scientists, climate researchers, social scientists, public health experts, and policymakers can support more holistic approaches to AI-enabled climate adaptation. Such collaboration can also facilitate the co-production of knowledge with affected communities, ensuring that research agendas are aligned with lived realities and practical needs.

Beyond these thematic priorities, future research must also pay closer attention to how AI-enabled climate adaptation research is designed, governed, and conducted in practice.

Future research should also prioritize context-sensitive and participatory methodological approaches, particularly in Global South settings where vulnerability is shaped by local social structures, informal institutions, and lived experiences.

Community-based and participatory research designs can help ensure that AI-enabled climate adaptation initiatives reflect local knowledge, cultural contexts, and differentiated needs, rather than relying solely on top-down or externally developed models (Adger, 2006; IPCC, 2022). Such approaches are especially important when working with older adults, women, and marginalized groups whose vulnerabilities may be poorly captured in conventional datasets.

There is also a need for greater integration of AI research with public health, social protection, and social work perspectives. Climate adaptation research has often focused on infrastructure and environmental systems, while underexamining human well-being, care needs, and psychosocial impacts. Future studies should explore how AI-enabled climate adaptation can be embedded within health systems, welfare programs, and community care structures to address both immediate climate risks and underlying social vulnerability (Haines & Ebi, 2019; Rafi, 2025). This includes examining how AI tools interact with caregiving arrangements, social pensions, and community health services in ageing societies.

In addition, future research should critically examine power relations and knowledge asymmetries in the development and deployment of AI technologies for climate adaptation. Many AI systems used in Global South contexts are designed, owned, or governed by actors based in the Global North, raising concerns about data ownership, accountability, and contextual relevance (Vinuesa et al., 2020). Research that foregrounds Global South leadership, locally grounded innovation, and equitable data governance is essential to avoid forms of technological dependency or data colonialism.

In addition, longitudinal and comparative research across regions and sectors can support learning about what works, for whom, and under what conditions. By combining technical evaluation with social, institutional, and ethical analysis, future scholarship can contribute to the development of AI-enabled climate adaptation strategies that are not only innovative but also inclusive, sustainable, and responsive to the realities of vulnerable populations in the Global South.

8. CONCLUSION

Climate change poses profound and persistent challenges for societies in the Global South, where climate risks intersect with structural inequalities, fragile institutions, and limited adaptive capacity. Vulnerable populations including older adults, women, informal workers, and low-income households face disproportionate exposure to climate hazards and often lack the resources necessary to cope with or recover from climate-induced shocks. Addressing these challenges requires adaptation strategies that are not only technologically innovative but also socially inclusive, ethically grounded, and institutionally robust.

This chapter has examined the role of artificial intelligence in enabling climate adaptation and resilience for vulnerable populations in the Global South. By situating AI within broader socio-technical and governance frameworks, the chapter has highlighted both the opportunities and limitations of AI-enabled climate adaptation. AI applications such as early warning systems, climate risk mapping, public health surveillance, and disaster response tools offer meaningful potential to enhance preparedness, improve targeting of resources, and strengthen resilience. However, these benefits are neither automatic nor guaranteed. This chapter contributes to existing scholarship by integrating AI-enabled climate adaptation with ageing, health, and social protection perspectives that remain underexamined in predominantly technology-driven climate discourse.

Equity, ethical, and governance challenges including digital divides, data bias, lack of transparency, institutional constraints, and environmental footprints pose significant risks to the sustainable and inclusive use of AI in climate adaptation. Without deliberate attention to these issues, AI-driven solutions may reinforce existing inequalities or produce new forms of vulnerability. The policy framework proposed in this chapter emphasizes human-centered design, integration with social protection and public health systems, ethical governance, environmental sustainability, and capacity building as essential pillars for inclusive AI-enabled climate adaptation.

Ultimately, sustaining climate action with AI requires a shift away from techno-solutionist narratives toward a more holistic understanding of AI as a supportive instrument within broader transformations toward resilience, equity, and sustainability. When embedded within inclusive governance structures and aligned with social justice principles, AI can contribute to long-term climate adaptation strategies that protect vulnerable populations and support sustainable development in the Global South. This perspective reinforces the central objective of sustaining climate action with AI by emphasizing long-term, inclusive, and governance-driven adaptation rather than short-term technological solutions.

AI can support sustained climate adaptation in the Global South only when technological innovation is embedded within inclusive institutions, accountable governance systems, and socially responsive policy frameworks.

List of Abbreviations

AI – Artificial Intelligence
IPCC – Intergovernmental Panel on Climate Change
ML – Machine Learning

REFERENCES

Adger, W. N. (2006). Vulnerability. *Global Environmental Change, 16*(3), 268–281. DOI: 10.1016/j.gloenvcha.2006.02.006

Alston, M. (2015). *Women and climate change in Bangladesh.* Routledge. DOI: 10.4324/9781315774589

Cutter, S. L., Boruff, B. J., & Shirley, W. L. (2012). Social vulnerability to environmental hazards. In *Hazards vulnerability and environmental justice* (pp. 115–132). Routledge. DOI: 10.4324/9781849771542

Dodman, D., Archer, D., & Satterthwaite, D. (2019). Responding to climate change in contexts of urban poverty and informality. *Environment and Urbanization, 31*(1), 3–12. DOI: 10.1177/0956247819830004

Floridi, L., Cowls, J., Beltrametti, M., Chatila, R., Chazerand, P., Dignum, V., Luetge, C., Madelin, R., Pagallo, U., Rossi, F., Schafer, B., Valcke, P., & Vayena, E. (2018). AI4People—An ethical framework for a good AI society: Opportunities, risks, principles, and recommendations. *Minds and Machines, 28*(4), 689–707. DOI: 10.1007/s11023-018-9482-5 PMID: 30930541

Haines, A., & Ebi, K. (2019). The imperative for climate action to protect health. *The New England Journal of Medicine, 380*(3), 263–273. DOI: 10.1056/NEJMra1807873 PMID: 30650330

IPCC. (2022). Impacts, adaptation and vulnerability. Contribution of Working Group II to the Sixth Assessment Report of the Intergovernmental Panel on Climate Change. O., Roberts, DC, Tignor, M., Poloczanska, ES, Mintenbeck, K., Alegría, A., Craig, M., Langsdorf, S., Löschke, S., Möller, V., et al., Eds, 3056.

Jobin, A., Ienca, M., & Vayena, E. (2019). The global landscape of AI ethics guidelines. *Nature Machine Intelligence, 1*(9), 389–399. DOI: 10.1038/s42256-019-0088-2

Krishnan, R., Sanjay, J., Gnanaseelan, C., Mujumdar, M., Kulkarni, A., & Chakraborty, S. (2020). Assessment of climate change over the Indian region. A Report of the Ministry of Earth Sciences (MoES), Government of India, 13-17.

Rafi, S. (2019). Critical analysis of the legal approach to protect the elderly population in India. *BSSS J Soc Work, 9*, 1–20.

Rafi, S. (2020). Age-friendly communities: creating a conducive environment for the elderly in India. Journal of Geriatric Medicine.

Rafi, S. (2026). Artificial intelligence and telemedicine in elderly healthcare: A mixed-methods study. *BMC Geriatrics*, *26*(1), 71. DOI: 10.1186/s12877-025-06872-y PMID: 41402744

Rafi, S. (2026). AI in Social Work, Counseling, and Psychology Education: Innovations, Ethics, and Educational Pathways. In AI-Powered Learning in Social Work, Counseling, and Psychology (pp. 1-24). IGI Global Scientific Publishing.

Rafi, S., & Rafi, S. (2022). Mental health and stress in the digital workplace. In *Managing the digital workplace in the post-pandemic* (pp. 26–40). Routledge. DOI: 10.4324/9781003283386-4

Rafi, S., & Saif, S. (2019). *Risky Health Behaviours among Elderly in India*. Today Publication.

Rafi, S., & Saif, S. (2021). Physical and mental health issues of elderly women in India: a review. Handbook of Research on Policies, Protocols, and Practices for Social Work in the Digital World, 376-392. DOI: 10.4018/978-1-7998-7772-1.ch021

Ren, X., Li, X., Ren, K., Song, J., Xu, Z., Deng, K., & Wang, X. (2021). Deep learning-based weather prediction: A survey. *Big Data Research*, *23*, 100178. DOI: 10.1016/j.bdr.2020.100178

Rolnick, D., Donti, P. L., Kaack, L. H., Kochanski, K., Lacoste, A., Sankaran, K., Ross, A. S., Milojevic-Dupont, N., Jaques, N., Waldman-Brown, A., Luccioni, A. S., Maharaj, T., Sherwin, E. D., Mukkavilli, S. K., Kording, K. P., Gomes, C. P., Ng, A. Y., Hassabis, D., Platt, J. C., & Bengio, Y. (2019). Tackling climate change with machine learning. *ACM Computing Surveys*, *55*(2), 1–96. DOI: 10.1145/3485128

Strubell, E., Ganesh, A., & McCallum, A. (2019, July). Energy and policy considerations for deep learning in NLP. In *Proceedings of the 57th annual meeting of the association for computational linguistics* (pp. 3645-3650). DOI: 10.18653/v1/P19-1355

Vicedo-Cabrera, A. M., Sera, F., Liu, C., Armstrong, B., Milojevic, A., Guo, Y., ... & Gasparrini, A. (2021). Short term association between ozone and mortality: global two stage time series study in 406 locations in 20 countries. bmj, 368.

Vinuesa, R., Azizpour, H., Leite, I., Balaam, M., Dignum, V., Domisch, S., Felländer, A., Langhans, S. D., Tegmark, M., & Fuso Nerini, F. (2020). The role of artificial intelligence in achieving the Sustainable Development Goals. *Nature Communications*, *11*(1), 233. DOI: 10.1038/s41467-019-14108-y PMID: 31932590

KEY TERMS AND DEFINITIONS

Artificial Intelligence (AI): Artificial intelligence refers to computational systems and techniques, including machine learning and predictive analytics, that enable machines to analyze large and complex datasets, identify patterns, and support decision-making processes relevant to climate risk assessment and adaptation planning.

Climate Adaptation: Climate adaptation involves adjustments in social, economic, and environmental systems in response to actual or expected climate impacts, aimed at reducing vulnerability, enhancing resilience, and minimizing harm to human and natural systems.

Climate Governance: Climate governance encompasses the institutional arrangements, policies, regulations, and decision-making processes through which climate mitigation and adaptation strategies are planned, implemented, and monitored at local, national, and global levels.

Climate Resilience: Climate resilience is the capacity of individuals, communities, institutions, and systems to anticipate, absorb, recover from, and adapt to climate-related shocks and stresses while maintaining essential functions and structures.

Digital Divide: The digital divide denotes disparities in access to digital technologies, connectivity, and digital literacy, which can limit the ability of marginalized populations to benefit from AI-enabled climate adaptation initiatives.

Early Warning Systems: Early warning systems are integrated mechanisms that use data, forecasting models, and communication channels to provide timely alerts about impending climate-related hazards, enabling preventive and protective actions.

Global South: The Global South encompasses regions in Africa, Asia, Latin America, and parts of the Middle East characterized by lower and middle-income economies, historical development inequalities, and heightened vulnerability to climate change impacts.

Responsible AI: Responsible AI refers to the design, development, and deployment of artificial intelligence systems in ways that ensure fairness, transparency, accountability, inclusivity, and environmental sustainability, particularly when addressing socially sensitive challenges such as climate adaptation.

Social Vulnerability: Social vulnerability describes the degree to which social, economic, and institutional factors influence the ability of individuals and communities to withstand, respond to, and recover from climate-related hazards.

Vulnerable Populations: Vulnerable populations refer to social groups that face heightened exposure and sensitivity to climate risks due to factors such as age, gender, health status, poverty, disability, and limited access to resources, services, or decision-making processes.

Chapter 11

An Integrated SOR-TAM Framework Using PLS-SEM Evidence for AI-Driven Climate Action in India:

From Prediction to Preservation

Piali Haldar
http://orcid.org/0000-0003-4729-6964
Brainware University, Kolkata, India

Dev Kumar Mandal
http://orcid.org/0009-0005-3740-6885
Brainware University, Kolkata, India

ABSTRACT

This chapter aims to address Artificial Intelligence and Climate Action, investigating the integration of the SOR model into the dissent technology acceptance model when applied to the Indian endeavours in the mentioned specific application of AI climate intelligence. AI as a stimulus is said to lead to impacts on cognitive and emotional states, especially perceived usefulness, ease of use, trust, and awareness of the environment in a way that would lead to more environmentally friendly behaviour. As a way of evidence for the above hypothesis, we evaluate the potential link between the variables in focus using a structural equation approach, which is known as Partial Least Squares Path Modelling (PLS-SEM) in survey data from 420 samples. The study findings indicate that both trust and sustainable culture

DOI: 10.4018/979-8-3373-9978-2.ch011

indeed influence the adoption of AI, while other factors, such as digital literacy and environmental orientation, predict more positive behavioural results and overall sustainability development recommendations.

1. INTRODUCTION

Climate change is undoubtedly one of the most urgent global issues of the 21st century for its tremendous impact on environmental conditions, economic stability, and quality of life. All countries these days have to deal more and more with the wrath of rising temperatures, unpredictable weather patterns, declining flora and fauna, and even the exhaustion of some very basic natural resources (Kushawaha et al., 2021). In the case of India, rapidly advancing developing countries in recent years pose certain challenges to the overall growth of the country. Due to its sizable population, the country is in the middle of a technological transformation and has a high dependency on sectors like agriculture that are vulnerable to the climate. India finds itself mired in very high levels of the risk parameters for climatic disaster (Lee et al., 2023). Thus, the adoption of new development models is of paramount importance, especially the ones that are most promising in terms of moderate greenhouse gas emissions.

In recent times, AI, a change-making technology on Earth, has caught the attention of leadership because it provides strategic solutions to tough environmental problems (Whig et al., 2025). It is worth noting that powered tools engage and process vast amounts of data, evaluate potential probability, and make informed decisions in many areas involving management of energy transfers, agriculture, hazards forecasting, and even staff scheduling. For example, it has been argued that artificial intelligence at a certain rate might improve while damaging floods and other extreme catastrophes are worse, leading to civil food insufficiency and social distress (Attia, 2025; Al Marzooqi, 2024; Rolnick et al., 2019). Such specific uses highlight the ability of AI to take on developmental responses in this context: instead of simply finding short-term solutions to disasters, optimal strategies may be put in place to avoid these problems altogether. However, AI's contribution in making outcomes sustainable is not confined to only technological advancements, but also the acceptance, trust, and implementation into the human decision-making process.

AI application for climate problem analysis and solution creation by AI is adverse due to the fact that society has not been able to pay much attention to how people view and engage with climate solutions that are AI-based. The implementation and use of persons are paramount to the realization of the benefits of these solutions, which should have people's trust and behaviour change behind. This motivates nationally and socially adjusting the approach by adding insights of socio-psychological character to views about the AI utilization in the name of ecology. The other name for this approach could be "Technology Acceptance Theory (TAM)". Current literature recognizes TAM as one of the most commonly applied frameworks in technology adoption research. It explains why some decreased once it was noted that it had not been delivered within the time provided. According to the model, perceived usefulness and perceived ease of use encompass the overall belief users have towards technology adoption, where usefulness is the extent to which an individual believes that using a particular system would enhance his/her performance. However, it was found that easier to use does not always correspond with usefulness, which changes the nature of control or relative advantage as perceived usefulness.

The difficulties outlined can be further explored through external perspectives, as the model of Stimulus – Organism – Response, in short SOR, attempts to focus on both the mental and emotional phenomena. This theory was first propagated in environmental studies and centres on the belief that external stimuli determine internal psychic predisposition, leading to response (Mehrabian & Russell, 1974). For instance, when considering the area of climate change and artificial intelligence, AI mechanisms, along with the specifications that boast accuracy and are user-friendly, like predictions, allowing for understanding, can be seen as thoughts put within the mind of the user that are not his/her own. According to these individuals, their feelings towards AI and how sensitive they are to the issue of climate change are a key factor that prevents them from getting involved in sustainable practices. By merging these two theories, TAM and SOR, a wise approach to the adoption of AI among individuals with robust analysis into the technology plus people's behaviour.

The convergence of SOR and TAM models becomes particularly relevant in the case of India, as it is often characterized by social and cultural heterogeneity, differences in the level of digital awareness, and exposure. The considerable effect on the acceptance and consequences of the technology in question. For instance, while the urbanites might much appreciate the help that comes with the adoption of artificial intelligence due to their exposure to IT, the rural and semi-urban population may not appreciate the change as much due to the unavailability, lack of trust, and ignorance. Understanding the factors that give rise to behaviour may also involve looking into the culture, in view of the fact that environmental values and practices are a female dimension in India, as described by Kumar et al (2021). For this reason, it is necessary to examine the use of AI for combating climate change with respect to levels of literacy as well as environmental conservation.

In addition to the core structural theories of SOR and TAM, mediation and moderation mechanisms are also considered in the explanation of the proposed model. Factors such as trust in the technology and environmental consciousness are treated as mediating factors that explain the gap between AI impetus and the behavioural consequences. One of the variables is trust, more of a concern in the context of AI's use; public concerns relating to the data acquisition, algorithm bias, and a lack of explainability can deter consumers. Other concerns, which are not directly related to technology the conventional environmental issue is also included. Putting it differently, users are more likely to use sustainable practices if such behaviour is encouraged by actively promoting environmentally friendly backdrops. They can help incorporate intelligence by simply providing information that changes over time for one to make current decisions.

In addition, the chapter analyses the moderator effects of levels of digital literacy and ecological dedication. Digital literacy, which is the depression, promotion, and use of AI technologies by individuals, influences the way in which individuals view a system in terms of its usefulness and ease of use (Tan et al., 2025). Conversely, environmental commitment and values refer to the level of commitment that each individual tends to have in the course of creating different returns from the environmental investment (Davis et al., 2009; Teng et al., 2014). Another very interesting and relevant variable was the Trust in AI (Afroogh et al., 2024). In the scenario of Trust in AI enabling climate mitigation activities in India, or Trust in AI driving climate change adaptation activities in India.

The given chapter employs Partial Least Squares Structural Equation Modelling (PLS-SEM) as an empirical tool for testing the hypothesized research model. PLS-SEM is noted for its suitability and applicability in handling complex models with various constructs entailing either mediation or moderation mechanisms (Hair et al., 2019). This is also congruent with the focus of the chapter, which is towards prediction, and this chapter endeavours to find out the main factors that bring out a sustainable behaviour. The application of PLS-SEM allows for the testing of both measurement and structural models, which assures that the constructs are reliable and valid while evaluating the presence and strength of the relationship.

The research uses an amalgam of principles from the sciences, environmental science, and inferring from behavioural psychology, and the SOR Framework. Integration of such perspectives vis-à-vis SOR and TAM in the current chapter helps us to place the study in the contemporary works on AI-assisted sustainability clearly and in a more effective manner, thereby providing a strategic context to the behavioural impacts of technological change. The output from the study has implications for practicing executives, technologists, and conservationists (i.e., policy, innovation research, and advocacy sectors). For example, how enhancing AI's comprehensibility improves trust and how the increase in digital literacy influences the acceptance of any AI solutions. There can also be effective use of AI in awareness campaigns of climate risks and how people can promote development for sustainability purposes, this time using AI.

Thus, the objective of this chapter is to bridge the chasm that separates technological possibilities and the actual conduct of people by looking at how AI can be most effectively harnessed for the sustenance of climate action in India. It goes beyond forecasting climate change and explains how AI is contingent upon, or rather compatible with, the human mind, trust, and societal principles in general. Researchers suggest that this comprehensive SOR–TAM theory, combined with data and measurements, via PLS-SEM methodology could be applied to advance AI-induced climate action imperatives globally. Given that India is progressing toward ecologically sustainable development, the use of AI within an ethical and inclusive framework is crucial to address the issues of increasing climate change and ensuring that progress does not break.

2. LITERATURE REVIEW

2.1 Artificial Intelligence in Climate Action

In Indian settings, AI is used in different areas such as agriculture, water management, and urban planning and design. The use of AI technology in farming helped the farmers maximize irrigation, use fertilizers precisely and practice no tillage promoting the efficient utilization of resources resulting in decrease of the beauty and productivity of the soil regarding the introduction of no till technology introduction (Adewusi et al., 2024). Another example of the use of AI technology in smart grid development which provides energy-saving technologies and incorporates the power generated by renewable sources into the grid (Ojadi et al., 2024). Although a variety of AI solutions are available in combating climate change, their application and effectiveness will largely depend on the acceptance and trust of the users and the technology that goes with it.

In addition, the ecological footprint of AI has started to undergo a critical discussion. Enormous AI models are rather costly to train and consume high quantities of computational power, thus causing substantial energy usage and consequently Carbon emission (Strubell et al., 2019). This is the paradox that is stimulating the idea of "Green AI", or green computing, an era that uses programs and computing devices that cause very little or no negative effect on the environment. In this context, a Green AI strategy would be using AI to accelerate work that achieves climate action and contributes to the achievement of sustainability.

2.2 Stimulus–Organism–Response (SOR) Framework

Regarding the Stimulus-Response (S-R) Model (launched by Hull in 1943), external environmental stimuli affect an individual his/her internal psychological state, and further, her/his behaviours in a certain way. Within this model, the term 'stimuli' indicates the outside forces that spark thought and feeling responses in the person (which functions as the organism), and this, in turn, results in a behaviour demonstrated by the person who does the action.

There are several research studies related to environmental as well as sustainability, where the SOR Model has seen much use concerning differentiation and behaviour directed towards protecting and enhancing the environment. In recent years, the term environmental stimuli have expanded to cover a variety of information, such as information, regulation, and technology, which management tends to use in energizing the person to act favourably in protecting sustainability (Jacoby, 2002). As a matter of fact, there a good number of recent research studies that have taken further the SOR framework, albeit not confined to the physical environment rather, to the area of human-computer interaction or digital spaces in which such technological features as the level of interactivity within an object, or the extent of personalization or clear presentation influence the usage of those objects (Islam et al., 2018).

Considering the SOR model in the efforts to determine how AI should be used to mitigate climate change, sustainability issues, and beyond, it could also view AI technologies as inputs that provide information and intelligence about environmental realities. These stimuli affect cognitive facets such as their perception of the benefits that would accrue to the use of the system and the corresponding emotional states, such as trust, environmental worry, or concern for the environment. People's predispositions help in explaining their actions, hence the different levels of utilization of AI and sustainable activities. As seen, the SOR paradigm entails a vast psychology of behaviour change with respect to AI-induced climate change.

2.3 Technology Acceptance Model (TAM)

The author suggested that the Technology Acceptance Model (TAM) be used for the purpose of correlating frameworks for technology use and growth at the time he developed it. The TAM theory (Davis, 1989) provides that there are two specific factors, such as the ease of use and the usefulness of a product, that lead to an individual's behaviour of using a product or not. Usefulness in this case includes factors that drive the performance above and beyond the changes in the efficiency of the user, while ease of use mainly deals with the user interaction factors.

Research has analysed the Technology Acceptance Model (S. Davis, 1989) in a variety of fields, such as online shopping, healthcare, e-learning, and ambient intelligence systems (Venkatesh & Davis, 2000). In the discussion of AI specifically, the use of AI to fight climate change may be affected by how helpful it is, and how easy it is to utilize the AI software, this is the ease of use of AI. Additionally, the wide availability of products, services, and skills contributes to advancing technology in this world of falsehood and fear.

Nevertheless, following the criticism that has been drawn to the Technology Acceptance Model (TAM) as the theory does not recognize emotional and environmental factors, there is a call for the application of other theoretical models. This limitation is especially poignant in climate change mitigation efforts, as scare, morality, and android user responsibility relational constructs are pertinent social and behavioural influences. As such, it is of high importance to juxtapose TAM with other theories, such as SOR, to gain an appreciation of the various dimensions of technology adoption in more sustainable settings.

2.4 Integration of SOR and TAM

The SOR-TAM Integration represents an Approaching, Open Perspective for Systematic Examination of the Interaction Technologies and Behavioural Patterns. TAM provides a realistic way in which the role of individuals in technology adoption can be understood in terms of perceptions of utility and ease of use, but SOR is broader and more extensive.

In the last couple of years, researchers have started including these two theoretical frameworks in their studies about digital behaviour in different situations. For instance, on the e-commerce and social networking topic, SOR and TAM have been used in one and the same study to find how the use of technological facilities either encourages user activity or helps the user make a decision (Islam et al., 2020). Artificial intelligence-induced behaviour modification, based on behaviour change, however, could be justified by the combined theoretical notion with regard to how stimuli created by technologies such as AI may or may not induce such change, but apply to reasoning and feelings too.

In this chapter, the AI features, such as two-dimensional interpretation of predictive accuracy and AI-related explainability issues, are considered as cues. It is these cues that actually have an effect on such cognitive elements as PU and PEOU and emotional elements such as trust and environmental awareness, which in turn bring about certain behaviours. Such a sophistication in the development of an organizational capacity for action necessitates the enhancement of both AI and psychology.

2.5 Role of Trust in AI

Clonability is a pivotal aspect in the advancement and use of AI technologies. At some point, users thought that AI structures are said to be sophisticated and self-governing, which might raise the issue of welfare, transparency, and morals. Trust in the severance of AI is understood as the faith that AI systems are valuable and beneficial and will be rendered as they are supposed to while promoting the user (Shin, 2021).

In the world of climate actions, trust plays a significant role, especially since most of the time these AI technologies are usually the ones coming in to provide important reporting and forecasting of climate-related aspects. For instance, farmers who base their weather predictions on AI need to trust that the predictions are true to prevent taking unnecessary actions. There have been explicit connections between the authenticity and conclusions about AI and trust, which can help ameliorate the adoption of entire AI systems in any given context (Rai, 2020).

Additionally, trust mediates the relationship between technological constructs and the following behavioural outcomes. The trust within AI users is likely to change with the perception of the ease-of-use and the perceived usefulness, or compatibility, being present with the AI. In other words, as the trust variable endogenizes, TAM and SOR models should be adjusted to accommodate this component.

2.6 Environmental Awareness and Sustainable Behaviour

Environmental awareness can be described as the concern of individuals regarding issues facing nature and their outcomes. It is of the essence as far as prompting activities that conserve energy, minimize waste, or promote sustainable continental economic development is concerned (Omer, 2008). AI can help increase environmental awareness by helping people access information about real-time climate conditions, as well as receiving suggestions for consumption modification.

For instance, AI-enabled services can estimate emissions, manage energy, and give suggestions about how to protect the environment and help individuals follow such tendencies. Such information can greatly affect awareness and the propensity of individuals to adopt more sustainable approaches. Levels of environmental awareness are directly proportional to the strength of pro-environmental behaviour intention (Kollmuss & Agyeman, 2002).

Integrating the concepts of TAM and SOR, environmental awareness is the agent that is responsible for transforming the environmental changes brought about by AI advancements into certain types of behaviour. Moreover, AI would help in modifying people's perspectives and drive, making it possible for a continuous change in the self-making of society against climate change.

2.7 Moderating Role of Digital Literacy and Environmental Values

There is evidence that suggests integrating digital literacy in education should focus on the qualitative aspects of the use of digital technologies rather than quantitative figures. Digital literacy is the ability to use digital technologies effectively. As for environmental values, they show how much people are concerned about promoting the cause of the environment.

Different levels of digital literacy persist in different parts of India and economic classes; this situation also affects the availability of AI technologies, including their utilization. People with better digital literacy see AI systems as more adoptable and easier to use than people with low digital skills, making it easier to adapt and use them (van Dijk, 2020). Environmental values, in a way, can also bridge the gap between trust and its manifestation in behaviour, for people who value environmental protection prefer to choose activities consistent with the preservation of the environment.

3. RESEARCH OBJECTIVES

1. To analyse the impact of AI stimuli on cognitive and emotional responses.
2. To examine the role of trust and environmental awareness as mediators.
3. To evaluate the moderating effects of digital literacy and environmental values.
4. To assess the influence of AI adoption on climate action behaviour.

4. CONCEPTUAL FRAMEWORK AND HYPOTHESIS DEVELOPMENT

In this chapter, two of the most popular theoretical concepts, that is, the Stimulus – Organism – Response Model and the Behavioural Intention Model, in order to understand factors of effective behavioural change in businesses in India, which are enacted through artificial intelligence. Thus, the framework examines the influence of environmental stimuli grounded in AI on the cognitive and emotional states of the users, thereby determining the sustainable changes in their behaviour. The framework is also designed with mediation and moderation mechanisms to provide further insight into the factors being studied and situational uniqueness.

4.1 Conceptual Framework

Figure 1. Conceptual framework

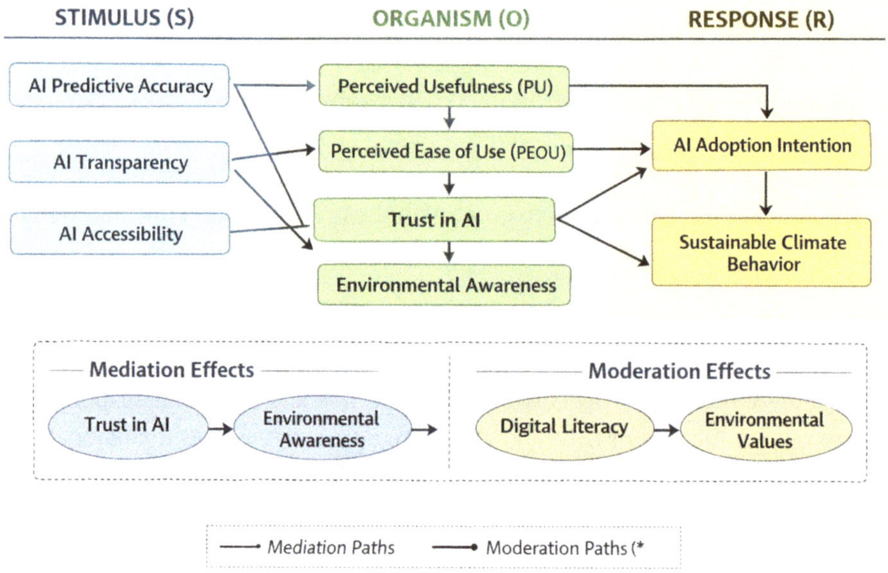

Source: Created by Author

The above Fig. 1 argues that external stimulation (S) leads to the evolution of the internal psychological state (O), which, as a result, prompts the action (R) (Mehrabian & Russell, 1974). The current dissertation aims at classifying AI-driven climate tools used for mitigation, adaptation, transparency, and precision. These tools are of significance because when applied well, they may address threats of climate change by, for example, providing a forecast of the weather, locating the user's carbon footprint, and even recommending how to cut on electricity use and so forth.

The organism component integrates TAM constructs, perceived usefulness (PU), and perceived ease of use (PEOU), as well as other affective variables such as trust in AI and environmental awareness. The systemic moderation provides that PU and PEOU are vital in the processes of technology adoption (Davis, 1989). Yet other aspects, such as emotional and ethical aspects, are also important when it comes to sustainability. Trust in AI is characterized as the degree of trust that users have in the reliability and openness of AI systems, while environmental awareness refers to an individual's knowledge and concerns for the environment.

The response portion of the project covers attitudes like intent in addressing AI in environmental integrity adherence, as well as other sustainable practices like energy efficiencies, green production, and even involvement in environmental activities. This code also moderates controls, where trust and environmental awareness mediate between the AI exposures and behaviours. In addition, they bring in interactions managing with digital literacy and environmental values, seeking to know the strength of the associations in the model.

4.2 Hypothesis Development

Stimulus → Organism Relationships

In simple terms, AI is at the core of integrating automation and agility to derive quick, efficient processes throughout knowledge creation. Though the overall perspective of AI firms in this research is not that of a specific application, the turn of the customer-service activity is the greatest art of the modern AI economy revolution. Back to the outlined symmetrical aspect of the research, "Theory of Reasoned Action" is the concept that has found its application in our study, given that it involves cognitive issues.

H1: AI predictive accuracy positively influences perceived usefulness.

Similarly, the degree to which people find AI challenging or accessible may be a major factor in their ability to use it. The two most important notions in technology acceptance are antecedents to it: perceived usefulness and perceived ease of use in the TAM (Venkatesh & Davis, 2000).

H2: AI accessibility positively influences perceived ease of use.

Transparency is of great interest in the public discourse on AI and has been connected to the notions of trust and understanding. Studies have suggested that Explainable AI might have a beneficial influence on user trust and acceptance (Rai, 2020).

H3: AI transparency positively influences trust in AI.

Organism → Response Relationships

It is established that the perceived usefulness and perceived ease of use significantly affect one's behavioural tendency to use a given technology (Davis, 1989). In the frame of environmentally related activities, people are willing to deploy machines based on AI technologies if they realize the benefits associated with their usability.

H4: Perceived usefulness positively influences AI adoption for climate action.
H5: Perceived ease of use positively influences AI adoption for climate action.

The level of confidence one possesses in native AI systems is directly linked to significant outcomes and behaviours, notably in critical areas like environmental conservation and climate change management. (Dwivedi et al 2021).

H6: Trust in AI positively influences sustainable climate behaviour.

Environmental awareness enhances individuals' motivation to engage in pro-environmental actions (Kollmuss & Agyeman, 2002).

H7: Environmental awareness positively influences climate action behaviour.

Mediation Effects

Belief and a wish to understand and preserve the environment are the middle links that engage AI information in generating actions. When AI components are open and supportive, they tend to encourage trust and concern as well, both of which are relevant to the extent of any change that occurs in human behaviour in view of a softer, sustainable attitude.

H8: Trust in AI mediates the relationship between AI transparency and climate action.
H9: Environmental awareness mediates the relationship between AI usefulness and climate action.

Moderation Effects

Digital literacy influences individuals' ability to use AI technologies effectively, thereby strengthening the relationship between perceived usefulness and adoption (van Dijk, 2020).

H10: Digital literacy moderates the relationship between perceived usefulness and AI adoption.

Environmental values reflect individuals' commitment to sustainability and can strengthen the impact of trust on behaviour.

H11: Environmental values moderate the relationship between trust in AI and climate action behaviour.

Climate change has grown into a major challenge across the globe. The use of data and artificial intelligence has driving factors including, but not limited to, technology, behaviour, emotion, and environment. Most notably, encouraging commitment and cognition for building trust and propagating the digital competencies and pro-environmental values as factors that can change or maintain these processes. Indeed, such a strategy would be effective for India, which has various socio-technological and cultural challenges that shape the intentions to use technology and engage in sustainable behaviour.

5. RESEARCH METHODOLOGY

5.1 Research Design and Data Collection

In order to collect the main data, it was decided to carry out a survey in which respondents were interviewed across urban and semi-urban areas of India. Recently, a convenient non-probability purposive sampling technique was used, meaning that the selection of the target population was based on those who are assumed to have some degree of exposure to digital technology and an understanding of AI software. An acceptable number of 420 focused samples was generated, which further ensured sufficiency in the pageant for this study.

The set questionnaire included two administrative sections: sociodemographic details and construct measurement assessments. Every other item in the measure was assessed on a 5-point Likert scale ranging from 1 (much disagree) to 5 (much agree). It was stated that, as either the modified individual items due to factors related to the application of climate change in the design or as scales, some items have been adapted with recognition to prior works in order for the content to be valid.

5.2 Measurement of Constructs

In the next section, we describe how the key concepts from the SOR-TAM theory will be measured in this study. Antecedent factors include AI predictive accuracy, transparency, and accessibility. Organism variables comprise perceived usefulness, perceived ease of use, trust in AI, and environmental awareness. Response variables include AI adoption intention and sustainable climate behaviour. Additionally, digital literacy and environmental values are measured as moderating variables.

5.3 Data Analysis Technique

The concrete application of the obtained data was done by using Structural Equation Modelling with Partial Least Squares and software wizard SmartPLS. The reason PLS-SEM is significantly important this week is that it can cope with intricate situations like multiple constructs and effects of mediation/moderation. It is also applicable to research bearing more on predictions and not to those where the sample data are assumed to be linear (Hair et al., 2019).

The data was subjected to the first stage-weighted analytical process. At this stage, the measurement model was assessed for reliability and validity. This was done by adopting a set of criteria that were duly satisfied by various statistics, reliability, and validity of the test based on the υ parameter (Cronbach's scale composites). This would help minimize common method bias. Second, the structural model was tested through t-tests and the use of bootstrapping for significance testing. The ability of the model to predict the presence or the absence of reliability in the model was determined by assessing the value of R^2 and Q^2.

5.4 Mediation and Moderation Analysis

To verify whether mediation is present, trust and environmental awareness underwent bootstrapping to estimate the confidence interval around the indirect effect. Meanwhile, the moderating role of digital literacy and environmental values was ascertained by the inclusion of interaction terms in the PLS-SEM model.

6. DATA ANALYSIS

Table 1. Descriptive Statistics

Variable	Category	Frequency (n)	Percentage (%)
Gender	Male	210	50.0
	Female	210	50.0
Age Group	18–25	150	35.7
	26–35	140	33.3
	36–45	80	19.0
	Above 45	50	11.9
Education Level	Undergraduate	180	42.9

continued on following page

Table 1. Continued

Variable	Category	Frequency (n)	Percentage (%)
	Postgraduate	170	40.5
	Others	70	16.6

The descriptive statistics of the study based on 420 respondents indicate a balanced gender distribution, with equal representation of male and female participants. The majority of respondents fall within the 18–35 age group, suggesting that the sample is relatively young and technologically active, as shown in Table 1.

6.1 Measurement Model Evaluation

Table 2. Reliability and Convergent Validity

Construct	Cronbach's Alpha	Composite Reliability (CR)	AVE
AI Predictive Accuracy	0.88	0.91	0.66
Transparency	0.86	0.90	0.64
Perceived Usefulness	0.89	0.92	0.68
Ease of Use	0.87	0.91	0.65
Trust in AI	0.88	0.92	0.67
Environmental Awareness	0.85	0.89	0.62
AI Adoption	0.90	0.93	0.70
Climate Action Behaviour	0.88	0.91	0.66

Examine the results in Table 2 and understand that the validity and reliability of all the variables are great. In other words, all the constructs of the framework meet the standards for reliability and validity. The values of Cronbach's alpha are established to be more than 0.70, suggesting that the internal consistency is consistent with the tested constructs. The composite reliability values of the instruments are above 0.90 for academic promotions and prenatal tests. The AVE values in all three constructs across the special education outmatch the 0.50 criterion, showing that these constructs are valid in that they explain over 50% of the variance of their indicators. These results, albeit in this preliminary empirical study, in their entirety make a strong case that the measurement model is solid and can be confidently used in further structural analysis. In view of the above, the constructs that are used in the chapter can be said to adequately represent the underlying theoretical concepts of the integrated SOR–TAM framework.

6.2 Discriminant Validity

Table 3. Discriminant Validity

Constructs	HTMT Value
PU – Trust	0.72
PU – Awareness	0.69
Trust – Behaviour	0.75
Awareness – Behaviour	0.71

Since every ratio in Table 3 is less than 0.85, it suggests that the constructs are distinct, confirming their independence. This implies that the functioning of each of the constructs holds a different term in the model. The HTMT values of moderate magnitude are a clear indication that there is a relationship between constructs such as perceived usefulness. This common practice of leaving out some of the constructs described by perceived usefulness, trust, and awareness still co-occurs. It is interesting to validate the SOR-TAM joint model since the overall structure does Concept of user fatigue and its central app sets in that each of the variables is causally related to the climate change behaviour determined by artificial intelligence. In short, there are indicators that the measure model used has excellent evidence of discriminant validity, and therefore, it is harmonious for further testing of hypotheses.

6.3 Structural Model Results

Table 4. Path Coefficients

Hypothesis	Path	β Value	T-value	P-value	Result
H1	AI Accuracy → PU	0.48	6.21	0.000	Supported
H2	Ease of Use → Adoption	0.42	5.89	0.000	Supported
H3	Transparency → Trust	0.51	6.75	0.000	Supported
H4	PU → Adoption	0.45	6.12	0.000	Supported
H5	Trust → Behaviour	0.38	5.45	0.000	Supported
H6	Awareness → Behaviour	0.29	4.98	0.000	Supported
H7	Adoption → Behaviour	0.41	5.78	0.000	Supported

Table 4 presents the model outcomes, highlighting that all anticipated connections are statistically significant, evidenced by high T-values (greater than 1.96) and P-values (less than 0.05). The influence of perceived usefulness on successful AI predictions is notably strong, and this pattern is consistent with the relationship between trust in AI and its perceived credibility. On the contrary, the two other points mentioned in the background are constructs of willingness to action syndrome, namely, ease of use and perceived usefulness, thereby significantly affecting AI adoption and rate. Moreover, there has been an advancement in this theory by including trust, green attitudes, and the adoption of one theory and influencing actions to adopt climate change mitigation practices. Additionally, β is determined on the basis of the strengths of the relationship in this study, which also strengthens the argument for the SOR–TAM's integration. These findings accentuate the imperative of both technological enhancements and the psychological factors in fostering environmentally friendly tech development in the domain of artificial intelligence in India.

6.4 Coefficient of Determination (R^2)

Table 5. R^2 Values

Endogenous Construct	R^2 Value
Perceived Usefulness	0.52
Trust	0.49
AI Adoption	0.58
Climate Behaviour	0.62

The R^2 values that are depicted in Table 5 do shed light on the explanatory capacity of the model. Particularly, the perceived usefulness (0.52) and the trust (0.49) variables show a moderate reliance significantly on AI stimuli to account for these constructs. In turn, the AI adoption (0.58) and the climate behaviour (0.62) variables indicate a significant extent of the modelling abilities in the sense that behavioural outcomes are captured by the model. As per the usual benchmarks, any value greater than 0.50 is considered to be quite high in social research. Hence, it is revealed by these findings that the Integrated SOR–TAM model has a very good capacity to explain AI-triggered climate actions, establishing that it can be successfully implemented in the perception of the Indian industry.

6.5 Mediation Analysis

Table 6. Indirect Effects

Path	Indirect Effect	T-value	Result
Transparency → Trust → Behaviour	0.19	4.87	Partial Mediation
PU → Awareness → Behaviour	0.17	4.45	Partial Mediation

The results of the mediation analysis shown in Table 6 reveal that trust and environmental awareness are powerful mediators in the impact of AI stimuli on environmentally friendly actions. Indirect effects are significant as presented by T-values exceeding 1.96. Nevertheless, the direct effects are still found to be significant, thereby rendering the mediation partial rather than complete. This thus highlights that the effects of AI features on behaviour are twofold: direct and through certain mental components. The obtained results further underline the necessity of building trust and enhancing environmental consciousness in order to make full use of AI functions regarding climate activities. Therefore, mediation has a critical function for the success of sustainable projects centred on AI.

6.6 Moderation Analysis

Table 7. Moderation Effects

Interaction Effect	β Value	T-value	Result
Digital Literacy × PU → Adoption	0.21	3.98	Significant
Environmental Values × Trust → Behaviour	0.24	4.12	Significant

Based on the results presented in Table 7, modulated impact has indeed occurred in these associations due to the values of digital literacy and environmental sustainability. It can be observed from the data that digital literacy carries an additional lever for the impact of perceived usefulness on AI germination, indicating that the trust in such tools will be more prominent in people with a high level of digital literacy. Also, environmental values contribute to a relationship between trust and climate change behaviour, meaning that those individuals who care a lot about sustainability issues tend to put their trust into action. The results specify the socio-cultural and circumstantial elements of employing AI in people's ecological practices, especially in the environment of a variety of India.

7. DISCUSSION

The main purpose of this chapter is to explore the ability of Artificial Intelligence (AI) to promote environmentally sustainable behaviour in India, within the confines of the SOR model and the Technology Acceptance Model (TAM), with the inclusion of mediating and moderating constructs. The sketch drawn here is powered by the actual empirical support; it explains completely how the defined stimuli introduced by AI are beneficial in driving the desired actions, such as the fight against climate change across diverse perspectives that include cognitive as well as emotional aspects.

The first goal was to understand how the impact of autonomous behaviour features affects the user's cognitive and emotional responses, specifically, predictability, interpretability, and controllability of an AI system. The findings prove AI predictability concerns as a big contributor to the perceived usefulness of AI technology (H1), in which people appreciate those AI systems that can provide helpful and local climate calls to action. The effectiveness of technological systems is a game; this is one reason why satisfaction and use are high among users for the machine equipment. Likewise, users were able to escalate the use of something because it is easy to comprehend (H2), which indicated the support that user-friendly systems encourage satisfaction with activities with them. In the context of India, which is extremely diversified in terms of socio-economic and educational levels of people, AI's profundity itself and the way it becomes easy for people in terms of implementation would be quite significant for adoption.

Researchers have observed that AI trust is influenced substantially by transparency (H3). This view is compatible with the perspective that the clarity of predictive and explanatory capacities of AI improves people's understanding and hence trust. This is especially critical in modifications aiming at climate change scenarios, as these activities are particularly high in tax and long-term consequences in nature. To sum up, this chapter examines the part played by the SOR framework, where AI characteristics are considered the main triggers. This is to help in accomplishing the first research objective.

The second hypothesis explored the role of non-rational and emotional factors in the formation and change of behaviour. The analysis of results indicates that it is the most influential factor in the process of AI diffusion (H4), providing evidence of the fundamental impact of performance expectancy in the context of technology acceptance. Moreover, attitudes towards deployment affect adoption favourably (H5), suggesting that decreased complexity leads to increased user propensity to use AI. The findings, the extended consumer attitude aspects related to AI adoption, align well with TAM, making it sufficient to explain the attachment.

It was additionally confirmed that trust in AI is closely connected with addressing the problems related to climate actions (H6). The role played by trust and how it enables or prevents the use of technology is also quite paramount in understanding the behaviour of consumers. It has been found necessary to include environmental consciousness as an intervening factor in the equation, especially in the decision process of mitigating climate change (H7). In other words, it is easier to accept that more educated individuals about the environment would behave more sustainably when it comes to action for climate change. All in all, it is important to point out that both cognitive appraisals and affective reactions are relevant for behaviour change in line with the application of SOR and TAM perspectives.

The third research question involved checking the effect of trust and environmental awareness as a mediating variable. Based on the outcomes, both the variables were able to partly pave the way for the association between AI stimuli and climate behaviour (H8 and H9). It is again observed that AI techniques promote sustainable behaviour through the individual only when the internal processes are involved. For example, even if everyone recognizes the accuracy and convenience of AI equipment, the use of it is best appreciated when there is trust in it, and the various actions in the environment are noticed. This result brings forward the urgency for the implementation of AI techniques that not only work well but also build infrastructure in terms of trust and awareness of AI systems. There is also mixed mediation because of the fact that the direct relationships are still strong and influencing behaviour, meaning AI influence lies on two paths.

Objective four investigated the two moderating variables of digital literacy and environmental values. The results remained that digital literacy significantly reinforced perceived usefulness and AI adoption, taking into consideration the relevant use theory (H10). This particular relationship suggests that the more advanced an individual's digital literacy is, the more easily they can identify and make use of AI technology. Given the different rates of digital literacy across India, this particular excerpt stresses the necessity of digital education as well as increasing capabilities following new technological trends.

To illustrate further, environmental priorities were noticed to have modulating effects on the bond existing between trust in AI and climate behaviour (H11). Environmental ethics would make someone more likely to enact green measures as opposed to just believing in AI. In turn, such an argument justifies the consideration of intrinsic values and cultural norms in the process of people's behavioural patterns. These moderation effects extend this understanding by revealing the complex interplay of environmental and other factors' effects in the context of AI implications that satisfy the fourth research aim.

The entire chapter demonstrates how acting to tackle climate change is no longer a matter of just having the required technology; it goes beyond technology. This could be a balance of norms and beliefs, biases, values, and attitudes, which one cannot easily quantify but has direct and indirect effects. The integrated SOR–TAM model does not just address stimulus and response relationships; it also describes the transformations between the stimuli and the responses. In explaining its cause-and-effect web, the model is highly plausible, which means no significant statistical scatter or lack of accuracy, especially in the case of data obtained from the relevant investigations conducted in Indian cities.

Taking into account all of the above, it may be said that all the hypotheses have been proven right. The findings suggest that if Artificial Intelligence is leveraged to change behaviours primarily by making it easy to use, perceived as useful if not also beneficial, and very clear and fair in how it works, it is likely to change all sorts of behaviours in a short while. So, it is how they are used, caused by the greater green-collar effects of AI-related industries that are brought into question. The way cognitive assistant agents are employed in society, the kind of systems with which they interact, and the social policies that ensure their completion are due for some consideration too. This chapter thus provides a tool for quantifying competing objectives and proposes an overlapping set of solution strategies for optimizing hierarchical optimization.

8. THEORETICAL AND PRACTICAL IMPLICATIONS

8.1 Theoretical Implications

In this chapter, the current research has sought to bridge the gap between two models in the area of technology: the Stimulus-Organism-Response (SOR) model and the Technology Acceptance Model (TAM), with controllable factors within the "green" paradigm. Over time, TAM has been targeted at factors that awakened pictures of utility and ease of use in nature, but the inclusion of such mechanisms as SOR enriches it with social and psychological aspects such as trust and eco concerns. This dilution of the understanding provides a more comprehensive perspective on how technology-related cues affect one's will and can be applied to both the gallant hypothetical principles and the environmental and AI environments, thus combining the perspective of both theories into one.

In addition, this section introduces 'mediation-moderation' mechanisms, developed within the framework for better explanation. The inclusion of external factors such as 'trust' and 'environmental awareness' emphasizes the importance of individual constructs in implementing AI actions. However, the additional dimensions of technological diffusion presented in terms of digital literacy and environmental values also imply a societal and cultural genesis. This approach effectively remedies a concern that has been expressed in the existing text of the paper - the development of models that reasonably represent social and human in the context of technology, adapted to sustainability.

In the future, the framework allows for the integration of further theories like the Unified Theory of Acceptance and Use of Technology (UTAUT), Value-Belief-Norm (VBN) theory, and Behavioural Economics models. In the same vein, research opportunities abound in inserting additive components into the existing model—that of the dynamic and longitudinal changes involving attitudes and behaviours towards AI, with the accumulation of experiences over time. Furthermore, the infusion of sophisticated analysis methodologies such as Artificial Neural Networks (ANN) or integrated SEM machine learning strategies will enrich predictability strength and scientific soundness.

8.2 Practical Implications

The chapter discusses some more possible areas and strategies for operators and regulators who view artificial intelligence as a possible tool for activities in climate protection in India. First of all, the results are clear that importance has to come while designing AI systems, which are also user-centric and explainable. It is because the perceived usefulness and trust are the main forces that affect adoption and usage. Devotees of artificial intelligence should make Explainable AI (XAI) their most immediate priority, as it enriches the level of predictability and reduces the level of uncertainty the users have. The imaginary yet vital function is to emphasize that not only does AI have to have high predictability, but high communication ability must be a standard.

Another angle to the matter of digital literacy, as a moderator, also brings up the essential role of the large-scale training efforts. Encouraged by this, leaders should launch major training interventions in these already technologically backward structures, such as rural and peri-urban areas. Overcoming the identified digital divide will be an essential part of realizing the potential of AI-based climate technologies in India. Digital literacy initiatives can be especially enhanced in Public-private partnerships, even in other fields than digital literacy and technology adoption, because they aim towards very good functionalities.

The third point underlined in the chapter is the key role which knowledge and attitudes towards the environment play in pertinent behavioural changes. In the scope of environmental protection behaviour, it is possible to use environmental policy such as computer systems, monitoring user activities, and providing user advice, as it can reach individuals offering personally designed information or urging them to care about the problem. For example, applications on mobile phones may allow users to look at the energy consumption or the amount of carbon emissions used at a particular time, which is likely to promote consciousness and change of behaviour. Further efforts can combine AI with behaviour change interventions applied as "nudges" with the purpose of enhancing sustainability initiatives and prolonging user engagement.

Moreover, the conclusions imply that it is essential for the legislation to encourage using technological advances in a way that fulfils ethical, in addition to sustainability, and effectiveness. It is highly recommended for every government to develop measures overseeing the proper use of AI in its climate applications, such as ensuring data protection, fairness, and transparency. Solutions to "Green AI" are inaction and even the hostility towards the development, which will entail negative environmental consequences on the one hand, and the complete abstention from any technological modification of human life on the other.

In the corporate sector, companies have the option to utilize smart solutions in the form of AI to enhance the Corporate Social Responsibility function. This would involve plans devised to promote environmental issues within the economic environment. Functions, market possibilities, and prospects will change, and these should be considered in a more detailed analysis.

9. CONCLUSION

Following the previous statements, we strive to clarify and further extend the AI, the other components of the TAM, together with the SOR's paradigm approach. The study will not only focus on Indian climate change mitigation but also resilience building in disaster areas, of which India is a part. With these improved concepts, it will be possible to assess more effectively the behaviour drivers of those who are advanced with respect to their age compared to others.

The findings of the study pointed out that AI-driven features, mainly accuracy and transparency, also majorly aid in shaping the users' mental as well as their emotional responses, such as perceived usefulness, ease of use, trust, and even environmental awareness. Can be argued that these states of mind are a result of the effective application of AI because when it comes to people and the environment, it is about AI technologies, which translates into climate adaptation as a cause. Also, the levels of trust in AI and confidence towards the environment intervention were identified as mediators necessary to assist in predicting behaviours across technology and impact levels, and expected behaviour only among outcomes. And then, the impacts on the users' experience of the technology were assessed in terms of digital literacy and environmental values, emphasizing further the necessity of social and technological preparedness in India.

Through the empirical validation of the SOR–TAM model integrated with AI using PLS-SEM, the thèsis further enhances what exists in scientific literature about the relationship between AI and changes in behaviour that result in sustainability. It brings attention to the fact that the effectiveness of AI in the solution of climate problems is not dependent only on its technical performance, but also on its fit with the human trust and holism of the AI–human cognitive dimension.

To summarize, the chapter concludes that in order to move from predict to preserve, there must be a synthesis of modern tools within a people-centred design, and it is supported policy-wise. This paper argues that as India endeavours to achieve environmentally friendly development, embracing AI inclusively, ethically, and equitably will assist, especially in building climate resilience in the long run, as well as responsible environmentalism.

10. LIMITATIONS AND FUTURE SCOPE

While it is true that the chapter does contain some interesting propositions and insights, the authors are, however, optimistic in suggesting that other research projects within the area of scholars' concern will lead to a new direction. To begin with, the chapter is of a cross-sectional nature, which would not allow any possible changes in the attitudes and behaviour of the respondents and the case study to be observed over a period of time. The use of a longitudinal approach is recommended for further research to investigate how the attitude towards AI and sustainable behaviours changes in the process of increased exposure and technological advancement. Secondly, the data used in this research are prone to social desirability and response biases. This is inevitably followed by the suggestion to include in the research behavioural or observational data.

Thirdly, since the study is related to the Indian market only, one can exhibit ethnographic distortion and distort the generalization of the information to different cultural and economic market structures across the globe. To illuminate the subject in a deeper context, a comparison study based on Western comparative advantage theory should be undertaken concerning AI-based climate control. Because the current chapter employs PLS-SEM as the tool for analysing the model, it is quite feasible to suggest the incorporation of advanced methods, such as Artificial Neural Networks (ANN) or possibly even some modified form of utilization, to make the projections more correct.

Possible scope of further research, one can consider additional factors such as policy actions, ethical considerations, maturity of the organization, or even applications regarding different industries for AI for climate issues that may include various interactions, which may be quite essential for AI because they enhance the prospects of AI sustainability and increase the adoption of the technology.

REFERENCES

Adewusi, A. O., Asuzu, O. F., Olorunsogo, T., Iwuanyanwu, C., Adaga, E., & Daraojimba, D. O. (2024). AI in precision agriculture: A review of technologies for sustainable farming practices. *World Journal of Advanced Research and Reviews*, *21*(1), 2276–2285. DOI: 10.30574/wjarr.2024.21.1.0314

Afroogh, S., Akbari, A., Malone, E., Kargar, M., & Alambeigi, H. (2024). Trust in AI: Progress, challenges, and future directions. *Humanities & Social Sciences Communications*, *11*(1), 1568. DOI: 10.1057/s41599-024-04044-8

Al Marzooqi, M. (2024). *The contribution of artificial intelligence (AI) to disaster response and management* (Doctoral dissertation, Anglia Ruskin Research Online (ARRO)).

Attia, F. (2025). AI in a Dual Challenge: Creative Remedies for the Hunger and Climate Disasters. *Emirati Journal of Business, Economics, &. Social Studies (Maynooth, Ireland)*, *4*(2), 254–268.

Davis, F. D. (1989). Perceived usefulness, perceived ease of use, and user acceptance of information technology. *Management Information Systems Quarterly*, *13*(3), 319–340. DOI: 10.2307/249008

Davis, J. L., Green, J. D., & Reed, A. (2009). Interdependence with the environment: Commitment, interconnectedness, and environmental behavior. *Journal of Environmental Psychology*, *29*(2), 173–180. DOI: 10.1016/j.jenvp.2008.11.001

Dwivedi, Y. K., Hughes, L., Ismagilova, E., Aarts, G., Coombs, C., Crick, T., Duan, Y., Dwivedi, R., Edwards, J., Eirug, A., Galanos, V., Ilavarasan, P. V., Janssen, M., Jones, P., Kar, A. K., Kizgin, H., Kronemann, B., Lal, B., Lucini, B., & Williams, M. D. (2021). Artificial Intelligence (AI): Multidisciplinary perspectives on emerging challenges, opportunities, and agenda for research, practice and policy. *International Journal of Information Management*, *57*, 101994. DOI: 10.1016/j.ijinfomgt.2019.08.002

Hair, J. F., Risher, J. J., Sarstedt, M., & Ringle, C. M. (2019). When to use and how to report the results of PLS-SEM. *European Business Review*, *31*(1), 2–24. DOI: 10.1108/EBR-11-2018-0203

Hull, C. L. (1943). The problem of intervening variables in molar behavior theory. *Psychological Review*, *50*(3), 273–291. DOI: 10.1037/h0057518

Islam, J. U., Rahman, Z., & Hollebeek, L. D. (2018). Consumer engagement in online brand communities: A solicitation of congruity theory. *Internet Research*, *28*(1), 23–45. DOI: 10.1108/IntR-09-2016-0279

Jacoby, J. (2002). Stimulus-organism-response reconsidered: An evolutionary step in modeling (consumer) behavior. *Journal of Consumer Psychology*, *12*(1), 51–57. DOI: 10.1207/S15327663JCP1201_05

Kollmuss, A., & Agyeman, J. (2002). Mind the gap: Why do people act environmentally and what are the barriers to pro-environmental behavior? *Environmental Education Research*, *8*(3), 239–260. DOI: 10.1080/13504620220145401

Kumar, S., Raut, R. D., Nayal, K., Kraus, S., Yadav, V. S., & Narkhede, B. E. (2021). To identify industry 4.0 and circular economy adoption barriers in the agriculture supply chain by using ISM-ANP. *Journal of Cleaner Production*, *293*, 126023. DOI: 10.1016/j.jclepro.2021.126023

Kushawaha, J., Borra, S., Kushawaha, A. K., Singh, G., & Singh, P. (2021). Climate change and its impact on natural resources. In *Water conservation in the era of global climate change* (pp. 333–346). Elsevier. DOI: 10.1016/B978-0-12-820200-5.00002-6

Lee, H., Calvin, K., Dasgupta, D., Krinner, G., Mukherji, A., Thorne, P., . . . Park, Y. (2023). *IPCC, 2023: Climate change 2023: Synthesis report, summary for policymakers. Contribution of working groups i, II and III to the sixth assessment report of the intergovernmental panel on climate change [core writing team, h. Lee and j. Romero (eds.)]. IPCC, geneva, Switzerland* (pp. 1-34). Intergovernmental Panel on Climate Change (IPCC).

Mehrabian, A., & Russell, J. A. (1974). *An approach to environmental psychology*. the MIT Press.

Ojadi, J. O., Odionu, C. S., Onukwulu, E. C., & Owulade, O. A. (2024). AI-enabled smart grid systems for energy efficiency and carbon footprint reduction in urban energy networks. *International Journal of Multidisciplinary Research and Growth Evaluation*, *5*(1), 1549–1566. DOI: 10.54660/.IJMRGE.2024.5.1.1549-1566

Omer, A. M. (2008). Energy, environment and sustainable development. *Renewable & Sustainable Energy Reviews*, *12*(9), 2265–2300. DOI: 10.1016/j.rser.2007.05.001

Rai, A. (2020). Explainable AI: From black box to glass box. *Journal of the Academy of Marketing Science*, *48*(1), 137–141. DOI: 10.1007/s11747-019-00710-5

Shin, D. (2021). The effects of explainability and causability on perception, trust, and acceptance: Implications for explainable AI. *International Journal of Human-Computer Studies*, *146*, 102551. DOI: 10.1016/j.ijhcs.2020.102551

Strubell, E., Ganesh, A., & McCallum, A. (2019, July). Energy and policy considerations for deep learning in NLP. In *Proceedings of the 57th annual meeting of the association for computational linguistics* (pp. 3645-3650). DOI: 10.18653/v1/P19-1355

Tan, P. H. P., Wijaya, S., Rahardja, U., Henry, B. N., & Anjani, A. (2025). Modeling the impact of digital literacy on ai based learning adoption through perceived usefulness and easeof use. *Sundara Advanced Research on Artificial Intelligence*, *1*(2), 56–64.

Teng, M. J., Wu, S. Y., & Chou, S. J. H. (2014). Environmental commitment and economic performance–short-term pain for long-term gain. *Environmental Policy and Governance*, *24*(1), 16–27. DOI: 10.1002/eet.1634

Van Dijk, J. (2020). *The digital divide*. John Wiley & Sons.

Venkatesh, V., & Davis, F. D. (2000). A theoretical extension of the technology acceptance model: Four longitudinal field studies. *Management Science*, *46*(2), 186–204. DOI: 10.1287/mnsc.46.2.186.11926

Whig, P., Gupta, S. K., Nadikattu, R. R., & Sharma, P. (2025). Application of AI in Environmental Sustainability. *Artificial Intelligence-Driven Models for Environmental Management*, 1-41.

Compilation of References

Aarthi, S., & Ravikumar, R. N. (2026). From regulation to transformation-economic and institutional pathways of environmental policy. In *The Ripple Effects of Environmental Policies* (pp. 49–80). IGI Global Scientific Publishing., DOI: 10.4018/979-8-3373-8998-1.ch002

Abatzoglou, J. T., & Williams, A. P. (2016). Impact of anthropogenic climate change on wildfire across western US forests. *Proceedings of the National Academy of Sciences of the United States of America*, *113*(42), 11770–11775. DOI: 10.1073/pnas.1607171113 PMID: 27791053

Abduljabbar, H. M., Abdullah, W. M., Bazool, S. D. S., Muhsin, A. I., Alfalahi, S. T. Y., & Temnikov, V. (2025). AI-Driven Environmental Impact Assessments: Ethical and Legal Considerations. *Environment and Social Psychology*, *10*(12). Advance online publication. DOI: 10.59429/esp.v10i12.3994

Abdul-Majid, M., Zahari, S. A., Othman, N., & Nadzri, S. (2024). Influence of technology adoption on farmers' well-being: Systematic literature review and bibliometric analysis. *Heliyon*, *10*(2), e24316. DOI: 10.1016/j.heliyon.2024.e24316 PMID: 38312653

Abed, N., Murugan, R., Deldari, A., Sankarannair, S., & Ramesh, M. V. (2025). IoT and AI-driven solutions for human-wildlife conflict: Advancing sustainable agriculture and biodiversity conservation. *Smart Agricultural Technology*, *10*, 100829. DOI: 10.1016/j.atech.2025.100829

Abo-Khalil, A. G. (2024). Integrating sustainability into higher education challenges and opportunities for universities worldwide. *Heliyon*, *10*(9), e29946. Advance online publication. DOI: 10.1016/j.heliyon.2024.e29946 PMID: 38707336

Abubakar, A. M., Zakarya, I. A., Hasnain, M., Sarkinbaka, Z. M., Mukwana, K. C., & Abdo, A. (2024). Potential breakthroughs in environmental monitoring and management. In *Advances in Geospatial Technologies* (pp. 239–282). IGI Global. DOI: 10.4018/979-8-3693-8104-5.ch011

Abyaneh, A. G., Ghanbari, H., Mohammadi, E., Amirsahami, A., & Khakbazan, M. (2025). An analytical review of artificial intelligence applications in sustainable supply chains. *Supply Chain Analytics*, *12*, 100173. DOI: 10.1016/j.sca.2025.100173

Achinas, S., & Quintero, M. (2025). Embodiment of machine learning in biogas power industry: An opinion. *Frontiers in Energy Research*, *13*, 1589782. Advance online publication. DOI: 10.3389/fenrg.2025.1589782

Adamakis, M., & Rachiotis, T. (2025). Artificial Intelligence in Higher Education: A State-of-the-Art Overview of Pedagogical Integrity, Artificial Intelligence Literacy, and Policy Integration. *Encyclopedia*, *5*(4), 180. DOI: 10.3390/encyclopedia5040180

Adanma, U. M., & Ogunbiyi, E. O. (2024). Artificial intelligence in environmental conservation: Evaluating cyber risks and opportunities for sustainable practices. *Computer Science & IT Research Journal*, *5*(5), 1178–1209. DOI: 10.51594/csitrj.v5i5.1156

Adewusi, A. O., Asuzu, O. F., Olorunsogo, T., Iwuanyanwu, C., Adaga, E., & Daraojimba, D. O. (2024). AI in precision agriculture: A review of technologies for sustainable farming practices. *World Journal of Advanced Research and Reviews*, *21*(1), 2276–2285. DOI: 10.30574/wjarr.2024.21.1.0314

Adger, W. N. (2006). Vulnerability. *Global Environmental Change*, *16*(3), 268–281. DOI: 10.1016/j.gloenvcha.2006.02.006

Afoma, U. M., Singh, S., Mishra, A. K., Sharma, C. K., Gupta, K., Mishra, M. K., Roy, B., Verma, V. V., & Sharma, V. K. (2025). Integrating Artificial Intelligence in Environmental Monitoring: A Paradigm Shift in Data-Driven Sustainability [Review of *Integrating Artificial Intelligence in Environmental Monitoring: A Paradigm Shift in Data-Driven Sustainability*]. *EcoHealth*. Springer Science+Business Media. DOI: 10.1007/s10393-025-01752-8

Afroogh, S., Akbari, A., Malone, E., Kargar, M., & Alambeigi, H. (2024). Trust in AI: Progress, challenges, and future directions. *Humanities & Social Sciences Communications*, *11*(1), 1568. DOI: 10.1057/s41599-024-04044-8

Afuwape, K. (2025). Legal and regulatory frameworks for blue carbon credit markets. *Journal of Foreign Legislation and Comparative Law*, *21*(3), 88. DOI: 10.61205/S199132220034876-9

Agrawal, J., & Arafat, M. Y. (2024). Transforming Farming: A Review of AI-Powered UAV Technologies in Precision Agriculture. *Drones (Basel)*, *8*(11), 664. DOI: 10.3390/drones8110664

Aijaz, N., Lan, H., Raza, T., Yaqub, M., Iqbal, R., & Pathan, M. S. (2025). Artificial intelligence in agriculture: Advancing crop productivity and sustainability. *Journal of Agriculture and Food Research*, *20*, 101762. DOI: 10.1016/j.jafr.2025.101762

Akanbi, M. B., Adedotun, K. J., Raji, A. K., & Banjoko, I. K. (2025). Optimizing water resource management in agriculture using ai-powered solar irrigation systems. *Journal of Science Innovation and Technology Research*. https://doi.org/DOI: 10.1016/j.sintl.2024.100292

Al Marzooqi, M. (2024). *The contribution of artificial intelligence (AI) to disaster response and management* (Doctoral dissertation, Anglia Ruskin Research Online (ARRO)).

Alduweib, E., Alromema, W., & Hezzam, E. (2025, April). Artificial intelligence for green agriculture and sustainable resource management: A comprehensive overview. In 2025 1st International Conference on Computational Intelligence Approaches and Applications (ICCIAA) (pp. 1-5). IEEE. https://doi.org/DOI: 10.1109/ICCIAA65327.2025.11013361

Ali, G., Mijwil, M. M., Adamopoulos, I., & Ayad, J. (2025). Leveraging the internet of things, remote sensing, and artificial intelligence for sustainable forest management. *Babylonian Journal of Internet of Things*, *2025*, 1–65. DOI: 10.58496/BJIoT/2025/001

Ali, Z., Muhammad, A., Lee, N., Waqar, M., & Lee, S. W. (2025). Artificial Intelligence for Sustainable Agriculture: A Comprehensive Review of AI-Driven Technologies in Crop Production. *Sustainability (Basel)*, *17*(5), 2281. DOI: 10.3390/su17052281

Alkhawaja, L., Idris, M., Al-Sayyed, S. W., & Jaber, A. R. (2025). Exploring the impact of Artificial Intelligence on students' skills for sustainable development in education. *Frontiers in Education*, *10*, 1691148. Advance online publication. DOI: 10.3389/feduc.2025.1691148

Allam, H., Gyamfi, B., & AlOmar, B. (2025). Sustainable Innovation: Harnessing AI and Living Intelligence to Transform Higher Education. *Education Sciences*, *15*(4), 398. DOI: 10.3390/educsci15040398

Alotaibi, E., & Nassif, N. (2024). Artificial intelligence in environmental monitoring: In-depth analysis. *Discover Artificial Intelligence*, *4*(1), 84. DOI: 10.1007/s44163-024-00198-1

Al-Raeei, M. (2024). Artificial intelligence for climate resilience: Advancing sustainable goals in SDGs 11 and 13 and its relationship to pandemics. *Discover Sustainability*, *5*(1), 513. Advance online publication. DOI: 10.1007/s43621-024-00775-5

Alston, M. (2015). *Women and climate change in Bangladesh*. Routledge. DOI: 10.4324/9781315774589

AlZubi, A. A., & Galyna, K. (2023). Artificial intelligence and internet of things for sustainable farming and smart agriculture. *IEEE Access : Practical Innovations, Open Solutions*, *11*, 78686–78692. DOI: 10.1109/ACCESS.2023.3298215

Amalan, A. A., & Aram, I. A. (2025). Artificial Intelligence adoption in Non-Chemical agriculture: An Integrated Mechanism for Sustainable Practices. *Sustainability (Basel)*, *17*(19), 8865. DOI: 10.3390/su17198865

Anjali, P. K., Sreerekha, S., & Monu, N. C. (2026). Sustainable Resource Management with AI: Innovations for a Greener Future. AI for Our Planet: How Artificial Intelligence can Solve. *Global Challenges (Hoboken, NJ)*, •••, 151.

Arab Momeni, M., Jain, V., & Bagheri, M. (2024). A multi-objective model for designing a sustainable closed-loop supply chain logistics network. *Logistics (Basel)*, *8*(1), 29. DOI: 10.3390/logistics8010029

Aragani, V. M., & Thirunagalingam, A. (2025). Leveraging advanced analytics for sustainable success. In *Advances in Business Strategy and Competitive Advantage* (pp. 229–248). IGI Global. DOI: 10.4018/979-8-3693-9750-3.ch012

Araújo, S. O., Peres, R. S., Barata, J., Lidon, F., & Ramalho, J. C. (2021). Characterising the Agriculture 4.0 Landscape—Emerging Trends, challenges and opportunities. *Agronomy (Basel)*, *11*(4), 667. DOI: 10.3390/agronomy11040667

Asorose, E. (2025). Integrating digital twins and AI-augmented predictive analytics for resilient, demand-driven global supply chain orchestration under volatility. *International Journal of Science and Research Archive*, *16*(2), 971–992. DOI: 10.30574/ijsra.2025.16.2.2430

Attia, F. (2025). AI in a Dual Challenge: Creative Remedies for the Hunger and Climate Disasters. *Emirati Journal of Business, Economics, &. Social Studies (Maynooth, Ireland)*, *4*(2), 254–268.

Awan, U., Sroufe, R., & Shahbaz, M. (2021). Industry 4.0 and the circular economy: A literature review and recommendations for future research. *Business Strategy and the Environment*, *30*(4), 2038–2060. DOI: 10.1002/bse.2731

Ayadi, R., Forouheshfar, Y., & Moghadas, O. (2025). Enhancing system resilience to climate change through artificial intelligence: A systematic literature review. *Frontiers in Climate*, *7*, 1585331. Advance online publication. DOI: 10.3389/fclim.2025.1585331

Ayim, C., Kassahun, A., Addison, C., & Tekinerdogan, B. (2022). Adoption of ICT innovations in the agriculture sector in Africa: A review of the literature. *Agriculture & Food Security*, *11*(1), 22. Advance online publication. DOI: 10.1186/s40066-022-00364-7

Ayoola, V. B., Idoko, I. P., Eromonsei, S. O., Afolabi, O., Apampa, A. R., & Oyebanji, O. S. (2024). The role of big data and AI in enhancing biodiversity conservation and resource management in the USA. *World Journal of Advanced Research and Reviews*, *23*(2), 1851–1873. DOI: 10.30574/wjarr.2024.23.2.2350

Ayre, M., Collum, V. M., Waters, W., Samson, P., Curro, A., Nettle, R., Paschen, J., King, B., & Reichelt, N. (2019). Supporting and practising digital innovation with advisers in smart farming. *NJAS Wageningen Journal of Life Sciences*, *90–91*(1), 1–12. DOI: 10.1016/j.njas.2019.05.001

Aziz, D., Rafiq, S., Saini, P., Ahad, I., Gonal, B., Rehman, S. A., Rashid, S., Saini, P., Rohela, G. K., Aalum, K., Singh, G., Gnanesh, B. N., & Nabila Iliya, M. (2025). Remote sensing and artificial intelligence: Revolutionizing pest management in agriculture. *Frontiers in Sustainable Food Systems*, *9*, 1551460. DOI: 10.3389/fsufs.2025.1551460

Bac, C. W., van Henten, E. J., Hemming, J., & Edan, Y. (2014). Harvesting robots for high-value crops: State-of-the-art review and challenges ahead. *Journal of Field Robotics*, *31*(6), 888–911. DOI: 10.1002/rob.21525

Bag, S., Routray, S., Rahman, M. S., & Gupta, S. (2025). Digital innovation for circular supply chain sustainability and resilience for achieving carbon neutrality: An empirical study. *Journal of Environmental Management*, *386*, 125665. DOI: 10.1016/j.jenvman.2025.125665 PMID: 40359861

Bahmutsky, S., Grassauer, F., Arulnathan, V., & Pelletier, N. (2024). A review of life cycle impacts and costs of precision agriculture for cultivation of field crops. *Sustainable Production and Consumption*, *52*, 347–362. DOI: 10.1016/j.spc.2024.11.010

Bains, R., Singh, N., & Singh, N. P. (2025). Artificial intelligence in predictive, personalized, and regenerative endodontics: A narrative review of emerging paradigms. *Journal of Healthcare Research and Education*, *1*(5), 5. DOI: 10.25259/JHRE_9_2025

Balafoutis, A. T.. (2017). Precision agriculture technologies positively contributing to greenhouse gas emissions mitigation. *Sustainability*, *9*(8), 1339. DOI: 10.3390/su9081339

Balakumar, S., Narayanan, M., Kamaraj, M., & Aravind, J. (2025). Harnessing artificial intelligence for sustainable environmental remediation a review [Review of *Harnessing artificial intelligence for sustainable environmental remediation a review*]. *International Journal of Environmental Science and Technology, 22*(13), 13189. Springer Science+Business Media. DOI: 10.1007/s13762-025-06528-9

Balaska, V., Adamidou, Z., Vryzas, Z., & Gasteratos, A. (2023). Sustainable crop protection via robotics and artificial intelligence solutions. *Machines (Basel)*, *11*(8), 774. DOI: 10.3390/machines11080774

Balasubramaniam, S., Nelson, S. G., Arishma, M., Rajan, A. S., & Satheesh Kumar, K. (2024). Machine Learning based Disease and Pest detection in Agricultural Crops. *EAI Endorsed Transactions on Internet of Things*, *10*. Advance online publication. DOI: 10.4108/eetiot.5049

Barroso-Barroso, C., Vega-Muñoz, A., Maradiaga-López, J., Salazar-Sepúlveda, G., & Carabantes-Silva, R. (2025). Smart Farming and the SDGs: Emerging research patterns and sustainability implications. *Agriculture*, *16*(1), 81. DOI: 10.3390/agriculture16010081

Baskara, F. X. R. (2024). Generative AI as an Enabler of Sustainable Education: Theoretical Perspectives and Future Directions. *British Journal of Teacher Education and Pedagogy*, *3*(3), 122–134. DOI: 10.32996/bjtep.2024.3.3.9

Bawa, S., & Asongu, S. A. (2026). Investigating How Firms in Emerging Economies Integrate AI-Driven Systems and ESG Metrics Into Biodiversity Conservation and Circular Economy Strategies. *Business Strategy and the Environment*, bse.70547. Advance online publication. DOI: 10.1002/bse.70547

Begho, T., Daubry, T. P., & Irabor, A. E. (2025). A Scoping Review of Artificial Intelligence and the Future of African Agriculture: Opportunities, Challenges and the Path Forward. *Advances in Agriculture, 2025*(1). DOI: 10.1155/aia/2862074

Ben-Daya, M., Hassini, E., & Bahroun, Z. (2019). Internet of things and supply chain management: A literature review. *International Journal of Production Research*, *57*(15–16), 4719–4742. DOI: 10.1080/00207543.2017.1402140

Bhattacharya, S., Govindan, K., Ghosh Dastidar, S., & Sharma, P. (2024). Applications of artificial intelligence in closed-loop supply chains: Systematic literature review and future research agenda. *Transportation Research Part E, Logistics and Transportation Review*, *184*, 103455. DOI: 10.1016/j.tre.2024.103455

Bibri, S. E., & Krogstie, J. (2017). Smart sustainable cities of the future: An extensive interdisciplinary literature review. *Sustainable Cities and Society*, *31*, 183–212. DOI: 10.1016/j.scs.2017.02.016

Birhane, A. (2021). Algorithmic injustice: A relational ethics approach. *Patterns (New York, N.Y.)*, *2*(2), 100205. DOI: 10.1016/j.patter.2021.100205 PMID: 33659914

Birner, R., Daum, T., & Pray, C. (2021). Who drives the digital revolution in agriculture? A review of supply-side trends, players and challenges. *Applied Economic Perspectives and Policy*, *43*(4), 1260–1285. DOI: 10.1002/aepp.13145

Bocken, N. M. P., de Pauw, I., Bakker, C., & van der Grinten, B. (2016). Product design and business model strategies for a circular economy. *Journal of Industrial and Production Engineering*, *33*(5), 308–320. DOI: 10.1080/21681015.2016.1172124

Boer, M. M., Resco de Dios, V., & Bradstock, R. A. (2020). Unprecedented burn area of Australian mega forest fires. *Nature Climate Change*, *10*(3), 171–172. DOI: 10.1038/s41558-020-0716-1

Bongarts Lebbe, T., Rey-Valette, H., Chaumillon, É., Camus, G., Almar, R., Cazenave, A., Claudet, J., Rocle, N., Meur-Férec, C., Viard, F., Mercier, D., Dupuy, C., Ménard, F., Rossel, B. A., Mullineaux, L., Sicre, M.-A., Zivian, A., Gaill, F., & Euzen, A. (2021). Designing coastal adaptation strategies to tackle sea level rise. *Frontiers in Marine Science*, *8*(740602), 740602. Advance online publication. DOI: 10.3389/fmars.2021.740602

Botero-Valencia, J., García-Pineda, V., Valencia-Arias, A., Valencia, J., Reyes-Vera, E., Mejia-Herrera, M., & Hernández-García, R. (2025). Machine learning in sustainable agriculture: Systematic review and research perspectives. *Agriculture*, *15*(4), 377. DOI: 10.3390/agriculture15040377

Bowman, D. M. J. S., Williamson, G. J., Abatzoglou, J. T., Kolden, C. A., Cochrane, M. A., & Smith, A. M. S. (2017). Human exposure and sensitivity to globally extreme wildfire events. Nature Ecology & Evolution, 1(3), 0058. DOI: 10.1038/s41559-016-0058

Breiman, L. (2001). Random forests. *Machine Learning*, *45*(1), 5–32. DOI: 10.1023/A:1010933404324

Bressanelli, G., Adrodegari, F., Pigosso, D. C. A., & Parida, V. (2022). Towards the smart circular economy paradigm: A definition, conceptualization, and research agenda. *Sustainability (Basel)*, *14*(9), 4960. DOI: 10.3390/su14094960

Bughin, J., Seong, J., Manyika, J., Chui, M., & Joshi, R. (2018). *Notes from the AI frontier: Modeling the impact of AI on the world economy*. McKinsey Global Institute._https://www.mckinsey.com/featured-insights/artificial-intelligence/notes-from-the-ai-frontier-modeling-the-impact-of-ai-on-the-world-economy

Buttigieg, P. L., Pafilis, E., Lewis, S. E., Schildhauer, M. P., Walls, R. L., & Mungall, C. J. (2016). The environment ontology in 2016: Bridging domains with increased scope, semantic density, and interoperation. *Journal of Biomedical Semantics*, *7*(1), 57. Advance online publication. DOI: 10.1186/s13326-016-0097-6 PMID: 27664130

Cagno, E., Neri, A., Negri, M., Bassani, C. A., & Lampertico, T. (2021). The role of digital technologies in operationalizing the circular economy transition: A systematic literature review. *Applied Sciences (Basel, Switzerland)*, *11*(8), 3328. DOI: 10.3390/app11083328

Candelieri, A. (2017). Clustering and support vector regression for water demand forecasting and anomaly detection. *Water (Basel)*, *9*(3), 224. Advance online publication. DOI: 10.3390/w9030224

Cannas, V. G., Ciano, M. P., Saltalamacchia, M., & Secchi, R. (2024). Artificial intelligence in supply chain and operations management: A multiple case study research. *International Journal of Production Research*, *62*(9), 3333–3360. DOI: 10.1080/00207543.2023.2232050

Carrer, M. J., De Souza Filho, H. M., De Mello Brandão Vinholis, M., & Mozambani, C. I. (2022). Precision agriculture adoption and technical efficiency: An analysis of sugarcane farms in Brazil. *Technological Forecasting and Social Change*, *177*, 121510. DOI: 10.1016/j.techfore.2022.121510

Cath, C., Wachter, S., Mittelstadt, B., Taddeo, M., & Floridi, L. (2018). Artificial intelligence and the 'good society': The US, EU, and UK approach. *Philosophy & Technology*, *31*(4), 505–528. PMID: 28353045

Ceccato, P., Flasse, S., & Gregoire, J. M. (2019). Designing a spectral index to estimate vegetation water content from remote sensing data. *Remote Sensing of Environment*, *82*(2–3), 188–197.

Cerjak, M., Medici, M., Faletar, I., Sundeep, J. V., & Canavari, M. (2025). Adoption of mobile-based agricultural extension services: Evidence from South India. *Journal of Rural Studies*, *120*, 103851. DOI: 10.1016/j.jrurstud.2025.103851

Chamorro-Padial, J., García, R., & Gil, R. (2024). A systematic review of open data in agriculture. *Computers and Electronics in Agriculture*, *219*, 108775. DOI: 10.1016/j.compag.2024.108775

Chan, C. K. Y. (2023). A comprehensive AI policy education framework for university teaching and learning. *International Journal of Educational Technology in Higher Education*, *20*(1), 38. Advance online publication. DOI: 10.1186/s41239-023-00408-3

Chang, F. J., Chang, L. C., & Chen, J. F. (2023). Artificial intelligence techniques in hydrology and water resources management. *Water (Basel)*, *15*(10), 1846. DOI: 10.3390/w15101846

Chaudhary, G. (2023). Environmental Sustainability: Can Artificial Intelligence be an Enabler for SDGs? *Nature Environment and Pollution Technology*, *22*(3), 1411–1420. DOI: 10.46488/NEPT.2023.v22i03.027

Chauhan, C., Parida, V., & Dhir, A. (2022). Linking circular economy and digitalisation technologies: A systematic literature review of past achievements and future promises. *Technological Forecasting and Social Change*, *177*, 121508. DOI: 10.1016/j.techfore.2022.121508

Chauhan, M., & Sahoo, D. R. (2024). Towards a Greener Tomorrow: Exploring the Potential of AI, Blockchain, and IoT in Sustainable Development. *Nature Environment and Pollution Technology*, *23*(2), 1105–1113. DOI: 10.46488/NEPT.2024.v23i02.044

Chisom, O. N., Biu, P. W., Umoh, A. A., Obaedo, B. O., Adegbite, A. O., & Abatan, A. (2024). Reviewing the role of AI in environmental monitoring and conservation: A data-driven revolution for our planet. *World Journal of Advanced Research and Reviews*, *21*(1), 161–171. DOI: 10.30574/wjarr.2024.21.1.2720

Chi, Z., & Liu, Z. (2023). Driving circular economy through digital technologies: A systematic literature review. *Sustainability*, *15*(24), 16608. DOI: 10.3390/su152416608

Choruma, D. J., Dirwai, T. L., Mutenje, M. J., Mustafa, M., Chimonyo, V. G. P., Jacobs-Mata, I., & Mabhaudhi, T. (2024). Digitalisation in agriculture: A scoping review of technologies in practice, challenges, and opportunities for smallholder farmers in sub-saharan africa. *Journal of Agriculture and Food Research*, *18*, 101286. DOI: 10.1016/j.jafr.2024.101286

Chun, Y., & Hwang, J. (2024). The Nexus of Artificial Intelligence and Green Innovation: A Cross-Density Analysis at the Country Level. *Journal of the Knowledge Economy*, *16*(1), 1688–1716. Advance online publication. DOI: 10.1007/s13132-024-02076-8

Chuvieco, E., Mouillot, F., van der Werf, G. R., San Miguel, J., Tanasse, M., Koutsias, N., & Mbow, C. (2020). Historical background and current developments for mapping burned area from satellite Earth observation. *Remote Sensing of Environment*, *225*, 45–64. DOI: 10.1016/j.rse.2019.02.013

Ciliberti, S., Frascarelli, A., Polenzani, B., Brunori, G., & Martino, G. (2024). Digitalisation strategies in the agri-food system: The case of PDO Parmigiano Reggiano. *Agricultural Systems*, *218*, 103996. DOI: 10.1016/j.agsy.2024.103996

Cova, T. J., Dennison, P. E., Kim, T. H., & Moritz, M. A. (2005). Setting wildfire evacuation trigger points using fire spread modeling and GIS. *Transactions in GIS*, *9*(4), 603–617. DOI: 10.1111/j.1467-9671.2005.00237.x

Cutter, S. L., Boruff, B. J., & Shirley, W. L. (2012). Social vulnerability to environmental hazards. In *Hazards vulnerability and environmental justice* (pp. 115–132). Routledge. DOI: 10.4324/9781849771542

D'amore, G., Di Vaio, A., Balsalobre-Lorente, D., & Boccia, F. (2022). Artificial intelligence in the water–energy–food model: A holistic approach towards sustainable development goals. *Sustainability (Basel)*, *14*(2), 867. DOI: 10.3390/su14020867

Daum, T., Scheiterle, L., Yameogo, V., Adegbola, Y. P., Mulinge, W., Kergna, A. O., Daudu, C., Angara, U. A., Zossou, R. C., Nientao, A., Fatunbi, O., Isuyi, L., & Birner, R. (2025). Moving beyond the productivity paradigm: Agricultural innovation systems and sustainable transformation in Africa. *Agricultural Systems*, *229*, 104445. DOI: 10.1016/j.agsy.2025.104445

Davis, F. D. (1989). Perceived usefulness, perceived ease of use, and user acceptance of information technology. *Management Information Systems Quarterly*, *13*(3), 319–340. DOI: 10.2307/249008

Davis, J. L., Green, J. D., & Reed, A. (2009). Interdependence with the environment: Commitment, interconnectedness, and environmental behavior. *Journal of Environmental Psychology*, *29*(2), 173–180. DOI: 10.1016/j.jenvp.2008.11.001

De Avila, F. R., & Barbosa, J. L. V. (2025). Smart environments in digital agriculture: A systematic review and taxonomy. *Computers and Electronics in Agriculture*, *236*, 110393. DOI: 10.1016/j.compag.2025.110393 de Groot, W. J., Flannigan, M. D., & Cantin, A. S. (2015). Climate change impacts on future boreal fire regimes. *Forest Ecology and Management*, *294*, 35–44. DOI: 10.1016/j.foreco.2012.09.027

Del-Coco, M., Leo, M., & Carcagnì, P. (2024). Machine learning for smart irrigation in agriculture: How far along are we? *Information (Basel)*, *15*(6), 306. DOI: 10.3390/info15060306

Dhakal, M., Banerjee, M., Pandey, K. K., & Seniya, C. (2026). Ai and IoT are no longer limited to high-tech labs but are improving nutrition-focused agricultural farming. *Discover Food*, *6*(1), 69. Advance online publication. DOI: 10.1007/s44187-025-00777-w

Dhanaraj, R. K., Maragatharajan, M., Sureshkumar, A., & Balakannan, S. P. (2025). On-device AI for climate-resilient farming with intelligent crop yield prediction using lightweight models on smart agricultural devices. *Scientific Reports*, *15*(1), 31195. DOI: 10.1038/s41598-025-16014-4 PMID: 40854942

Dhanaraju, M., Chenniappan, P., Ramalingam, K., Pazhanivelan, S., & Kaliaperumal, R. (2022). Smart Farming: Internet of Things (IoT)-Based Sustainable agriculture. *Agriculture*, *12*(10), 1745. DOI: 10.3390/agriculture12101745

Dignum, V. (2019). *Responsible artificial intelligence: How to develop and use AI in a responsible way*. Springer., DOI: 10.1007/978-3-030-30371-6

Dixit, K., Aashish, K., & Dwivedi, A. K. (2023). Antecedents of smart farming adoption to mitigate the digital divide – extended innovation diffusion model. *Technology in Society*, *75*, 102348. DOI: 10.1016/j.techsoc.2023.102348

Dodman, D., Archer, D., & Satterthwaite, D. (2019). Responding to climate change in contexts of urban poverty and informality. *Environment and Urbanization*, *31*(1), 3–12. DOI: 10.1177/0956247819830004

Dolci, V., Bigliardi, B., Petroni, A., Pini, B., Filippelli, S., & Tagliente, L. (2024). Integrating Industry 4.0 and circular economy: A conceptual framework for sustainable manufacturing. *Procedia Computer Science*, *232*, 1711–1720. DOI: 10.1016/j.procs.2024.01.169

Duan, C., Xiu, G., & Yao, F. (2019). Multi-period E-closed-loop supply chain network considering consumers' preference for products and AI-push. *Sustainability (Basel)*, *11*(17), 4571. DOI: 10.3390/su11174571

Dueben, P. D., & Bauer, P. (2018). Challenges and design choices for global weather and climate models based on machine learning. *Geoscientific Model Development*, *11*(10), 3999–4009. DOI: 10.5194/gmd-11-3999-2018

Duguma, A. L., & Bai, X. (2024). How the internet of things technology improves agricultural efficiency. *Artificial Intelligence Review, 58*(2), 63. Advance online publication. DOI: 10.1007/s10462-024-11046-0

Duncan, E., Ruder, S., Kc, K. B., Wittman, H., & Fraser, E. (2026). The political economy of agricultural data governance: Farmer perspectives on trust in data sharing. *Journal of Rural Studies, 123*, 104076. DOI: 10.1016/j.jrurstud.2026.104076

Dwivedi, Y. K., Hughes, L., Ismagilova, E., Aarts, G., Coombs, C., Crick, T., Duan, Y., Dwivedi, R., Edwards, J., Eirug, A., Galanos, V., Ilavarasan, P. V., Janssen, M., Jones, P., Kar, A. K., Kizgin, H., Kronemann, B., Lal, B., Lucini, B., & Williams, M. D. (2021). Artificial Intelligence (AI): Multidisciplinary perspectives on emerging challenges, opportunities, and agenda for research, practice and policy. *International Journal of Information Management, 57*, 101994. DOI: 10.1016/j.ijinfomgt.2019.08.002

Eaton, B., Gregory, J., Drach, B., Taylor, K., Hankin, S., Caron, J., Signell, R., Bentley, P., Rappa, G., Höck, H., Pamment, A., Juckes, M., & Raspaud, M. (2017). *NetCDF climate and forecast (CF) metadata conventions* (Version 1.7). CF Conventions Committee. https://cfconventions.org/cf-conventions/cf-conventions.html

Eccles, R. G., & Klimenko, S. (2019). The investor revolution. *Harvard Business Review, 97*(3), 106–116.

Efremova, N., Foley, J. C., Unagaev, A., & Karimi, R. (2023). AI for sustainable agriculture and rangeland monitoring. In *The ethics of artificial intelligence for the sustainable development goals* (pp. 399–422). Springer International Publishing., DOI: 10.1007/978-3-031-21147-8_22

El Jarroudi, M., Kouadio, L., Delfosse, P., Bock, C. H., Mahlein, A. K., Fettweis, X., Mercatoris, B., Adams, F., Lenné, J. M., & Hamdioui, S. (2024). Leveraging edge artificial intelligence for sustainable agriculture. *Nature Sustainability, 7*(7), 846–854. DOI: 10.1038/s41893-024-01352-4

Elbasi, E., Mostafa, N., AlArnaout, Z., Zreikat, A. I., Cina, E., Varghese, G., Shdefat, A., Topcu, A. E., Abdelbaki, W., Mathew, S., & Zaki, C. (2022). Artificial intelligence technology in the agricultural sector: A systematic literature review. *IEEE Access : Practical Innovations, Open Solutions, 11*, 171–202. DOI: 10.1109/ACCESS.2022.3232485

Elith, J., Graham, C. H., Anderson, R. P., Dudík, M., Ferrier, S., Guisan, A., Hijmans, R. J., Huettmann, F., Leathwick, J. R., Lehmann, A., Li, J., Lohmann, L. G., Loiselle, B. A., Manion, G., Moritz, C., Nakamura, M., Nakazawa, Y., Overton, J. M., Peterson, A. T., & Zimmermann, N. E. (2006). Novel methods improve prediction of species' distributions from occurrence data. *Ecography*, *29*(2), 129–151. DOI: 10.1111/j.2006.0906-7590.04596.x

Elliott, C. P., Commander, L. E., Merino-Martín, L., Golos, P. J., Stevens, J., & Miller, B. P. (2022). An approach to defining and achieving restoration targets for a threatened plant community. *Ecological Applications : A Publication of the Ecological Society of America*, *32*(6), e2613. DOI: 10.1002/eap.2613 PMID: 35366034

Erike, A., Ikerionwu, C., Azubogu, A., & Obodoagwu, V. (2025). Is AI for illiterate farmers? A systematic literature review of AI and machine learning applications and challenges for precision agriculture. *Discover Artificial Intelligence*, *5*(1), 204. Advance online publication. DOI: 10.1007/s44163-025-00457-9

Esiri, A. E., Sofoluwe, O. O., & Ukato, A. (2024). Digital twin technology in oil and gas infrastructure: Policy requirements and implementation strategies. *Engineering Science & Technology Journal*, *5*(6), 2039–2049. DOI: 10.51594/estj.v5i6.1221

Espinel, R., Herrera-Franco, G., Rivadeneira García, J. L., & Escandón-Panchana, P. (2024). Artificial intelligence in agricultural mapping. *Agriculture*, *14*(7), 1071. DOI: 10.3390/agriculture14071071

Estopinan, J., Servajean, M., Bonnet, P., Munoz, F., & Joly, A. (2022). Deep species distribution modeling from sentinel-2 image time-series: A global scale analysis on the orchid family. *Frontiers in Plant Science*, *13*, 839327. DOI: 10.3389/fpls.2022.839327 PMID: 35528931

Fairbairn, M., & Kish, Z. (2023). Setting data free: The politics of open data for food and agriculture. *New Media & Society*, *25*(8), 1935–1959. DOI: 10.1177/14614448231174520

Farasin, A., Colomba, L., & Garza, P. (2021). Double-step U-Net: A deep learning-based approach for the estimation of wildfire damage severity through Sentinel-2 satellite data. *Applied Sciences (Basel, Switzerland)*, *10*(12), 4332. DOI: 10.3390/app10124332

Farshadfar, Z., Mucha, T., & Tanskanen, K. (2024). Leveraging machine learning for advancing circular supply chains: A systematic literature review. *Logistics (Basel)*, *8*(4), 108. DOI: 10.3390/logistics8040108

Feng, C., Ye, G., Zeng, J., Zeng, J., Jiang, Q., He, L., Zhang, Y., & Xu, Z. (2023). Sustainably developing global blue carbon for climate change mitigation and economic benefits through international cooperation. *Nature Communications*, *14*(1), 6144. DOI: 10.1038/s41467-023-41870-x PMID: 37783692

Ferraris, S., Meo, R., Pinardi, S., Salis, M., & Sartor, G. (2023). Machine learning as a strategic tool for helping cocoa farmers in Côte D'Ivoire. *Sensors (Basel)*, *23*(17), 7632. DOI: 10.3390/s23177632 PMID: 37688090

Filho, W. L., Kim, E., Borsatto, J. M. L. S., & Marcolin, C. B. (2025). Using artificial intelligence in sustainability teaching and learning. *Environmental Sciences Europe*, *37*(1), 124. Advance online publication. DOI: 10.1186/s12302-025-01159-w

Fiske, I., & Chandler, R. (2011). unmarked: An R package for fitting hierarchical models of wildlife occurrence and abundance. *Journal of Statistical Software*, *43*(10), 1–23. DOI: 10.18637/jss.v043.i10

Flannigan, M., Cantin, A. S., de Groot, W. J., Wotton, M., Newbery, A., & Gowman, L. M. (2013). Global wildland fire season severity in the 21st century. *Forest Ecology and Management*, *294*, 54–61. DOI: 10.1016/j.foreco.2012.10.022

Floridi, L., Cowls, J., Beltrametti, M., Chatila, R., Chazerand, P., Dignum, V., & Vayena, E. (2018). AI4People—an ethical framework for a good AI society: Opportunities, risks, principles, and recommendations. *Minds and Machines*, *28*(4), 689–707. DOI: 10.1007/s11023-018-9482-5 PMID: 30930541

Fountas, S., Mylonas, N., Malounas, I., Rodias, E., Hellmann Santos, C., & Pekkeriet, E. (2020). Agricultural robotics for field operations. *Sensors (Basel)*, *20*(9), 2672. Advance online publication. DOI: 10.3390/s20092672 PMID: 32392872

Frazier, A. E., & Song, L. (2024). Artificial intelligence in landscape ecology: Recent advances, perspectives, and opportunities. *Current Landscape Ecology Reports*, *10*(1), 1. Advance online publication. DOI: 10.1007/s40823-024-00103-7

Frija, A. (2024). A system readiness approach to support the packaging and scaling of innovation bundles for farming systems transformation. *Agricultural Systems*, *221*, 104148. DOI: 10.1016/j.agsy.2024.104148

Frikha, M. A., & Mrad, M. (2025). AI-Driven Supply Chain Decarbonization: Strategies for Sustainable Carbon Reduction. *Sustainability (Basel)*, *17*(21), 9642. DOI: 10.3390/su17219642

Frost, J., Gambacorta, L., Huang, Y., Shin, H. S., & Zbinden, P. (2019). BigTech and the changing structure of financial intermediation. *Economic Policy*, *34*(100), 761–799. DOI: 10.1093/epolic/eiaa003

Garske, B., Bau, A., & Ekardt, F. (2021). Digitalization and AI in European agriculture: A strategy for achieving climate and biodiversity targets? *Sustainability (Basel), 13*(9), 4652. DOI: 10.3390/su13094652

Gaur, A., & Kumar, M. (2020). A comprehensive study on machine learning and deep learning techniques for wildfire prediction. *International Journal of Environmental Research and Public Health, 17*(2), 602. DOI: 10.3390/ijerph17020602 PMID: 31963490

Geels, F. W. (2004). From sectoral systems of innovation to socio-technical systems: Insights about dynamics and change from sociology and institutional theory. *Research Policy, 33*(6–7), 897–920. DOI: 10.1016/j.respol.2004.01.015

Geodetic Parameter Dataset, E. P. S. G. (2023). *EPSG registry*. International Association of Oil & Gas Producers. https://epsg.org

George, B. (2023). Artificial intelligence for Circular Economy 2.0: Opportunities, challenges, and outlook. *Clean Technologies and Recycling, 3*(3), 166–171. DOI: 10.3934/ctr.2023011

Ghisellini, P., Cialani, C., & Ulgiati, S. (2016). A review on circular economy: The expected transition to a balanced interplay of environmental and economic systems. *Journal of Cleaner Production, 114*, 11–32. DOI: 10.1016/j.jclepro.2015.09.007

Giarratano, J. C., & Riley, G. D. (2005). *Expert systems: Principles and programming* (4th ed.). Thomson Course Technology.

Giglio, L., Schroeder, W., & Justice, C. O. (2016). The collection 6 MODIS active fire detection algorithm and fire products. *Remote Sensing of Environment, 178*, 31–41. DOI: 10.1016/j.rse.2016.02.054 PMID: 30158718

Gimenez, O., Kery, M., & Schaub, M. (2021). Integrated population models: Theory and ecological applications with R and JAGS. *Methods in Ecology and Evolution, 12*(3), 388–401. DOI: 10.1111/2041-210X.13524

Godara, S., Begam, S., Bana, R. S., Bedi, J., Jain, R., Haque, M. A., Parsad, R., Marwaha, S., Patial, M., Shirzad, S., & Nirmal, R. (2024). TPTC: Topic-wise problems' trend clusters for smart agricultural insights extraction and forecasting of farmer's information demand. *Scientific Reports, 14*(1), 29272. DOI: 10.1038/s41598-024-80488-x PMID: 39587214

Goldberg, D. E. (1989). *Genetic algorithms in search, optimization, and machine learning*. Addison-Wesley.

Goodfellow, I., Bengio, Y., & Courville, A. (2016). *Deep learning*. The MIT Press.

Govindan, K., & Soleimani, H. (2017). A review of reverse logistics and closed-loop supply chains: A Journal of Cleaner Production focus. *Journal of Cleaner Production*, *142*(Part 1), 371–384. DOI: 10.1016/j.jclepro.2016.03.126

Greif, L., Kimmig, A., Bobbou, S. E., Jurisch, P., & Ovtcharova, J. (2024). Strategic view on the current role of AI in advancing environmental sustainability: A SWOT analysis. *Discover Artificial Intelligence*, *4*(1), 45. Advance online publication. DOI: 10.1007/s44163-024-00146-z

Greiner, R. (2025). Conservation fencing for coastal wetland restoration: Technical requirements and financial viability as a nature-based climate solution. *Sustainability (Basel)*, *17*(16), 7295. DOI: 10.3390/su17167295

Gryshova, I., Balian, A., Antonik, I., Miniailo, V., Nehodenko, V., & Nyzhnychenko, Y. (2024). Artificial intelligence in climate smart in agricultural: Toward a sustainable farming future. *Access J*, *5*(1), 125–140. DOI: 10.46656/access.2024.5.1(8)

Guidotti, R., Monreale, A., Ruggieri, S., Turini, F., Giannotti, F., & Pedreschi, D. (2018). A survey of methods for explaining black box models. *ACM Computing Surveys, 51*(5). *ACM Computing Surveys*, *93*(5), 1–42. DOI: 10.1145/3236009

Gul, D., & Banday, R. U. Z. (2024). *Transforming crop management through advanced AI and machine learning: Insights into innovative strategies for sustainable agriculture*. AI, Computer Science and Robotics Technology., DOI: 10.5772/acrt.20240030

Habila, M. A., Ouladsmane, M., & Alothman, Z. A. (2023). Role of artificial intelligence in environmental sustainability. In *Visualization techniques for climate change with machine learning and artificial intelligence* (pp. 449–469). Elsevier., DOI: 10.1016/B978-0-323-99714-0.00009-1

Hackfort, S. (2023). Unlocking sustainability? The power of corporate lock-ins and how they shape digital agriculture in Germany. *Journal of Rural Studies*, *101*, 103065. DOI: 10.1016/j.jrurstud.2023.103065

Haines, A., & Ebi, K. (2019). The imperative for climate action to protect health. *The New England Journal of Medicine*, *380*(3), 263–273. DOI: 10.1056/NEJMra1807873 PMID: 30650330

Hair, J. F., Risher, J. J., Sarstedt, M., & Ringle, C. M. (2019). When to use and how to report the results of PLS-SEM. *European Business Review*, *31*(1), 2–24. DOI: 10.1108/EBR-11-2018-0203

Haleem, A., Javaid, M., Khan, I. H., & Mohan, S. (2023). Significant applications of artificial intelligence towards attaining sustainability. *Journal of Industrial Integration and Management*, *8*(04), 489–520. DOI: 10.1142/S2424862223500331

Hamdan, A., Ibekwe, K. I., Etukudoh, E. A., Umoh, A. A., & Ilojianya, V. I. (2024). AI and machine learning in climate change research: A review of predictive models and environmental impact [Review of *AI and machine learning in climate change research: A review of predictive models and environmental impact*]. *World Journal of Advanced Research and Reviews, 21*(1), 1999. GSC Online Press. DOI: 10.30574/wjarr.2024.21.1.0257

Ham, Y. G., Kim, J. H., & Luo, J. J. (2019). Deep learning for ENSO forecasts. *Nature*, *573*(7775), 568–572. DOI: 10.1038/s41586-019-1559-7 PMID: 31534218

Hangl, J., Behrens, V. J., & Krause, S. (2022). Barriers, drivers, and social considerations for AI adoption in supply chain management: A tertiary study. *Logistics (Basel)*, *6*(3), 63. DOI: 10.3390/logistics6030063

Han, H., Liu, Z., Li, J., & Zeng, Z. (2024). Challenges in remote sensing based climate and crop monitoring: Navigating the complexities using AI. *Journal of Cloud Computing (Heidelberg, Germany)*, *13*(1), 34. Advance online publication. DOI: 10.1186/s13677-023-00583-8

Haque, A. B., Islam, A. N., & Mikalef, P. (2022). Explainable Artificial Intelligence (XAI) from a user perspective: A synthesis of prior literature and problematizing avenues for future research. *Technological Forecasting and Social Change*, *186*, 122120. DOI: 10.1016/j.techfore.2022.122120

Hashem, I. A. T., Yaqoob, I., Anuar, N. B., Mokhtar, S., Gani, A., & Samee, U. (2015). The rise of "big data" on cloud computing: Review and open research issues. *Information Systems*, *47*, 98–115. DOI: 10.1016/j.is.2014.07.006

Henderson, P., Hu, J., Romoff, J., Brunskill, E., Jurafsky, D., & Pineau, J. (2020). Towards the systematic reporting of energy and carbon footprints of machine learning. *Journal of Machine Learning Research*, *21*(248), 1–43.

Hochreiter, S., & Schmidhuber, J. (1997). Long short-term memory. *Neural Computation*, *9*(8), 1735–1780. DOI: 10.1162/neco.1997.9.8.1735 PMID: 9377276

Hong, T. T. M., Tung, N. T. T., & Thanh, N. T. P. (2025). Mapping artificial intelligence research in higher education toward sustainable development. *Discover Sustainability*, *6*(1), 1240. Advance online publication. DOI: 10.1007/s43621-025-02162-0

Hull, C. L. (1943). The problem of intervening variables in molar behavior theory. *Psychological Review*, *50*(3), 273–291. DOI: 10.1037/h0057518

Hunecke, C., Engler, A., Jara-Rojas, R., & Poortvliet, P. M. (2017). Understanding the role of social capital in adoption decisions: An application to irrigation technology. *Agricultural Systems*, *153*, 221–231. DOI: 10.1016/j.agsy.2017.02.002

Huntingford, C., Jeffers, E. S., Bonsall, M. B., Christensen, H. M., Lees, T., & Yang, H. (2019). Machine learning and artificial intelligence to aid climate change research and preparedness. *Environmental Research Letters*, *14*(12), 124007. DOI: 10.1088/1748-9326/ab4e55

Huot, F., Hu, R. L., Goyal, N., Sankar, T., Ihme, M., & Chen, Y. F. (2022). Next day wildfire spread: A machine learning dataset to predict wildfire spreading from remote-sensing data. *IEEE Transactions on Geoscience and Remote Sensing*, *60*, 1–13. DOI: 10.1109/TGRS.2022.3192974

INSPIRE. (2007). *Directive 2007/2/EC of the European Parliament and of the Council establishing an infrastructure for spatial information in the European Community (INSPIRE)*. Official Journal of the European Union. https://inspire.ec.europa.eu

Intergovernmental Panel on Climate Change. (2021). *Climate change 2021: The physical science basis. Contribution of Working Group I to the Sixth Assessment Report of the Intergovernmental Panel on Climate Change*. Cambridge University Press., DOI: 10.1017/9781009157896

International Energy Agency. (2021). *Artificial intelligence and energy*. https://www.iea.org/reports/artificial-intelligence-and-energy

IPCC. (2022). Impacts, adaptation and vulnerability. Contribution of Working Group II to the Sixth Assessment Report of the Intergovernmental Panel on Climate Change. O., Roberts, DC, Tignor, M., Poloczanska, ES, Mintenbeck, K., Alegría, A., Craig, M., Langsdorf, S., Löschke, S., Möller, V., et al., Eds, 3056.

Islam, J. U., Rahman, Z., & Hollebeek, L. D. (2018). Consumer engagement in online brand communities: A solicitation of congruity theory. *Internet Research*, *28*(1), 23–45. DOI: 10.1108/IntR-09-2016-0279

ISO. (2019). *ISO 8601-1:2019: Date and time — Representations for information interchange — Part 1: Basic rules*. International Organization for Standardization. https://www.iso.org/standard/70907.html

Jaafari, A., Zenner, E. K., Panahi, M., & Shahabi, H. (2019). Hybrid artificial intelligence models based on a neuro-fuzzy system and metaheuristic optimization algorithms for spatial prediction of wildfire probability. *Agricultural and Forest Meteorology*, *266*, 198–207. DOI: 10.1016/j.agrformet.2018.12.015

Jabed, M. A., & Murad, M. A. (2024). Crop yield prediction in agriculture: A comprehensive review of machine learning and deep learning approaches, with insights for future research and sustainability. *Heliyon*, *10*(24), e40836. DOI: 10.1016/j.heliyon.2024.e40836 PMID: 39720079

Jacoby, J. (2002). Stimulus-organism-response reconsidered: An evolutionary step in modeling (consumer) behavior. *Journal of Consumer Psychology*, *12*(1), 51–57. DOI: 10.1207/S15327663JCP1201_05

Jain, P., Coogan, S. C., Subramanian, S. G., Crowley, M., Taylor, S., & Flannigan, M. D. (2020). A review of machine learning applications in wildfire science and management. *Environmental Reviews*, *28*(4), 478–505. DOI: 10.1139/er-2020-0019

Janssen, M., & Kuk, G. (2016). Big and open linked data (BOLD) in government: A challenge to transparency and privacy? *Government Information Quarterly*, *33*(2), 363–368.

Janssen, M., Van Der Voort, H., & Wahyudi, A. (2017). Factors influencing big data decision-making quality. *Journal of Business Research*, *70*, 338–345. DOI: 10.1016/j.jbusres.2016.08.007

Jobin, A., Ienca, M., & Vayena, E. (2019). The global landscape of AI ethics guidelines. *Nature Machine Intelligence*, *1*(9), 389–399. DOI: 10.1038/s42256-019-0088-2

Johnston, F. H., Henderson, S. B., Chen, Y., Randerson, J. T., Marlier, M., DeFries, R. S., Kinney, P., Bowman, D. M. J. S., & Brauer, M. (2012). Estimated global mortality attributable to smoke from landscape fires. *Environmental Health Perspectives*, *120*(5), 695–701. DOI: 10.1289/ehp.1104422 PMID: 22456494

Jolly, W. M., Cochrane, M. A., Freeborn, P. H., Holden, Z. A., Brown, T. J., Williamson, G. J., & Bowman, D. M. J. S. (2015). Climate-induced variations in global wildfire danger from 1979 to 2013. *Nature Communications*, *6*(1), 7537. DOI: 10.1038/ncomms8537 PMID: 26172867

Jones, J., Harris, E., Febriansah, Y., Adiwijaya, A., & Hikam, I. N. (2024). Ai for sustainable development: Applications in natural resource management, agriculture, and waste management. *International Transactions on Artificial Intelligence*, *2*(2), 143–149. DOI: 10.33050/italic.v2i2.549

Jordan, A., Dockens, A. L., Pierson, N. A., & Ren, X. (2025). Editorial: AI's impact on higher education: transforming research, teaching, and learning. *Frontiers in Education, 10*, 1682901. Advance online publication. DOI: 10.3389/feduc.2025.1682901

Kamilaris, A., & Francesc, X. Prenafeta-Boldú. (2018). Deep learning in agriculture: A survey. *Agricultural Systems, 162*, 70–90. DOI: 10.1016/j.agsy.2018.02.016

Kamilaris, A., & Prenafeta-Boldú, F. X. (2018). Deep learning in agriculture: A survey. *Computers and Electronics in Agriculture, 147*, 70–90. DOI: 10.1016/j.compag.2018.02.016

Kasischke, E. S., & Turetsky, M. R. (2006). Recent changes in the fire regime across the North American boreal region. *Geophysical Research Letters, 33*(9), 2006GL025677. Advance online publication. DOI: 10.1029/2006GL025677

Kaymaz Mühling, Ş. M. (2023). Utilizing artificial intelligence (AI) for the identification and management of marine protected areas (MPAs): A review. *Journal of Geoscience and Environment Protection, 11*(09), 118–132. DOI: 10.4236/gep.2023.119008

Khan, S., Mazhar, T., Shahzad, T., Khan, M. A., Rehman, A. U., Saeed, M. M., & Hamam, H. (2025). Harnessing AI for sustainable higher education: Ethical considerations, operational efficiency, and future directions. *Discover Sustainability, 6*(1), 403. Advance online publication. DOI: 10.1007/s43621-025-00938-y

Khatri, B. B., & Karki, P. D. (2023). Artificial Intelligence (AI) in Higher Education: Growing Academic Integrity and Ethical Concerns. *Nepalese Journal of Development and Rural Studies, 20*(1), 1–7. DOI: 10.3126/njdrs.v20i01.64134

Kirchherr, J., Reike, D., & Hekkert, M. (2017). Conceptualizing the circular economy: An analysis of 114 definitions. *Resources, Conservation and Recycling, 127*, 221–232. DOI: 10.1016/j.resconrec.2017.09.005

Klerkx, L., Jakku, E., & Labarthe, P. (2019). A review of social science on digital agriculture. *NJAS Wageningen Journal of Life Sciences, 90–91*(1), 100315. DOI: 10.1016/j.njas.2019.100315

Knutti, R. (2008). Should we believe model predictions of future climate change? *Philosophical Transactions of the Royal Society A: Mathematical, Physical and Engineering Sciences, 366*(1885), 4647–4664. https://doi.org/DOI: 10.1098/rsta.2008.0169

Kollmuss, A., & Agyeman, J. (2002). Mind the gap: Why do people act environmentally and what are the barriers to pro-environmental behavior? *Environmental Education Research, 8*(3), 239–260. DOI: 10.1080/13504620220145401

Korzeb, Z., Alonso-Fariñas, B., Diéguez, A. I. I., Naharro, F. J., Kobylińska, U., di Pietro, F., Palacín-Sánchez, M.-J., Rollnik-Sadowska, E., Szpilko, D., Szydło, J., & Gallegos, A. de la T. (2024). The future of European universities on the path to sustainable development. *Engineering Management in Production and Services*, *16*(2), 68–89. DOI: 10.2478/emj-2024-0014

Kose, U., Prasath, V. B., Mondal, M., Podder, P., & Bharati, S. (Eds.). (2022). *Artificial intelligence and smart agriculture technology*. CRC Press., DOI: 10.1201/9781003299059

Kotwal, V. S.Vasifa S. Kotwal. (2025). A Digital Twin framework for intelligent water treatment, quality monitoring, and autonomous control. *International Journal of Applied Mathematics*, *38*(7s), 565–580. DOI: 10.12732/ijam.v38i7s.500

Kratzert, F., Klotz, D., Brenner, C., Schulz, K., & Herrnegger, M. (2018). Rainfall–runoff modelling using Long Short-Term Memory (LSTM) networks. *Hydrology and Earth System Sciences*, *22*(11), 6005–6022. DOI: 10.5194/hess-22-6005-2018

Krishnan, R., Sanjay, J., Gnanaseelan, C., Mujumdar, M., Kulkarni, A., & Chakraborty, S. (2020). Assessment of climate change over the Indian region. A Report of the Ministry of Earth Sciences (MoES), Government of India, 13-17.

Kumari, K., Mirzakhani Nafchi, A., Mirzaee, S., & Abdalla, A. (2025). AI-driven future farming: Achieving climate-smart and sustainable agriculture. *AgriEngineering*, *7*(3), 89. DOI: 10.3390/agriengineering7030089

Kumar, P., & Leonardi, N. (2025). Predicting morphological changes along a macrotidal coastline using a two-stage machine learning model. *Water Resources Research*, *61*(4), e2024WR037523. Advance online publication. DOI: 10.1029/2024WR037523

Kumar, S., Raut, R. D., Nayal, K., Kraus, S., Yadav, V. S., & Narkhede, B. E. (2021). To identify industry 4.0 and circular economy adoption barriers in the agriculture supply chain by using ISM-ANP. *Journal of Cleaner Production*, *293*, 126023. DOI: 10.1016/j.jclepro.2021.126023

Kumar, V., Sharma, K. V., Kedam, N., Patel, A., Kate, T. R., & Rathnayake, U. (2024). A comprehensive review on smart and sustainable agriculture using IoT technologies. *Smart Agricultural Technology*, *8*, 100487. DOI: 10.1016/j.atech.2024.100487

Kuppusamy, P. (2025). Artificial Intelligence-powered digital twin predictive analytics model for smart healthcare system. In *Advances in Computational Intelligence and Robotics* (pp. 271–324). IGI Global. DOI: 10.4018/979-8-3373-0538-7.ch009

Kushawaha, J., Borra, S., Kushawaha, A. K., Singh, G., & Singh, P. (2021). Climate change and its impact on natural resources. In *Water conservation in the era of global climate change* (pp. 333–346). Elsevier. DOI: 10.1016/B978-0-12-820200-5.00002-6

Kyathanahally, S. P., Hardeman, T., Reyes, M., Merz, E., Bulas, T., Brun, P., Pomati, F., & Baity-Jesi, M. (2022). Ensembles of data-efficient vision transformers as a new paradigm for automated classification in ecology. *Scientific Reports*, *12*(1), 18590. Advance online publication. DOI: 10.1038/s41598-022-21910-0 PMID: 36329061

Lan, J., & Ban, Q. (2025). The Farm-Level Economic and Environmental Benefits of Precision Agriculture Technology Adoption: A Meta-Analysis of Global Evidence. *Sustainability (Basel)*, *17*(24), 11223. DOI: 10.3390/su172411223

Lary, D. J., Alavi, A. H., Gandomi, A. H., & Walker, A. L. (2016). Machine learning in geosciences and remote sensing. *Geoscience Frontiers*, *7*(1), 3–10. DOI: 10.1016/j.gsf.2015.07.003

Laskurain-Iturbe, I., Arana-Landín, G., Landeta-Manzano, B., & Uriarte-Gallastegi, N. (2021). Exploring the influence of Industry 4.0 technologies on the circular economy. *Journal of Cleaner Production*, *321*, 128944. DOI: 10.1016/j.jclepro.2021.128944

Lebrasse, M. C., Schaeffer, B. A., Coffer, M. M., Whitman, P. J., Zimmerman, R. C., Hill, V. J., Islam, K. A., Li, J., & Osburn, C. L. (2022). Temporal stability of seagrass extent, leaf area, and carbon storage in St. Joseph Bay, Florida: A semi-automated remote sensing analysis. *Estuaries and Coasts*, *45*(7), 2082–2101. DOI: 10.1007/s12237-022-01050-4 PMID: 37009415

LeCun, Y., Bengio, Y., & Hinton, G. (2015). Deep learning. *Nature*, *521*(7553), 436–444. DOI: 10.1038/nature14539 PMID: 26017442

Lee, H., Calvin, K., Dasgupta, D., Krinner, G., Mukherji, A., Thorne, P., . . . Park, Y. (2023). *IPCC, 2023: Climate change 2023: Synthesis report, summary for policymakers. Contribution of working groups i, II and III to the sixth assessment report of the intergovernmental panel on climate change [core writing team, h. Lee and j. Romero (eds.)]. IPCC, geneva, Switzerland* (pp. 1-34). Intergovernmental Panel on Climate Change (IPCC).

Lehuedé, S. (2024). An elemental ethics for artificial intelligence: Water as resistance within AI's value chain. *AI & Society*. Advance online publication. DOI: 10.1007/s00146-024-01922-2

Lestari, S. L., Mayaguezz, H., & Febryano, I. G. (2025). Artificial intelligence in mangrove management policy in Indonesia: A systematic literature review. *Jurnal Perikanan Unram*, *15*(6), 3560–3569. DOI: 10.29303/jp.v15i6.2035

Liakos, K. G., Busato, P., Moshou, D., Pearson, S., & Bochtis, D. (2018). Machine learning in agriculture: A review. *Sensors (Basel)*, *18*(8), 2674. DOI: 10.3390/s18082674 PMID: 30110960

Lindsey, D. T., Schmit, T. J., MacKenzie, W. M., Pavolonis, M. J., Seaman, C. J., & Movva, L. (2020). 10.35 µm GOES-16 ABI channel application for fires. *Journal of Applied Remote Sensing*, *12*(4), 046016. DOI: 10.1117/1.JRS.12.046016

Liu, T., Sun, Y., Wang, C., Zhang, Y., Qiu, Z., Gong, W., Lei, S., Tong, X., & Duan, X. (2021). Unmanned aerial vehicle and artificial intelligence revolutionizing efficient and precision sustainable forest management. *Journal of Cleaner Production*, *311*, 127546. DOI: 10.1016/j.jclepro.2021.127546

Li, X., Zhang, Y., Li, M., & Ouyang, Y. (2023). A transformer-based approach for wildfire smoke detection using remote sensing imagery. *International Journal of Wildland Fire*, *32*(4), 501–516. DOI: 10.1071/WF22130

Longley, P. A., Goodchild, M. F., Maguire, D. J., & Rhind, D. W. (2015). *Geographic information science and systems* (4th ed.). Wiley.

Lopes de Sousa Jabbour, A. B., Jabbour, C. J. C., Godinho Filho, M., & Roubaud, D. (2018). Industry 4.0 and the circular economy: A proposed research agenda and original roadmap for sustainable operations. *Annals of Operations Research*, *270*(1), 273–286. DOI: 10.1007/s10479-018-2772-8

Maddipatla, S. (2026). Green AI for sustainable Big Data platforms. *Journal of Information Systems Engineering & Management*, *11*(1s), 228–241. DOI: 10.52783/jisem.v11i1s.14065

Malik, F. (2025). Carbon-conscious intelligence: Life Cycle Assessment and Green standards for Generative AI. *International Journal of Innovative Science and Research Technology*, (IJISRT25AUG585), 1030–1037. DOI: 10.38124/ijisrt/25aug585

Mallinger, K., Corpaci, L., Neubauer, T., Tikasz, I. E., Goldenits, G., & Banhazi, T. (2024). Breaking the barriers of technology adoption: Explainable AI for requirement analysis and technology design in smart farming. *Smart Agricultural Technology*, *9*, 100658. DOI: 10.1016/j.atech.2024.100658

Mambile, C., & Mwogosi, A. (2024). Transforming higher education in Tanzania: Unleashing the true potential of AI as a transformative learning tool. *Technological Sustainability*, *4*(1), 51–76. DOI: 10.1108/TECHS-03-2024-0014

Mana, A., Allouhi, A., Hamrani, A., Rehman, S., Jamaoui, I. E., & Jayachandran, K. (2024). Sustainable AI-based production agriculture: Exploring AI applications and implications in agricultural practices. *Smart Agricultural Technology*, *7*, 100416. DOI: 10.1016/j.atech.2024.100416

Mansour, M., Zobi, M. A., & Alomair, M. (2025). Artificial Intelligence, ESG Governance, and Green Innovation Efficiency in Emerging Economies. *Economies (Basel)*, *14*(1), 11. DOI: 10.3390/economies14010011

Margules, C. R., & Pressey, R. L. (2000). Systematic conservation planning. *Nature*, *405*(6783), 243–253. DOI: 10.1038/35012251 PMID: 10821285

Martos, V., Ahmad, A., Cartujo, P., & Ordoñez, J. (2021). Ensuring agricultural sustainability through remote sensing in the era of agriculture 5.0. *Applied Sciences (Basel, Switzerland)*, *11*(13), 5911. DOI: 10.3390/app11135911

Ma, W., & Rahut, D. B. (2024). Climate-smart agriculture: Adoption, impacts, and implications for sustainable development. *Mitigation and Adaptation Strategies for Global Change*, *29*(5), 44. Advance online publication. DOI: 10.1007/s11027-024-10139-z

Mazaherian, A., & Nourbakhsh, E. (2025). *Beyond the Hype: Critical Analysis of Student Motivations and Ethical Boundaries in Educational AI Use in Higher Education.* DOI: 10.48550/ARXIV.2511.11369

Mehrabian, A., & Russell, J. A. (1974). *An approach to environmental psychology.* the MIT Press.

Mehrabi, Z., McDowell, M. J., Ricciardi, V., Levers, C., Martinez, J. D., Mehrabi, N., Wittman, H., Ramankutty, N., & Jarvis, A. (2020). The global divide in data-driven farming. *Nature Sustainability*, *4*(2), 154–160. DOI: 10.1038/s41893-020-00631-0

Mienye, I. D., Sun, Y., & Ileberi, E. (2024). Artificial intelligence and sustainable development in Africa: A comprehensive review [Review of *Artificial intelligence and sustainable development in Africa: A comprehensive review*]. *Machine Learning with Applications, 18*, 100591. Elsevier BV. DOI: 10.1016/j.mlwa.2024.100591

Miller, T., Durlik, I., Kostecka, E., Kozlovska, P., Łobodzińska, A., Sokołowska, S., & Nowy, A. (2025). Integrating artificial intelligence agents with the internet of things for enhanced environmental monitoring: Applications in water quality and climate data. *Electronics (Basel)*, *14*(4), 696. DOI: 10.3390/electronics14040696

Mishra, H. (2025). Artificial Intelligence Technologies in Agriculture. In AgriTech Revolution: Next-Gen Solutions and Challenges in Modern Agriculture (pp. 371-396). Singapore: Springer Nature Singapore. https://doi.org/DOI: 10.1007/978-981-95-1268-3_19

Mishra, H., & Dwivedi, S. Remote Sensing for Sustainable Agriculture: A Machine Learning Approach to Optimizing Farm Yield and Economic Returns. In Artificial Intelligence and Computer Vision for Ecological Informatics (pp. 116-150). CRC Press.

Mishra, A., Dutta, P., Jayasankar, S., Jain, P., & Mathiyazhagan, K. (2023). A review of reverse logistics and closed-loop supply chains in the perspective of circular economy. *Benchmarking*, *30*(3), 975–1020. DOI: 10.1108/BIJ-11-2021-0669

Mishra, H., & Mishra, D. (2023). Artificial intelligence and machine learning in agriculture: Transforming farming systems. *Res. Trends Agric. Sci*, *1*, 1–16.

Mishra, H., & Mishra, D. (2024). AI for data-driven decision-making in smart agriculture: from field to farm management. In *Artificial intelligence techniques in smart agriculture* (pp. 173–193). Springer Nature Singapore., DOI: 10.1007/978-981-97-5878-4_11

Mishra, H., & Mishra, D. (2025). Robotics, drones, remote sensing, GIS, and IoT tools for agricultural operations and water management. In *Integrated Land and Water Resource Management for Sustainable Agriculture* (Vol. 2, pp. 21–49). Springer Nature Singapore., DOI: 10.1007/978-981-97-9800-1_2

Mishra, H., Mishra, R., Olayinka, O. T., Ghaib, A. A., Raza, M. Y., & Maroof, K. (2025). *The Role of Digital Agriculture in Mitigating Climate Change and Ensuring Food Security: An Overview. Climate Resilient and Sustainable Agriculture* (Vol. 2). Social and Transformative Strategies., DOI: 10.1007/978-3-032-04141-8_11

Mnih, V., Kavukcuoglu, K., Silver, D., Rusu, A. A., Veness, J., Bellemare, M. G., Graves, A., Riedmiller, M., Fidjeland, A. K., Ostrovski, G., Petersen, S., Beattie, C., Sadik, A., Antonoglou, I., King, H., Kumaran, D., Wierstra, D., Legg, S., & Hassabis, D. (2015). Human-level control through deep reinforcement learning. *Nature*, *518*(7540), 529–533. DOI: 10.1038/nature14236 PMID: 25719670

Mohan, R. N. V. J., Rayanoothala, P. S., & Sree, R. P. (2025). Next-gen agriculture: Integrating AI and XAI for precision crop yield predictions. *Frontiers in Plant Science*, *15*, 1451607. DOI: 10.3389/fpls.2024.1451607 PMID: 39845494

Mohanty, S. P., Hughes, D. P., & Salathé, M. (2016). Using deep learning for image-based plant disease detection. *Frontiers in Plant Science*, *7*, 1419. Advance online publication. DOI: 10.3389/fpls.2016.01419 PMID: 27713752

Moilanen, A., Wilson, K. A., & Possingham, H. P. (Eds.). (2009). *Spatial conservation prioritization: Quantitative methods and computational tools*. Oxford University Press. DOI: 10.1093/oso/9780199547760.001.0001

Molnar, C. (2020). Interpretable machine learning: A guide for making black box models explainable. Leanpub. https://christophm.github.io/interpretable-ml-book/

Molnar, C. (2020). *Interpretable machine learning*. Lulu. Com.

Moosavi, V. (2017). Delineating wildfire risk areas using linked weights-of-evidence and GIS—Application in the Zagros forests, western Iran. *International Journal of Wildland Fire*, *26*(6), 556–565.

Morash, E. A., & Clinton, S. R. (1998). Supply chain integration: Customer value through collaborative closeness versus operational excellence. *Journal of Marketing Theory and Practice*, *6*(4), 104–120. DOI: 10.1080/10696679.1998.11501814

Moritz, M. A., Batllori, E., Bradstock, R. A., Gill, A. M., Handmer, J., Hessburg, P. F., & Sugihara, N. G. (2012). Learning to coexist with wildfire. *Nature*, *515*(7525), 58–66. DOI: 10.1038/nature13946 PMID: 25373675

Morseletto, P. (2020). Targets for a circular economy. *Resources, Conservation and Recycling*, *153*, 104553. DOI: 10.1016/j.resconrec.2019.104553

Morshedi, M., Hargaden, V., Papakostas, N., & Ghadimi, P. (2026). The adoption of digital technologies in circular supply chains: From theoretical developments to practical applications. *Logistics (Basel)*, *10*(1), 18. DOI: 10.3390/logistics10010018

Mossie, W. A. (2022). The Impact of Climate-Smart Agriculture Technology on Productivity: Does Row Planting Matter? Evidence from Southern Ethiopia. *TheScientificWorldJournal*, *2022*, 1–11. DOI: 10.1155/2022/3218287 PMID: 35991088

Mudenda, S. (2025). Impact of Artificial Intelligence on College and University Students: A Global Transformation of the Education System. *Creative Education*, *16*(11), 1801–1828. DOI: 10.4236/ce.2025.1611112

Mumtaz, N., Izhar, T., Pandey, G., & Labhasetwar, P. K. (2022). Utilizing artificial intelligence for environmental sustainability. In *Artificial intelligence for renewable energy systems* (pp. 259–279). Woodhead Publishing., DOI: 10.1016/B978-0-323-90396-7.00010-9

Munz, J., Maurmann, I., Schuele, H., & Doluschitz, R. (2024). Digital transformation at what cost? A case study from Germany estimating the adoption potential of precision farming technologies under different scenarios. *Smart Agricultural Technology*, *9*, 100585. DOI: 10.1016/j.atech.2024.100585

Nadeem, W., Ashraf, A. R., Khan, H., & Kumar, V. (2025). Impact of AI strategies on climate-change performance: Responsible AI and crisis management perspectives. *Technovation*, *150*, 103390. DOI: 10.1016/j.technovation.2025.103390

Nagaraj, B., & Kalaivani, A. (2023). The Emerging Role of Artificial Intelligence in STEM Higher Education: A Critical Review [Review of The Emerging Role of Artificial Intelligence in STEM Higher Education: A Critical Review]. *International Research Journal of Multidisciplinary Technovation*, *1*, 1–19. Advance online publication. DOI: 10.54392/irjmt2351

Naghibi, S. A., Pourghasemi, H. R., & Dixon, B. (2016). GIS-based groundwater potential mapping using boosted regression tree, classification and regression tree, and random forest machine learning models in Iran. *Environmental Monitoring and Assessment*, *188*(1), 44. Advance online publication. DOI: 10.1007/s10661-015-5049-6 PMID: 26687087

Najah, A., El-Shafie, A., Karim, O. A., & El-Shafie, A. H. (2013). Application of artificial neural networks for water quality prediction. *Neural Computing & Applications*, *22*(S1), 187–201. DOI: 10.1007/s00521-012-0940-3

Nasirahmadi, A., & Hensel, O. (2022). Toward the next generation of digitalization in agriculture based on digital twin paradigm. *Sensors (Basel)*, *22*(2), 498. DOI: 10.3390/s22020498 PMID: 35062459

Nasruddin, B., Baderan, D. W. K., Rahim, S., Rauf, A., & Hamidun, M. S. (2025). Toward regenerative coastal seascapes: Integrating blue carbon, restoration ecology, and spatial planning across marine and coastal ecosystems. *International Journal of Research and Review*, *401*, 401. Advance online publication. DOI: 10.52403/ijrr.20251245

Nauslar, N. J., Abatzoglou, J. T., & Marsh, P. T. (2018). Structure, meteorology, and climatology of the Diablo and Santa Ana wind events of northern and southern California. *Natural Hazards*, *91*(3), 1475–1490. DOI: 10.1007/s11069-018-3141-y

Nautiyal, C. T., Nautiyal, P., Papnai, G., Mittal, H., Agrawal, K., & Nandini, R. (2024). Importance of smart agriculture and use of artificial intelligence in shaping the future of agriculture. *Journal of Scientific Research and Reports*, *30*(3), 129–138. DOI: 10.9734/jsrr/2024/v30i31864

Nawaz, M., & Babar, M. I. K. (2024). IoT and AI: A panacea for climate change-resilient smart agriculture. *Discover Applied Sciences*, *6*(10), 517. Advance online publication. DOI: 10.1007/s42452-024-06228-y

Ndiaye, S. M. (2025). *Building Trustworthiness as a Requirement for AI in Africa: Challenges*. Stakeholders and Perspectives., DOI: 10.1007/978-3-031-75674-0_3

Ngulube, P. (2025). Leveraging information and communication technologies for sustainable agriculture and environmental protection among smallholder farmers in tropical Africa. *Discover Environment*, *3*(1), 9. DOI: 10.1007/s44274-025-00190-1

Norouzzadeh, M. S., Nguyen, A., Kosmala, M., Swanson, A., Palmer, M. S., Packer, C., & Clune, J. (2018). Automatically identifying, counting, and describing wild animals in camera-trap images with deep learning. *Proceedings of the National Academy of Sciences of the United States of America*, *115*(25), E5716–E5725. DOI: 10.1073/pnas.1719367115 PMID: 29871948

Nti, E. K., Cobbina, S. J., Attafuah, E. E., Opoku, E., & Gyan, M. A. (2022). Environmental sustainability technologies in biodiversity, energy, transportation and water management using artificial intelligence: A systematic review. *Sustainable Futures : An Applied Journal of Technology, Environment and Society*, *4*, 100068. DOI: 10.1016/j.sftr.2022.100068

Nti, E. K., Cobbina, S. J., Attafuah, E. E., Senanu, L. D., Amenyeku, G., Gyan, M. A., Forson, D., & Safo, A. R. (2023). Water pollution control and revitalization using advanced technologies: Uncovering artificial intelligence options towards environmental health protection, sustainability and water security. *Heliyon*, *9*(7), e18170. Advance online publication. DOI: 10.1016/j.heliyon.2023.e18170 PMID: 37496916

Nugroho, M. A., Suyanto, U. Y., Hariyanto, D., & Arianisari, S. (2025). Uneven But Accelerating: AI Adoption in Higher Education. *International Journal of Advanced Computer Science and Applications*, *16*(10). Advance online publication. DOI: 10.14569/IJACSA.2025.0161041

Nxumalo, G. S., & Chauke, H. (2025). Challenges and opportunities in smallholder agriculture digitization in South Africa. *Frontiers in Sustainable Food Systems*, *9*, 1583224. Advance online publication. DOI: 10.3389/fsufs.2025.1583224

Ojadi, J. O., Odionu, C. S., Onukwulu, E. C., & Owulade, O. A. (2024). AI-enabled smart grid systems for energy efficiency and carbon footprint reduction in urban energy networks. *International Journal of Multidisciplinary Research and Growth Evaluation*, *5*(1), 1549–1566. DOI: 10.54660/.IJMRGE.2024.5.1.1549-1566

Ojija, F., Ogwu, M. C., Ally, J. S., John, J. P., Stephano, A., Felix, N., & Tekka, R. S. (2025). Artificial intelligence-driven solutions for mitigating human–wildlife conflict in biodiversity hotspots. *Science Progress*, *108*(4), 00368504251394584. Advance online publication. DOI: 10.1177/00368504251394584 PMID: 41259221

Olawade, D. B., Wada, O. Z., David-Olawade, A. C., Fapohunda, O., Ige, A. O., & Ling, J. (2024). Artificial intelligence potential for net zero sustainability: Current evidence and prospects. *Next Sustainability*, *4*, 100041. DOI: 10.1016/j.nxsust.2024.100041

Oloruntoba, O. (2024). Green cloud computing: AI for sustainable database management. *World Journal of Advanced Research and Reviews*, *23*(3), 3242–3257. DOI: 10.30574/wjarr.2024.23.3.2611

Omer, A. M. (2008). Energy, environment and sustainable development. *Renewable & Sustainable Energy Reviews*, *12*(9), 2265–2300. DOI: 10.1016/j.rser.2007.05.001

Open Geospatial Consortium. (2022). *OGC API standards*. OGC. https://ogcapi.ogc.org

Opesemowo, O. A. G., & Adekomaya, V. (2024). Harnessing Artificial Intelligence for Advancing Sustainable Development Goals in South Africa's Higher Education System: A Qualitative Study. *International Journal of Learning Teaching and Educational Research*, *23*(3), 67–86. DOI: 10.26803/ijlter.23.3.4

Osrof, H. Y., Tan, C. L., Angappa, G., Yeo, S. F., & Tan, K. H. (2023). Adoption of smart farming technologies in field operations: A systematic review and future research agenda. *Technology in Society*, *75*, 102400. DOI: 10.1016/j.techsoc.2023.102400

Padhiary, M., Hoque, A., Prasad, G., Kumar, K., & Sahu, B. (2025). Precision agriculture and AI-driven resource optimization for sustainable land and resource management. In *Smart water technology for sustainable management in modern cities* (pp. 197–232). IGI Global Scientific Publishing., DOI: 10.4018/979-8-3693-8074-1.ch009

Pagell, M., & Wu, Z. (2009). Building a more complete theory of sustainable supply chain management using case studies of 10 exemplars. *The Journal of Supply Chain Management*, *45*(2), 37–56. DOI: 10.1111/j.1745-493X.2009.03162.x

Page, M. J., McKenzie, J. E., Bossuyt, P. M., Boutron, I., Hoffmann, T. C., Mulrow, C. D., Shamseer, L., Tetzlaff, J. M., Akl, E. A., Brennan, S. E., & Chou, R. (2021). The PRISMA 2020 statement: An updated guideline. *BMJ (Clinical Research Ed.)*, •••, 372. PMID: 33782057

Pagoropoulos, A., Pigosso, D. C. A., & McAloone, T. C. (2017). The emergent role of digital technologies in the circular economy: A review. *Procedia CIRP*, *64*, 19–24. DOI: 10.1016/j.procir.2017.02.047

Pantazi, X. E., Moshou, D., Alexandridis, T., Whetton, R. L., & Mouazen, A. M. (2016). Wheat yield prediction using machine learning and advanced sensing techniques. *Computers and Electronics in Agriculture*, *121*, 57–65. DOI: 10.1016/j.compag.2015.11.018

Pasha Mohammed, S., Deepika, J., Sritharan, N., Ravichandran, V., Prasanthrajan, M., & Kannan, P. (2025). A systematic literature review on artificial intelligence in transforming precision agriculture for sustainable farming: Current status and future directions. *Plant Science Today*, *12*(2), 1–13. DOI: 10.14719/pst.6175

Pasupuleti, M. K. (2024). *EcoInformatics: Advancing climate and biodiversity solutions with geospatial data*. National Education Services., DOI: 10.62311/nesx/rb978-81-971242-6-6

Pawlak, K., & Kołodziejczak, M. (2020). The role of agriculture in ensuring food security in developing countries: Considerations in the context of the problem of sustainable food production. *Sustainability (Basel)*, *12*(13), 5488. DOI: 10.3390/su12135488

Phadikar, S., Sil, J., & Das, A. K. (2013). Rice diseases classification using feature selection and rule generation techniques. *Computers and Electronics in Agriculture*, *90*, 76–85. DOI: 10.1016/j.compag.2012.11.001

Phillips, S. J., Anderson, R. P., & Schapire, R. E. (2006). Maximum entropy modeling of species geographic distributions. *Ecological Modelling*, *190*(3–4), 231–259. DOI: 10.1016/j.ecolmodel.2005.03.026

Pimenow, S., Pimenowa, O., Prus, P., & Niklas, A. (2025). The impact of artificial intelligence on the sustainability of regional ecosystems: Current challenges and future prospects. *Sustainability (Basel)*, *17*(11), 4795. DOI: 10.3390/su17114795

Platon, V., Pavelescu, F.-M., Antonescu, D., Constantinescu, A., Frone, S., Surugiu, M., Mazilescu, R., & Popa, F. (2024). New evidence about artificial intelligence and eco-investment as boosters of the circular economy. *Environmental Technology & Innovation*, *35*, 103685. DOI: 10.1016/j.eti.2024.103685

Poenaru, M. M., Manta, L. F., Gher escu, C., & Manta, A. G. (2025). Shaping the future of horticulture: Innovative technologies, artificial intelligence, and robotic automation through a bibliometric lens. *Horticulturae*, *11*(5), 449. DOI: 10.3390/horticulturae11050449

Pournader, M., Ghaderi, H., Hassanzadegan, A., & Fahimnia, B. (2021). Artificial intelligence applications in supply chain management. *International Journal of Production Economics*, *241*, 108250. DOI: 10.1016/j.ijpe.2021.108250

Prapas, I., Bountos, N. I., Kondylatos, S., Michail, D., Camps-Valls, G., & Papoutsis, I. (2021). Deep learning for global wildfire forecasting. arXiv preprint arXiv:2211.00534. https://doi.org//arXiv.2211.00534DOI: 10.48550

Programme, U. N. E. (2023). *Strengthening transparency of non-state actors: How national experiences and new digital technologies can strengthen the Transparency efforts of non-state actors*. United Nations Environment Programme., DOI: 10.59117/20.500.11822/43573

Qwaid, M. A., Sarker, M. T., Shawon, S. M., & Zubair, H. (2026). AI-enabled smart farming framework for sustainable date palm cultivation in arid regions using machine learning and IoT integration. *Scientific Reports*, *16*(1), 5125. DOI: 10.1038/s41598-026-36106-z PMID: 41526481

Rafi, S. (2020). Age-friendly communities: creating a conducive environment for the elderly in India. Journal of Geriatric Medicine.

Rafi, S. (2026). AI in Social Work, Counseling, and Psychology Education: Innovations, Ethics, and Educational Pathways. In AI-Powered Learning in Social Work, Counseling, and Psychology (pp. 1-24). IGI Global Scientific Publishing.

Rafi, S., & Saif, S. (2021). Physical and mental health issues of elderly women in India: a review. Handbook of Research on Policies, Protocols, and Practices for Social Work in the Digital World, 376-392. DOI: 10.4018/978-1-7998-7772-1.ch021

Rafi, S. (2019). Critical analysis of the legal approach to protect the elderly population in India. *BSSS J Soc Work*, *9*, 1–20.

Rafi, S. (2026). Artificial intelligence and telemedicine in elderly healthcare: A mixed-methods study. *BMC Geriatrics*, *26*(1), 71. DOI: 10.1186/s12877-025-06872-y PMID: 41402744

Rafi, S., & Rafi, S. (2022). Mental health and stress in the digital workplace. In *Managing the digital workplace in the post-pandemic* (pp. 26–40). Routledge. DOI: 10.4324/9781003283386-4

Rafi, S., & Saif, S. (2019). *Risky Health Behaviours among Elderly in India*. Today Publication.

Rai, A. (2020). Explainable AI: From black box to glass box. *Journal of the Academy of Marketing Science*, *48*(1), 137–141. DOI: 10.1007/s11747-019-00710-5

Raihan, A., Paul, A., & Rahman, M. (2024). Artificial Intelligence (AI) for Environmental Sustainability: A Concise Review of Technology Innovations in Energy, Transportation, Biodiversity, and Water Management. *Journal of Technology Innovations and Energy*, *3*(2), 64–73. DOI: 10.56556/jtie.v3i2.953

Rajbongshi, A., Johora, F. T., Hossain, A., Sarker, M. S., Rahman, M. H., Rahman, M. W., Alotaibi, F. T., & Moni, M. A. (2026). Leveraging explainable AI for sustainable agriculture: A comprehensive review of recent advances. *Artificial Intelligence Review*, *59*(3), 105. Advance online publication. DOI: 10.1007/s10462-025-11459-5

Raman, R., Pattnaik, D., Lathabai, H. H., Kumar, C., Govindan, K., & Nedungadi, P. (2024). Green and sustainable AI research: An integrated thematic and topic modeling analysis. *Journal of Big Data*, *11*(1), 55. Advance online publication. DOI: 10.1186/s40537-024-00920-x

Ramzan, S., Ali, B., Raza, A., Hussain, I., Fitriyani, N. L., Gu, Y., & Syafrudin, M. (2024). An innovative artificial neural network model for smart crop prediction using sensory network-based soil data. *PeerJ. Computer Science*, *10*, e2478. DOI: 10.7717/peerj-cs.2478 PMID: 39650399

Rane, N. (2023). Roles and challenges of ChatGPT and similar generative artificial intelligence for achieving the sustainable development goals (SDGs). Available at *SSRN* 4603244. https://doi.org/DOI: 10.2139/ssrn.4603244

Rane, N. L., Rane, J., & Paramesha, M. (2024). Artificial Intelligence and business intelligence to enhance Environmental, Social, and Governance (ESG) strategies: Internet of things, machine learning, and big data analytics in financial services and investment sectors. In *Trustworthy Artificial Intelligence in Industry and Society*. Deep Science Publishing., DOI: 10.70593/978-81-981367-4-9_3

Rathinavel, P., Ravikumar, R. N., Aarthi, S., Habib, M., Salaeva, N., & Pardaev, J. (2025). Augmenting human capabilities. In *Advances in Computational Intelligence and Robotics* (pp. 1–30). IGI Global Scientific Publishing. DOI: 10.4018/979-8-3373-2372-5.ch001

Rathoure, A. K., & Ram, B. L. G. (2024). Unveiling the marvels of biodiversity: Recent advancements in conservation efforts. *Biodiversity International Journal*, *7*(2), 51–61. DOI: 10.15406/bij.2024.07.00211

Ravikumar, A. S., Kurbanova, N., Sevdim, K. E., Zaynitdinovna, B. D., & Jumaniyazova, N. (2026). Bridging knowledge and policy. In *Advances in Computational Intelligence and Robotics* (pp. 119–150). IGI Global Scientific Publishing. DOI: 10.4018/979-8-3373-7620-2.ch005

Ravikumar, R. N., & Aarthi, S. (2026). Energy-aware Federated Learning for climate monitoring. In *Advances in Computational Intelligence and Robotics* (pp. 113–146). IGI Global Scientific Publishing. DOI: 10.4018/979-8-3373-7554-0.ch005

Ravikumar, R. N., Aarthi, S., Nadarajan, S., Masharipova, G., Mamadiyarov, Z., & Tamilselvi, S. (2025). Big data and cloud computing for climate-oriented remote sensing. In *Advances in Geospatial Technologies* (pp. 165–196). IGI Global Scientific Publishing. DOI: 10.4018/979-8-3373-3501-8.ch006

Ravikumar, R. N., & Aarthi, S. (2025). Climate data without borders. In *Cross-Border Cooperation in the Age of Climate Change* (pp. 61–90). IGI Global Scientific Publishing., DOI: 10.4018/979-8-3373-3972-6.ch003

Ray, D. K., Mueller, N. D., West, P. C., & Foley, J. A. (2013). Yield trends are insufficient to double global crop production by 2050. *PLoS One*, *8*(6), e66428. DOI: 10.1371/journal.pone.0066428 PMID: 23840465

Razak, A., Yogarayan, S., Sayeed, M. S., & Derafi, M. I. F. M. (2024). Agriculture 5.0 and Explainable AI for Smart Agriculture: A Scoping Review [Review of Agriculture 5.0 and Explainable AI for Smart Agriculture: A Scoping Review]. *Emerging Science Journal*, *8*(2), 744–760. DOI: 10.28991/ESJ-2024-08-02-024

Reichstein, M., Camps-Valls, G., Stevens, B., Jung, M., Denzler, J., Carvalhais, N., & Prabhat, . (2019). Deep learning and process understanding for data-driven Earth system science. *Nature*, *566*(7743), 195–204. DOI: 10.1038/s41586-019-0912-1 PMID: 30760912

Reitano, M., Segovia, M. S., & Nayga, R. M.Jr. (2025). A systematic review on the impact of Artificial Intelligence in the agri-food supply chain. *Food Policy*, *137*, 102983. DOI: 10.1016/j.foodpol.2025.102983

Ren, X., Li, X., Ren, K., Song, J., Xu, Z., Deng, K., & Wang, X. (2021). Deep learning-based weather prediction: A survey. *Big Data Research*, *23*, 100178. DOI: 10.1016/j.bdr.2020.100178

Reynolds, S. A., Beery, S., Burgess, N. D., Burgman, M. A., Butchart, S. H. M., Cooke, S. J., Coomes, D. A., Danielsen, F., Minin, E. D., Durán, A. P., Gassert, F., Hinsley, A., Jaffer, S., Jones, J. P. G., Li, B. V., Aodha, O. M., Madhavapeddy, A., O'Donnell, S., Oxbury, W. M., … Sutherland, W. J. (2024). The potential for AI to revolutionize conservation: a horizon scan [Review of *The potential for AI to revolutionize conservation: a horizon scan*]. *Trends in Ecology & Evolution, 40*(2), 191. Elsevier BV. DOI: 10.1016/j.tree.2024.11.013

Riahi, Y., Saikouk, T., Gunasekaran, A., & Badraoui, I. (2021). Artificial intelligence applications in supply chain: A descriptive bibliometric analysis and future research directions. *Expert Systems with Applications*, *173*, 114702. DOI: 10.1016/j.eswa.2021.114702

Ricart, S., Gandolfi, C., & Castelletti, A. (2023). Climate change awareness, perceived impacts, and adaptation from farmers' experience and behavior: A triple-loop review. *Regional Environmental Change*, *23*(3), 82. Advance online publication. DOI: 10.1007/s10113-023-02078-3

Roberts, H., Zhang, J., Bariach, B., Cowls, J., Gilburt, B., Juneja, P., Tsamados, A., Ziosi, M., Taddeo, M., & Floridi, L. (2024). Artificial intelligence in support of the circular economy: Ethical considerations and a path forward. *AI & Society*, *39*(3), 1405–1418. DOI: 10.1007/s00146-022-01596-8

Rolnick, D., Donti, P. L., Kaack, L. H., Kochanski, K., Lacoste, A., Sankaran, K., & Bengio, Y. (2019). Tackling climate change with machine learning. *ACM Computing Surveys*, *55*(2), 1–96. DOI: 10.1145/3485128

Romera, A., Sharifi, M., & Charters, S. (2024). Digitalization in agriculture. Towards an integrative approach. *Computers and Electronics in Agriculture*, *219*, 108817. DOI: 10.1016/j.compag.2024.108817

Rotz, S., Gravely, E., Mosby, I., Duncan, E., Finnis, E., Horgan, M., LeBlanc, J., Martin, R., Neufeld, H. T., Nixon, A., Pant, L., Shalla, V., & Fraser, E. (2019). Automated pastures and the digital divide: How agricultural technologies are shaping labour and rural communities. *Journal of Rural Studies*, *68*, 112–122. DOI: 10.1016/j.jrurstud.2019.01.023

Roy, T., Garza-Reyes, J. A., Kumar, V., Kumar, A., & Agrawal, R. (2022). Redesigning traditional linear supply chains into circular supply chains: A study into its challenges. *Sustainable Production and Consumption*, *31*, 113–126. DOI: 10.1016/j.spc.2022.02.004

Ruiz, I., Pompeu, J., Ruano, A., Franco, P., Balbi, S., & Sanz, M. J. (2023). Combined artificial intelligence, sustainable land management, and stakeholder engagement for integrated landscape management in Mediterranean watersheds. *Environmental Science & Policy*, *145*, 217–227. DOI: 10.1016/j.envsci.2023.04.011

Russell, S., & Norvig, P. (2021). *Artificial intelligence: A modern approach* (4th ed.). Pearson.

Ryu, J., Cha, D., & Oh, J. (2020). Development of irrigation water management system using deep reinforcement learning. *Applied Sciences (Basel, Switzerland)*, *10*(22), 8085. Advance online publication. DOI: 10.3390/app10228085

Saha, S., Kucher, O. D., Utkina, A. O., & Rebouh, N. Y. (2025). Precision agriculture for improving crop yield predictions: A literature review. *Frontiers in Agronomy*, *7*, 1566201. DOI: 10.3389/fagro.2025.1566201

Saheb, T., Dehghani, M., & Saheb, T. (2022). Artificial intelligence for sustainable energy: A contextual topic modeling and content analysis. *Sustainable Computing : Informatics and Systems*, *35*, 100699. DOI: 10.1016/j.suscom.2022.100699

Saikanth, K., Singh, B. V., Sachan, D. S., & Singh, B. (2023). Advancing sustainable agriculture: A comprehensive review for optimizing food production and environmental conservation. *International Journal of Plant and Soil Science*, *35*(16), 417–425. DOI: 10.9734/ijpss/2023/v35i163169

Saiz-Rubio, V., & Rovira-Más, F. (2020). From Smart Farming towards Agriculture 5.0: A Review on Crop Data Management. *Agronomy (Basel)*, *10*(2), 207. DOI: 10.3390/agronomy10020207

Sánchez, J. M., Rodríguez, J. P., & Espitia, H. E. (2022). Bibliometric analysis of publications discussing the use of the artificial intelligence technique agent-based models in sustainable agriculture. *Heliyon*, *8*(12), e12005. Advance online publication. DOI: 10.1016/j.heliyon.2022.e12005 PMID: 36478806

Sandbrook, C. (2024). Beyond the Hype: Navigating the Conservation Implications of Artificial Intelligence. *Conservation Letters*, *18*(1), e13076. Advance online publication. DOI: 10.1111/conl.13076

Sanders, C. E., Mayfield-Smith, K. A., & Lamm, A. J. (2021). Exploring Twitter discourse around the use of artificial intelligence to advance agricultural sustainability. *Sustainability (Basel)*, *13*(21), 12033. DOI: 10.3390/su132112033

Sang, L. T., Ravikumar, R. N., Aarthi, S., & Ben, T. L. (2025). Optimizing sales and inventory performance using AI and business intelligence in POS environments. In *Advances in Computational Intelligence and Robotics* (pp. 259–290). IGI Global Scientific Publishing. DOI: 10.4018/979-8-3373-4392-1.ch011

Santos, M. R. C., Carvalho, L. C., & Francisco, E. (2025). A Capability-Based framework for Knowledge-Driven AI innovation and sustainability. *Information (Basel)*, *16*(11), 987. DOI: 10.3390/info16110987

Sarkis, J., Zhu, Q., & Lai, K. (2011). An organizational theoretic review of green supply chain management literature. *International Journal of Production Economics*, *130*(1), 1–15. DOI: 10.1016/j.ijpe.2010.11.010

Sauvagerd, M., Mayer, M., & Hartmann, M. (2024). Digital platforms in the agricultural sector: Dynamics of oligopolistic platformisation. *Big Data & Society*, *11*(4), 20539517241306365. Advance online publication. DOI: 10.1177/20539517241306365

Savec, V. F., & Jedrinović, S. (2024). The Role of AI Implementation in Higher Education in Achieving the Sustainable Development Goals: A Case Study from Slovenia. *Sustainability (Basel)*, *17*(1), 183. DOI: 10.3390/su17010183

Schmidhuber, J. (2015). Deep learning in neural networks: An overview. *Neural Networks : The Official Journal of the International Neural Network Society*, *61*, 85–117. DOI: 10.1016/j.neunet.2014.09.003 PMID: 25462637

Schwaeke, J., Gerlich, C., Nguyen, H. L., Kanbach, D. K., & Gast, J. (2025). Artificial intelligence (AI) for good? Enabling organizational change towards sustainability. *Review of Managerial Science*, *19*(10), 3013–3038. DOI: 10.1007/s11846-025-00840-x

Screpnik, C. R., Zamudio, E., & Gimenez, L. I. (2025). Artificial Intelligence in Agriculture: A Systematic Review of crop yield Prediction and Optimization. *IEEE Access : Practical Innovations, Open Solutions*, *13*, 70691–70697. DOI: 10.1109/ACCESS.2025.3560631

Sebihi, A., Schoelen, L., & Uwamwezi, B. (2025). Unveiling AI's impact in Rwandan higher education: Navigating trends, challenges, and imperatives at the University of Rwanda. *Discover Education*, *4*(1), 166. Advance online publication. DOI: 10.1007/s44217-025-00576-6

Sehgal, D., Singh, R., & Dwivedi, S. (2021). Deep learning-based hybrid model for environmental forecasting: A review. *Journal of Environmental Informatics*, *37*(2), 89–104. DOI: 10.3808/jei.202100456

Selvakumar, A., & Natarajan, R. (2026). Neuro-environmental AI orchestration for climate-adaptive cross-border logistics systems. In *Advances in Computational Intelligence and Robotics* (pp. 267–298). IGI Global Scientific Publishing. DOI: 10.4018/979-8-3373-7847-3.ch010

Senoo, E. E. K., Anggraini, L., Kumi, J. A., Karolina, L. B., Akansah, E., Sulyman, H. A., Mendonça, I., & Aritsugi, M. (2024). IoT solutions with artificial intelligence technologies for precision agriculture: Definitions, applications, challenges, and opportunities. *Electronics (Basel)*, *13*(10), 1894. DOI: 10.3390/electronics13101894

Seuring, S., & Müller, M. (2008). From a literature review to a conceptual framework for sustainable supply chain management. *Journal of Cleaner Production*, *16*(15), 1699–1710. DOI: 10.1016/j.jclepro.2008.04.020

Shah, E. A., Raiz, A., & Rashid, M. (2024). *Ethical Considerations in Use of Artificial Intelligence in Digital Marketing*. https://doi.org/DOI: 10.36968/JPDC-V08-I02-25

Shahzadi, G., Jia, F., Chen, L., & John, A. (2024). AI adoption in supply chain management: A systematic literature review. *Journal of Manufacturing Technology Management*, *35*(6), 1125–1150. DOI: 10.1108/JMTM-09-2023-0431

Shaikh, T. A., Mir, W. A., Rasool, T., & Sofi, S. (2022). Machine Learning for smart agriculture and Precision Farming: Towards making the fields talk. *Archives of Computational Methods in Engineering*, *29*(7), 4557–4597. DOI: 10.1007/s11831-022-09761-4

Shang, M., & Xie, J. (2024). Agricultural sustainable development: Soil, water resources, biodiversity, climate change, and technological innovation. *Advances in Resources Research*, *4*(2), 181–204.

Sharma, A., Jain, A., Gupta, P., & Chowdary, V. (2021). Machine learning applications for precision agriculture. *Agricultural Water Management*, *241*, 106–123. DOI: 10.1016/j.agwat.2020.106547

Sharma, K., Verma, N., Batra, N., & Bhatia, S. (2025). A strategic management framework for sustainable machine learning with Eco-AI and digital innovation. In *Enhancing Operational Efficiency and Predictive Maintenance Through Digital Innovation* (pp. 1–14). IGI Global Scientific Publishing., DOI: 10.4018/979-8-3373-2474-6.ch001

Sharma, R., Shishodia, A., Gunasekaran, A., Min, H., & Munim, Z. H. (2022). The role of artificial intelligence in supply chain management: Mapping the territory. *International Journal of Production Research*, *60*(24), 7527–7550. DOI: 10.1080/00207543.2022.2029611

Sharma, S., Jeet, V., Misha, S., & Pansari, J. K. (2023). Prediction of forest fire using deep learning-based approaches. *Computers & Electrical Engineering*, *108*, 108719. DOI: 10.1016/j.compeleceng.2023.108719

Shekarian, E., Das, S., & Krishnan, R. (2021). Analyzing the structure of closed-loop supply chains: A game theoretic approach. *Sustainability (Basel)*, *13*(3), 1397. DOI: 10.3390/su13031397

Shen, C. (2018). A transdisciplinary review of deep learning research and its relevance for water resources scientists. *Water Resources Research*, *54*(11), 8558–8593. DOI: 10.1029/2018WR022643

Shepherd, M., Turner, J. A., Small, B., & Wheeler, D. (2018). Priorities for science to overcome hurdles thwarting the full promise of the 'digital agriculture' revolution. *Journal of the Science of Food and Agriculture*, *100*(14), 5083–5092. DOI: 10.1002/jsfa.9346 PMID: 30191570

Shim, J. P., Warkentin, M., Courtney, J. F., Power, D. J., Sharda, R., & Carlsson, C. (2002). Past, present, and future of decision support technology. *Decision Support Systems*, *33*(2), 111–126. DOI: 10.1016/S0167-9236(01)00139-7

Shin, D. (2021). The effects of explainability and causability on perception, trust, and acceptance: Implications for explainable AI. *International Journal of Human-Computer Studies*, *146*, 102551. DOI: 10.1016/j.ijhcs.2020.102551

Shivaprakash, K. N., Swami, N., Mysorekar, S., Arora, R., Gangadharan, A., Vohra, K., Jadeyegowda, M., & Kiesecker, J. M. (2022). Potential for artificial intelligence (AI) and machine learning (ML) applications in biodiversity conservation, managing forests, and related services in India. *Sustainability (Basel)*, *14*(12), 7154. DOI: 10.3390/su14127154

Shreha, S. A., Ravikumar, R. N., & Aarthi, S. (2025). Neuro-symbolic AI and Behavioral Intelligence. In *Advances in Computational Intelligence and Robotics* (pp. 121–152). IGI Global Scientific Publishing. DOI: 10.4018/979-8-3373-5012-7.ch005

Shuliang, Z., Brentuo, E., Horsey, E. M., & Osei, B. A. (2025). Farmers' perceptions and government-farmer value co-creation: The roles of artificial intelligence adoption and government support. *International Journal of Innovation Studies*. https://doi.org/DOI: 10.1016/j.ijis.2025.10.004

Siano, P. (2014). Demand response and smart grids—A survey. *Renewable & Sustainable Energy Reviews*, *30*, 461–478. DOI: 10.1016/j.rser.2013.10.022

Siirila-Woodburn, E., Gerlein-Safdi, C., Mueller, J., Dennedy-Frank, J., Negron-Juarez, R., & (OSTI), O. of S. and T. I. (2021). *AI predicted shifts in watershed hydrodynamics driven by extreme weather and fire*. Atmospheric Radiation Measurement (ARM) Archive. DOI: 10.2172/1769660

Silva, T. C. C., & Hämäläinen, M. (2024). Innovating for the Future: AI and HRM Capabilities for Sustainability in Higher Education. *Academy of Management Proceedings, 2024*(1). DOI: 10.5465/AMPROC.2024.19249abstract

Silva, T. H. H., & Sehnem, S. (2022). Industry 4.0 and the circular economy: Integration opportunities generated by startups. *Logistics (Basel)*, *6*(1), 14. DOI: 10.3390/logistics6010014

Silvestro, D., Goria, S., Sterner, T., & Antonelli, A. (2022). Improving biodiversity protection through artificial intelligence. *Nature Sustainability*, *5*(5), 415–424. DOI: 10.1038/s41893-022-00851-6 PMID: 35614933

Singhal, N., Vardhan, H., Jain, R., Vashistha, P., Pandey, A., Wagri, N. K., & Gaur, A. (2025). Algorithms for nature: Integrating technology, ecology, and society for sustainable conservation. *Environmental Systems Research*, *14*(1), 30. Advance online publication. DOI: 10.1186/s40068-025-00431-5

Singh, S., & Jain, P. (2022). Applications of artificial intelligence for the development of sustainable agriculture. In *Agro-biodiversity and Agri-ecosystem Management* (pp. 303–322). Springer Nature Singapore., DOI: 10.1007/978-981-19-0928-3_16

Sinha, G. (2024). Artificial intelligence in water management for sustainable farming: A review. *Journal of Scientific Research and Reports*, *30*(6), 511–525. DOI: 10.9734/jsrr/2024/v30i62068

Slimani, S., Omri, A., & Jabeur, S. B. (2025). When and how does artificial intelligence impact environmental performance? *Energy Economics*, *148*, 108643. DOI: 10.1016/j.eneco.2025.108643

Sreeram, R., Krishna, S. A., Kumar, A. S., Remya, S., & Cho, Y. Y. (2025). Soil moisture monitoring technologies in smart agriculture: A comprehensive review. *Farming System*, *4*(2), 100189. DOI: 10.1016/j.farsys.2025.100189

Srivastava, S., Shardeo, V., Dwivedi, A., & Paul, S. K. (2026). Analyzing enablers of artificial intelligence for decarbonization: Implications for circular supply chains. *Annals of Operations Research*, *359*(1), 953–980. DOI: 10.1007/s10479-025-06843-x

Stellinga, L., Korenhof, P., & Blok, V. (2025). Bio-Centered Artificial Intelligence: Towards a Progressive AI-biosphere Relation Through the Concepts of Poiesis and Mimesis. *Topoi*. Advance online publication. DOI: 10.1007/s11245-025-10270-9

Stephens, S. L., Agee, J. K., Fulé, P. Z., North, M. P., Romme, W. H., Swetnam, T. W., & Turner, M. G. (2013). Managing forests and fire in changing climates. *Science*, *342*(6154), 41–42. DOI: 10.1126/science.1240294 PMID: 24092714

Strubell, E., Ganesh, A., & McCallum, A. (2019, July). Energy and policy considerations for deep learning in NLP. In *Proceedings of the 57th annual meeting of the association for computational linguistics* (pp. 3645-3650). DOI: 10.18653/v1/P19-1355

Sutton, R. S., & Barto, A. G. (2018). *Reinforcement learning: An introduction* (2nd ed.). MIT Press., https://mitpress.mit.edu/9780262039246/reinforcement-learning/

Tabak, M. A., Norouzzadeh, M. S., Wolfson, D. W., Sweeney, S. J., VerCauteren, K. C., Snow, N. P., Halseth, J. M., Di Salvo, P. A., Lewis, J. S., White, M. D., Teton, B., Beasley, J. C., Schlichting, P. E., Boughton, R. K., Wight, B., Newkirk, E. S., Ivan, J. S., Odell, E. A., Brook, R. K., & Miller, R. S. (2019). Machine learning to classify animal species in camera trap images: Applications in ecology. *Methods in Ecology and Evolution*, *10*(4), 585–590. DOI: 10.1111/2041-210X.13120

Tan, P. H. P., Wijaya, S., Rahardja, U., Henry, B. N., & Anjani, A. (2025). Modeling the impact of digital literacy on ai based learning adoption through perceived usefulness and easeof use. *Sundara Advanced Research on Artificial Intelligence*, *1*(2), 56–64.

Teng, M. J., Wu, S. Y., & Chou, S. J. H. (2014). Environmental commitment and economic performance–short-term pain for long-term gain. *Environmental Policy and Governance*, *24*(1), 16–27. DOI: 10.1002/eet.1634

Tian, Y., Yu, J., & Gu, J. (2024). Overview of digital twins in mining equipment. In *Advances in Transdisciplinary Engineering*. IOS Press., DOI: 10.3233/ATDE240242

Toorajipour, R., Sohrabpour, V., Nazarpour, A., Oghazi, P., & Fischl, M. (2021). Artificial intelligence in supply chain management. *Journal of Business Research*, *122*, 502–517. DOI: 10.1016/j.jbusres.2020.09.009

Twabu, K. (2025). Investigating schumpeter's innovation theory in the context of AI in higher education research. *Discover Education*, *4*(1), 389. Advance online publication. DOI: 10.1007/s44217-025-00855-2

Ukoba, K., Onisuru, O. R., Jen, T., Madyira, D. M., & Olatunji, K. O. (2025). Predictive modeling of climate change impacts using Artificial Intelligence: A review for equitable governance and sustainable outcome. *Environmental Science and Pollution Research International*, *32*(17), 10705–10724. DOI: 10.1007/s11356-025-36356-w PMID: 40183866

Ullah, F., Saqib, S., & Xiong, Y. (2024). Integrating artificial intelligence in biodiversity conservation: Bridging classical and modern approaches. *Biodiversity and Conservation*, *34*(1), 45–65. DOI: 10.1007/s10531-024-02977-9

UNDRR. (2017). Terminology on disaster risk reduction. United Nations Office for Disaster Risk Reduction. https://www.undrr.org/terminology

Upadhyay, A., Mukhuty, S., Kumar, V., & Kazancoglu, Y. (2021). Blockchain technology and the circular economy: Implications for sustainability and social responsibility. *Journal of Cleaner Production*, *293*, 126130. DOI: 10.1016/j.jclepro.2021.126130

Upadhyay, A., Yadav, A., Aggarwal, G., & Maan, V. (2024). Comparative analysis of sustainable AI techniques used to reduce the Carbon Footprints in the healthcare sector. *International Journal of Scientific Research in Engineering and Management*, *08*(11), 1–7. DOI: 10.55041/IJSREM38583

Uztürk, D., & Büyüközkan, G. (2024). Industry 4.0 technologies in Smart Agriculture: A review and a Technology Assessment Model proposition. *Technological Forecasting and Social Change*, *208*, 123640. DOI: 10.1016/j.techfore.2024.123640

van der Werf, G. R., Randerson, J. T., Giglio, L., van Leeuwen, T. T., Chen, Y., Rogers, B. M., Mu, M., van Marle, M. J. E., Morton, D. C., Collatz, G. J., Yokelson, R. J., & Kasibhatla, P. S. (2017). Global fire emissions estimates during 1997–2016. *Earth System Science Data*, *9*(2), 697–720. DOI: 10.5194/essd-9-697-2017

Van Dijk, J. (2020). *The digital divide*. John Wiley & Sons.

Van Klompenburg, T., Kassahun, A., & Catal, C. (2020). Crop yield prediction using machine learning: A systematic literature review. *Computers and Electronics in Agriculture*, *177*, 105709. DOI: 10.1016/j.compag.2020.105709

Vapnik, V. N. (1995). *The nature of statistical learning theory*. Springer., DOI: 10.1007/978-1-4757-2440-0

Varghese, A. (2022). AI-Driven Solutions for Energy Optimization and Environmental Conservation in Digital Business Environments. *Asia Pacific Journal of Energy and Environment*, *9*(1), 49–60. DOI: 10.18034/apjee.v9i1.736

Vemana, V. C. (2025). Digital twins in remote monitoring and telehealth: Bridging the gap with Ai. In *Digital Twins: Building Trust and Security in the Digital Age (Unlocking Innovation in Education, Humanities, Management, and Sciences)* (pp. 20–34). Iterative International Publishers (IIP), Selfypage Developers Pvt Ltd. DOI: 10.58532/nbennurDTBCSW2

Venkatesh, V., & Davis, F. D. (2000). A theoretical extension of the technology acceptance model: Four longitudinal field studies. *Management Science*, *46*(2), 186–204. DOI: 10.1287/mnsc.46.2.186.11926

Vicedo-Cabrera, A. M., Sera, F., Liu, C., Armstrong, B., Milojevic, A., Guo, Y., ... & Gasparrini, A. (2021). Short term association between ozone and mortality: global two stage time series study in 406 locations in 20 countries. bmj, 368.

Vijayakumar, S., Murugaiyan, V., Ilakkiya, S., Kumar, V., Sundaram, R. M., & Kumar, R. M. (2025). Opportunities, challenges, and interventions for agriculture 4.0 adoption. *Discover Food*, *5*(1), 265. Advance online publication. DOI: 10.1007/s44187-025-00576-3

Vilar, L., Camia, A., & San-Miguel-Ayanz, J. (2016). A comparison of remote sensing products and forest fire statistics for improving fire information in Mediterranean Europe. *European Journal of Remote Sensing*, *49*(1), 185–201.

Vinoth, S., Chinnasamy, G., & Madbouly, A. (2025). Evaluating the integration and impact of artificial intelligence (AI) tools on academic learning, assessment, and research practices in higher education. *Education and Information Technologies*, *30*(18), 25655–25682. DOI: 10.1007/s10639-025-13719-x

Vinuesa, R., Azizpour, H., Leite, I., Balaam, M., Dignum, V., Domisch, S., & Fuso Nerini, F. (2020). The role of artificial intelligence in achieving the Sustainable Development Goals. *Nature Communications*, *11*(1), 233. DOI: 10.1038/s41467-019-14108-y PMID: 31932590

Voulgaridis, K., Lagkas, T., Angelopoulos, C. M., & Nikoletseas, S. E. (2022). IoT and digital circular economy: Principles, applications, and challenges. *Computer Networks*, *219*, 109456. DOI: 10.1016/j.comnet.2022.109456

Wainaina, P., & Sun, Y. (2025). Educators' perceptions and willingness to integrate Generative Artificial Intelligence in teaching and research: Evidence from Kenyan higher education. *Discover Education*, *4*(1), 347. Advance online publication. DOI: 10.1007/s44217-025-00820-z

Wang, R., Sun, Y., Zong, J., Wang, Y., Cao, X., Wang, Y., Cheng, X., & Zhang, W. (2024). Remote sensing application in ecological restoration monitoring: A systematic review. *Remote Sensing (Basel)*, *16*(12), 2204. DOI: 10.3390/rs16122204

Wang, T., Zuo, Y., Manda, T., Hwarari, D., & Yang, L. (2025). Harnessing artificial intelligence, machine learning and deep learning for sustainable forestry management and conservation: Transformative potential and future perspectives. *Plants*, *14*(7), 998. DOI: 10.3390/plants14070998 PMID: 40219066

Wang, Y., Qi, Y., Li, J., Huan, L., Li, Y., Xie, B., & Wang, Y. (2023). The wind and photovoltaic power forecasting method based on digital twins. *Applied Sciences (Basel, Switzerland)*, *13*(14), 8374. DOI: 10.3390/app13148374

Wani, A. K., Rahayu, F., Ben Amor, I., Quadir, M., Murianingrum, M., Parnidi, P., Ayub, A., Supriyadi, S., Sakiroh, S., Saefudin, S., Kumar, A., & Latifah, E. (2024). Environmental resilience through artificial intelligence: Innovations in monitoring and management. *Environmental Science and Pollution Research International*, *31*(12), 18379–18395. DOI: 10.1007/s11356-024-32404-z PMID: 38358626

Whig, P., Gupta, S. K., Nadikattu, R. R., & Sharma, P. (2025). Application of AI in Environmental Sustainability. *Artificial Intelligence-Driven Models for Environmental Management*, 1-41.

Whittaker, J., Blanchi, R., Haynes, K., Leonard, J., & Opie, K. (2020). Experiences of sheltering during the Black Saturday bushfires: Implications for policy and research. *International Journal of Disaster Risk Reduction*, *23*, 119–127. DOI: 10.1016/j.ijdrr.2017.05.002

Wieczorek, J., Bloom, D., Guralnick, R., Blum, S., Döring, M., Giovanni, R., Robertson, T., & Vieglais, D. (2012). Darwin Core: An evolving community-developed biodiversity data standard. *PLoS One*, *7*(1), e29715. Advance online publication. DOI: 10.1371/journal.pone.0029715 PMID: 22238640

Wilkinson, M. D., Dumontier, M., Aalbersberg, I. J., Appleton, G., Axton, M., Baak, A., Blomberg, N., Boiten, J.-W., da Silva Santos, L. B., Bourne, P. E., Bouwman, J., Brookes, A. J., Clark, T., Crosas, M., Dillo, I., Dumon, O., Edmunds, S., Evelo, C. T., Finkers, R., & Mons, B. (2016). The FAIR Guiding Principles for scientific data management and stewardship. *Scientific Data*, *3*(1), 160018. Advance online publication. DOI: 10.1038/sdata.2016.18 PMID: 26978244

Wirtz, B. W., Weyerer, J. C., & Geyer, C. (2019). Artificial intelligence and the public sector—Applications and challenges. *Public Management Review*, *21*(4), 596–615.

Wolfert, S., Ge, L., Verdouw, C., & Bogaardt, M. J. (2017). Big data in smart farming: A review. *Agricultural Systems*, *153*, 69–80. DOI: 10.1016/j.agsy.2017.01.023

Wong, J., Hoong, P., Teo, E., & Lin, A. (2022). Digital twin: A conceptualization of the task-technology fit for individual users in the building maintenance sector. *IOP Conference Series. Earth and Environmental Science*, *1101*(9), 092041. DOI: 10.1088/1755-1315/1101/9/092041

Wukich, C. (2016). Social media use in emergency management. *Journal of Emergency Management (Weston, Mass.)*, *14*(4), 271–285. DOI: 10.5055/jem.2016.0263 PMID: 26312654

Wu, P., & Zhong, Y. (2025). Artificial intelligence in sustainable agriculture: Towards a socio-technical roadmap. *Smart Agricultural Technology*, *12*, 101578. DOI: 10.1016/j.atech.2025.101578

Wynn, M., & Jones, P. (2022). Digital technology deployment and the circular economy. *Sustainability (Basel)*, *14*(15), 9077. DOI: 10.3390/su14159077

Xia, H., Chen, Z., Milisavljevic-Syed, J., & Salonitis, K. (2024). Uncertain programming model for designing multi-objective reverse logistics networks. *Cleaner Logistics and Supply Chain*, *12*, 100155. DOI: 10.1016/j.clscn.2024.100155

Xiang, X., Li, Q., Khan, S., & Khalaf, O. I. (2021). Urban water resource management for sustainable environment planning using artificial intelligence techniques. *Environmental Impact Assessment Review*, *86*, 106515. DOI: 10.1016/j.eiar.2020.106515

Xing, Y., & Wang, X. (2024). Precision agriculture and water conservation strategies for sustainable crop production in arid regions. *Plants*, *13*(22), 3184. DOI: 10.3390/plants13223184 PMID: 39599396

Xu, R., Yu, P., Abramson, M. J., Johnston, F. H., Samet, J. M., Bell, M. L., Haines, A., Ebi, K. L., Li, S., & Guo, Y. (2020). Wildfires, global climate change, and human health. *The New England Journal of Medicine*, *383*(22), 2173–2181. DOI: 10.1056/NEJMsr2028985 PMID: 33034960

Yadav, G., Mangla, S. K., Bhattacharya, A., & Luthra, S. (2020). Exploring indicators of circular economy adoption framework through a hybrid decision support approach. *Journal of Cleaner Production*, *277*, 124186. DOI: 10.1016/j.jclepro.2020.124186

Yadav, M., & Singh, G. (2023). Environmental sustainability with artificial intelligence. [IJMR]. *EPRA International Journal of Multidisciplinary Research*, *9*(5), 213–217.

Yang, Y., Chen, L., Diao, Z., Gao, P., Zhang, B., & Zhao, C. (2026). Recent advances in crop pest detection, forecasting and early warning: A review. *Artificial Intelligence in Agriculture*. https://doi.org/DOI: 10.1016/j.aiia.2026.03.009

Yang, J., Tian, H., Tao, B., Ren, W., Kush, J., Liu, Y., & Wang, Y. (2014). Rare and extreme events: The case of Canadian forests. *Environmental Research Letters*, *9*(7), 075005.

Yang, M., Fu, M., & Zhang, Z. (2021). The adoption of digital technologies in supply chains: Drivers, process and impact. *Technological Forecasting and Social Change*, *169*, 120795. DOI: 10.1016/j.techfore.2021.120795

Ye, G., Wang, Z., Yin, X., Tang, B., & Shen, H. (2022). A robust anomaly detection algorithm for water quality data based on autoencoder. *Journal of Hydrology (Amsterdam), 609*, 127686. Advance online publication. DOI: 10.1016/j.jhydrol.2022.127686

Yi, S., & Yi, J. (2024). Reservoir-based flood forecasting and warning: Deep learning versus machine learning. *Applied Water Science, 14*(11), 237. DOI: 10.1007/s13201-024-02298-w

Yousaf, H. (2024). AI for environmental sustainability: Applications in resource management and conservation. *Journal of AI Range, 1*(2), 13–26.

Yuan, S., & Pan, X. (2023). The effects of digital technology application and supply chain management on corporate circular economy: A dynamic capability view. *Journal of Environmental Management, 341*, 118082. DOI: 10.1016/j.jenvman.2023.118082 PMID: 37141715

Yue, Y., Wang, L., Brandt, M., Zhang, X., & Wang, K. (2024). A social-ecological framework to enhance sustainable reforestation under geological constraints. *Earth's Future, 12*(5), e2023EF004335. Advance online publication. DOI: 10.1029/2023EF004335

Zakarneh, B., Aljabr, F., Mahmoud, D. A. M., Annamalai, A. N., Chakraoui, R., & Yakoob, A. (2025). Impact of Sustainability Awareness on AI Perception, Engagement, and Motivation among University-Level Literature Students. *World Journal of English Language, 16*(2), 295. DOI: 10.5430/wjel.v16n2p295

Zehouani, N., Ababou, M., & Faquir, S. (2024, May). Harnessing artificial intelligence for environmental sustainability: a comprehensive review of its applications in biodiversity, energy, transportation, and water management. In International Conference on Digital Technologies and Applications (pp. 410-419). Cham: Springer Nature Switzerland. DOI: 10.1007/978-3-031-68650-4_39

Zhai, Z., Martínez, J. F., Beltran, V., & Martínez, N. L. (2020). Decision support systems for agriculture 4.0: Survey and challenges. *Computers and Electronics in Agriculture, 170*, 105256. DOI: 10.1016/j.compag.2020.105256

Zhang, C., & Kovacs, J. M. (2012). The application of small unmanned aerial systems for precision agriculture. *Precision Agriculture, 13*(6), 693–712. DOI: 10.1007/s11119-012-9274-5

Zhang, G., Wang, M., & Liu, K. (2019). Forest fire susceptibility modeling using a convolutional neural network for Yunnan province of China. *International Journal of Disaster Risk Science, 10*(3), 386–403. DOI: 10.1007/s13753-019-00233-1

Zhang, P., Guo, Z., Ullah, S., Melagraki, G., Afantitis, A., & Lynch, I. (2021). Nanotechnology and artificial intelligence to enable sustainable and precision agriculture. *Nature Plants*, *7*(7), 864–876. DOI: 10.1038/s41477-021-00946-6 PMID: 34168318

Zhang, P., Sutley, E. J., & Hamideh, S. (2023). Wildfire risk and social vulnerability: Mapping compound risks in rural communities of the American West. *International Journal of Disaster Risk Reduction*, *88*, 103614.

Zhang, Y., Wang, G., & Li, Y. (2022). Artificial intelligence in smart agriculture: A systematic review. *Computers and Electronics in Agriculture*, *195*, 106800. DOI: 10.1016/j.compag.2022.106800

Zhu, X. X., Tuia, D., Mou, L., Xia, G. S., Zhang, L., Xu, F., & Fraundorfer, F. (2017). Deep learning in remote sensing: A comprehensive review and list of resources. *IEEE Geoscience and Remote Sensing Magazine*, *5*(4), 8–36. DOI: 10.1109/MGRS.2017.2762307

Ziemba, E., Duong, C. D., Ejdys, J., González-Pérez, M. A., Kazlauskaitė, R., Korzyński, P., Mazurek, G., Paliszkiewicz, J., Stankevičienė, J., & Wach, K. (2024). Leveraging artificial intelligence to meet the sustainable development goals. *Journal of Economics and Management*, *46*, 508–583. DOI: 10.22367/jem.2024.46.19

About the Contributors

Mohieddine Rahmouni is an Associate Professor of Economics and Quantitative Methods at King Faisal University and the University of Tunis, specializing in the econometrics of innovation, industrial dynamics, and technological change. He holds a joint Ph.D. from the University of Bordeaux and the University of Tunis and has built an international academic career spanning France, Tunisia, and Saudi Arabia. His research has been published in well-established peer-reviewed journals such as Structural Change and Economic Dynamics, the International Journal of Innovation and Technology Management, and the International Journal of Technological Learning, Innovation and Development, as well as leading journals in education economics. His work focuses on the relationships between R&D intensity, firm productivity, institutional quality, and sustainable development, with recent contributions addressing climate-related challenges, including the economic effects of climate shocks, decarbonization, and the role of emerging technologies such as AI in green growth and resilience in emerging economies.

***

Senthil Kumar A.V. is working as Principal, Nehru Institute of Information Technology and Management, Coimbatore, India. He has worked as Professor and Director, PG and Research Department of Computer Applications, Hindusthan College of Arts & Science, Coimbatore, Tamilnadu for 15 years and as Senior Grade Lecturer in CMS College of Science and Commerce for 14 years. To his credit he has industrial experience for five years and teaching experience of 29 years. He has also received his Doctor of Science (D.Sc in Computer Science). He has to his credit 86 Book Chapters, 234 papers in International and National Journals, 85 papers in International Conferences in International and National Conferences, and edited 17 books and 3 Text books. He is as Associate Editor of IEEE Access. He is an Editor-in-Chief for many journals and Key Member for India, Machine

Intelligence Research Lab (MIR Labs). He is an Editorial Board Member and Reviewer for various International Journals. He is also a Committee member for various International Conferences. He has guided 16 Ph.D scholars and guiding 3 Ph. D scholars now.

Ioannis Adamopoulos, are: Associated Teaching, Coordinator, and Research Staff at Hellenic Open University, School of Social Science, of MSc Program, Public Health & Policies. Postdoc Researcher in Department of Public Health Policy, Sector of Occupational & Environmental Health, School of Public Health, University of West Attica. Public Health and Environmental Inspector at the Region of Attica, Department of Environmental Hygiene and Public Health Inspections. Member and scientific associate researcher at the Global Burden Diseases GBD Collaborator Network, which contributes significantly to the dynamic community of global experts in the field of public health at the Institute for Health Metrics and Evaluation, Research Center for Population Health (IHME) at the University of Washington. Member of the International Association: ICOH, EUPHA, APA, UK & Ireland Occupational and Environmental Epidemiology Society, MEDICHEM, and Humanitarian Network of Experts. STUDIES (PhD, MPH, BSc in Public Occupational and Environmental Health). Specialist on: Statistical and meta-analysis, systematic reviews, and network meta-analysis research. S P S S - R E S E A R C H. Working as an Inspector of Public Health and Safety in the Ministry of Health & Unified Food Control Agency (EFET) & Prefectures local governance in Greece until now in the Region of Attica. Assigned registered main teaching staff member of the National School of Public Administration. Editor in: BMC Public Health journal, Springer Nature

Thompson Xavier Ananth B.B.A., M.S.W., M.Phil., Ph.D. President - University Council DMI St. Joseph University In Tanzania, Dar es Salaam, Tanzania, txaa@sjuit.ac.tz

K. Aravinthan is an accomplished academician serving as an Associate Professor in the Department of Master of Computer Applications (MCA) at Anjalai Ammal Mahalingam Engineering College, Kovilvenni, Thiruvarur District, Tamil Nadu. With extensive experience in teaching, research, and academic administration, he has contributed significantly to the advancement of computer science education. His areas of expertise include data analytics, artificial intelligence, software engineering, and emerging computing technologies. Dr. Aravinthan is actively involved in guiding undergraduate and postgraduate students, fostering innovation, and promoting research-oriented learning. He has authored several research papers in reputed journals and conferences and has participated in various faculty development programs, workshops, and seminars. His commitment to academic excellence and continuous learning has played a key role in enhancing the quality of education and research within the institution.

Anorgul Ashirova is a distinguished academic affiliated with the Department of General Professional Sciences at Mamun University, Khiva, Uzbekistan. She is committed to advancing higher education through quality teaching, academic mentorship, and interdisciplinary learning. Her academic role reflects a strong dedication to developing students' foundational and professional competencies essential for academic and career growth. Dr. Ashirova's interests lie in strengthening general professional education, promoting knowledge integration, and supporting innovative teaching practices that enhance student engagement and understanding. She contributes to the academic environment by encouraging critical thinking, ethical values, and lifelong learning among students. Through her scholarly and educational efforts, she supports institutional excellence and the preparation of capable graduates who can respond effectively to contemporary professional challenges. Her work continues to reflect a deep commitment to educational development and intellectual progress in Uzbekistan's higher education sector.

Aditya Bala is currently working as an Assistant Professor in the School of Education at Rayat Bahra University, Mohali, Punjab. She specializes in the Teaching of Science and holds a Master's degree in Education (M.Ed.) and a Master's degree in Botany (M.Sc.). Dr. Bala has actively contributed to the field of education through her research, with several papers published in peer-reviewed journals. Her academic interests include science education, educational technology, and innovative teaching-learning practices. She is committed to enhancing the quality of teacher education and fostering effective pedagogical approaches among future educators.

Babamuratov Bekzod is an academic professional in the Department of Medicine at Termez University of Economics and Service, Uzbekistan. He is recognized for his commitment to medical education, academic development, and the advancement of healthcare knowledge through teaching and scholarly engagement. His work focuses on supporting students in building strong theoretical understanding and practical awareness in the medical field. He is actively involved in promoting quality education and fostering a learning environment that encourages critical thinking, ethical values, and professional responsibility among future healthcare practitioners. His academic interests include medical sciences, healthcare improvement, and interdisciplinary approaches that connect education with real-world health challenges. Through his teaching and academic contributions, Dr. Babamuratov Bekzod continues to support institutional growth and the preparation of skilled and socially responsible medical graduates. His dedication to excellence in education reflects his broader commitment to strengthening healthcare knowledge and academic progress in Uzbekistan.

Monica C is a full-time Research Scholar at the School of Commerce & Management Studies, Dayananda Sagar University, Bengaluru. Her research focuses on Artificial Intelligence, Financial Inclusion, and Sustainability.

Kajul Chauhan is pursuing PhD. from Department of Geography,Panjab University Chandigarh. My research area of interest is Apple cultivation, Horticulture, Agriculture, Climate change. Currently Iam working with Rayat Bahra University as an Assistant Professor of Geography.

Nisha Goyal is an Associate Professor at the School of Commerce and Management Studies, Dayananda Sagar University, Bengaluru, with over 15 years of teaching and research experience and a PhD from MNIT Jaipur (UGC NET-JRF qualified). Her teaching includes Financial Management, M&A, International Finance, Research Methods, and Statistics, and she is skilled in SPSS, AMOS, Smart-PLS, SAS-JMP, and Advanced Excel. Her research focuses on Corporate & Behavioural Finance, FinTech, Sustainable Finance, Financial Literacy, and AI in Finance, with publications in ABDC- A listed and Scopus/WoS journals.

Piali Haldar earned her MBA from the Institute of Rural Management Anand and completed her PhD at Birla Institute of Technology, Mesra. Her core areas of specialization include rural marketing, rural development, branding strategy, and consumer behaviour. With over 20 years of experience in academia and research, she has published extensively in renowned journals listed in ABDC and Scopus. She has authored one textbook on Marketing and edited three academic books. Dr. Haldar has presented her research at numerous national and international conferences and has contributed more than 50 papers to reputed journals, edited volumes, and conference proceedings.

Rajiv Kalebar is an Associate Professor at Christ (Deemed to be University), Bangalore, I bring over 23 years of rich, multifaceted experience spanning Corporate leadership, Entrepreneurship, and Academic research. My journey bridges real-world industry expertise and innovative interdisciplinary scholarship — enabling me to deliver profound insights and impactful learning to my students and professional peers alike. Before embarking on my academic career, I gained invaluable experience working with leading organizations such as Berger Paints, Hindustan Unilever Ltd, and Hutchison Essar (Vodafone Idea Ltd), where I developed a deep understanding of corporate finance, strategic management, and operational excellence. As a cofounder of Vishal Financial School and an award-winning entrepreneur (Best B-Plan Competition, TiE-Hubli), I understand the dynamics of financial markets and the entrepreneurial mindset firsthand. In academia, my passion lies in pioneering **interdisciplinary research** at the intersection of **Finance, Behavioral Science, and Therapeutic Practices**. I am deeply involved in exploring innovative concepts like **Financial Somatics** — investigating how embodied experiences and therapies such as **Dance Movement Therapy (DMT)** can transform financial decision-making and emotional well-being. My research has been recognized with Best Paper Awards and featured by premier institutions like the Indian Institute of Management Ahmedabad (IIMA). I actively contribute to the professional community through collaborations with organizations such as PRMIA India, LIC of India, SBI Life, and IIFL, where I conduct specialized training on financial risk management and market strategies. I also engage with industry and academia via conferences like the International Conference on Business & Finance and Finance Conference 2025, where I present cutting-edge research and foster collaborative learning. Holding a Ph.D. from Mysore University and an MBA from Karnatak University, I am committed to academic excellence and continuous innovation in teaching methodologies. My approach integrates real-world relevance with emerging research trends, empowering students and professionals to navigate complex financial landscapes with confidence and insight.

Preet Kanwal is an Associate Professor at the Mittal School of Business, Lovely Professional University, with over 23 years of academic and research experience in Human Resource Management. A Post-Doctoral Fellow from the University of Quebec in Montreal, she holds a PhD in HRM from I.K. Gujral Punjab Technical University. Her research interests include quality of work life, labour laws, industrial relations, organisational behaviour, and workforce well-being. She has published widely in Scopus-indexed and UGC-listed journals, authored books and numerous book chapters, and presented at several national and international conferences. She has taught internationally for Victoria University (Online), Australia, and Sunway University (Online), Malaysia. She recently completed an ICSSR-funded project as Co-Principal Investigator on empowering women artisans for Vision Viksit Bharat@ 2047 and is currently a Co-Investigator on an ongoing ICSSR-funded project on enhancing Social-Emotional Learning (SEL) in government schools across Punjab.

Fredrick Kayusi is a Kenyan academician and researcher currently working as a Adjunct Lecturer of Geosciences (Earth Sciences) in the Department of Environmental Studies, Geography & Planning at Maasai Mara University. He also works as a Lecturer of Research Skills, Social studies, Geography and Mathematics at Narok Teachers Training University College (NTTC). He was born on April 20,

1991, in Kisii County. His specific specialization falls under Earth Sciences with major focus on geomorphology and climatology. Current Education: PhD Scholar Geosciences Pwani University (since August 2024). M.Sc. Geography (Geosciences/ Earth Sciences) Pwani University from 2021-2023. B.Ed.(Arts) in Geography and Mathematics at Masinde Muliro University of Science and Technology (MMUST) from 2011 to 2015. Professional Certifications: Advanced Diploma in Geographic Information Systems (GIS) and a certificate in GIS from Alison, Ireland. Professional and Research Interests: Fredrick has more than 60 publications as an author and/or co-author. He is also a peer reviewer for some international journals. His research interests are very broad but mostly lie within the intersection of environmental, geomorphological and climatological science, with modern technology: Geosciences: Slope stability analysis (notably in soapstone mining sites), sediment dynamics, and geohazards. Application of Artificial Intelligence (AI) and Machine Learning towards climate resilience, precision agronomy as well as healthcare governance. Sustainable transformations in post-mining regions, adaptation to climate change, plus biodiversity within urban green spaces. He enjoys recognition as a research scientist at Pwani University where he has been over five years now partaking in projects that involve remote sensing concerning coastal geomorphology and the blue economy.

Uzma Malik holds a Ph.D. in Public Administration from Aligarh Muslim University, where she also completed her post-graduation. She has qualified both the UGC-NET and the Jammu & Kashmir SET examinations. Her academic work primarily focuses on poverty and sustainable development. She has actively participated in numerous national and international conferences, presenting research papers and attending academic workshops. She has over three years of teaching experience as a Lecturer in various government colleges in Jammu and Kashmir. In addition to her academic engagements, she has contributed to newspaper writing and maintains a strong interest in scholarly and creative writing. During the COVID-19 pandemic, she was associated with an NGO, contributing to community-oriented initiatives. She has published several research papers in UGC-listed and peer-reviewed journals and has also contributed book chapters to academic publications.

Dev Kumar Mandal currently serves as an Assistant Professor in the Department of Management at Brainware University, Kolkata. Before this, he worked at ICRA Analytics Ltd and Shriram Finance Ltd. He holds an MBA in Finance and a Bachelor of Commerce degree from Vidyasagar University. Additionally, he has successfully cleared the UGC NET and WB SET examinations in Management.

Sajad Ahmad Mir Assistant Professor I Public Administration I Political Science I Police Administration I Research Methodology Introduction I am a dedicated academician and researcher with a Ph.D. in Public Administration, committed to advancing knowledge through rigorous research and scholarly publications in esteemed Scopus-indexed journals. My academic journey is enriched by active participation in international conferences, contributing valuable insights to the field. My passion for education and research has driven me to create meaningful contributions in Public Administration, Political Science, Police Administration, and Research Methodology. Academic and Professional Journey My professional journey in academia began with a strong inclination toward teaching and research. Over the years, I have refined my pedagogical approaches, integrating innovative teaching methodologies with research-driven insights to enhance student learning outcomes. My academic career has been shaped by various teaching and research experiences, which have provided me with a comprehensive understanding of higher education's dynamic landscape.

Harshit Mishra completed his Ph.D. in Agricultural Economics in 2025 from the Department of Agricultural Economics, Acharya Narendra Deva University of Agriculture and Technology, Kumarganj, Ayodhya, Uttar Pradesh. He completed his Master's degree from the same university in 2022. During his academic journey, he has been recognized with several prestigious awards, including the Research Excellence Award, Young Researcher Award, Best M.Sc. (Ag) Thesis Award, Best Oral Presentation Award, and Best Poster Presentation Award conferred by esteemed academic societies. He is the author of two books: "Fundamentals of Agri-Business Management & Intellectual Property Rights" and "Manual cum Workbook of Finance and Cooperation." In addition, he has published over 24 research and review articles in reputed NAAS-rated, Scopus-indexed, and Web of Science-indexed peer-reviewed journals. His scholarly contributions also include more than 48 popular articles published in both Hindi and English across various reputed magazines, and over 81 book chapters in edited volumes published by renowned international publishers such as Springer Nature, CRC Press, Apple Academic Press, Routledge (Taylor & Francis Group), IGI Global Scientific Publishing, Nova Science Publishers and many other reputed academic publishing houses.

Ravikumar R. Natarajan is an experienced academician with over 13+ years of teaching and research expertise across institutions in India, the UAE, and the Maldives. He currently serves as an Assistant Professor at Marwadi University, Rajkot, specializing in Machine Learning, Recommender Systems, and AI-driven applications in healthcare and security. He holds a Ph.D. in Machine Learning and Recommender Systems from Amity University, Jaipur, a Master of Engineering in

CSE from VMRF University, Salem, and a Bachelor of Engineering in CSE from Anna University. He also has a MiniMBA from the Swiss eLearning Institute. Dr. Ravikumar has authored more than 8+ multiple SCIE/Scopus-indexed journal articles, 25+ IEEE/Springer conference papers, and co-authored a book on Artificial Intelligence. He has also published 9 patents and contributed 85+ book chapters in emerging domains such as Federated Learning, Blockchain, and Digital Twins for IGI Global, Wiley and Springer publisher. He actively mentors students through initiatives like the Research Activities Club (RAC) and Student Research Internship Training (SRIT), promoting innovation and research culture.

Sameen Rafi holds a PhD in Social Work with a focus on aging and health. Her research centers on elderly care, women's well-being, and social support systems. She has published several papers in peer-reviewed journals and is committed to interdisciplinary and evidence-based social welfare research.

Rashmi Mishra is a faculty member at the University of Technology and Applied Sciences, Oman, with over 22 years of academic and industry experience. Her work focuses on economics, human resource management, and quality assurance, with publications in Scopus-indexed journals and active roles as reviewer and session chair at international conferences.

Aarthi S is a dedicated Assistant Professor in Computer Science with over six years of teaching experience. She is currently pursuing her Ph.D. at Marwadi University, Gujarat, and holds an M.Phil. and MCA from Bharathiar University, Tamil Nadu. Throughout her academic career, Aarthi has specialized in programming languages such as C++, Java, and C, and contributed to research in recommender systems and AI-related data privacy. She has published journal articles, contributed to book chapters, and actively participated in technical workshops. Known for her student mentorship and engagement in academic events, Aarthi remains committed to fostering innovation and excellence in computer science education.

Navneet Saini is an Associate Professor with more than a decade of experience in education. I am passionate about guiding students and had the honor of supervising a Ph.D. candidate at Panjab University, Chandigarh, who successfully completed her doctorate under my guidance. My work focuses on critical social issues, with a specific emphasis on gender and women's studies. I have authored three books and edited five others, all of which explore these important topics. My research has also been published in various papers in respected journals and edited books, contributing to the ongoing academic discourse. Through my writing and teaching, I aim to create meaningful dialogue and promote positive social change.

Prabhleen Saini is an Assistant Professor at the University School of Education, Rayat Bahra University, Mohali, and a Research Scholar in the Department of Education at Panjab University, Chandigarh. She is actively engaged in academic research in the field of education, with a focus on emerging issues of education. Her work reflects a strong commitment to advancing educational research and contributing to innovative teaching-learning processes. Through her academic and research pursuits, she aims to bridge theoretical insights with practical applications in the education sector.

Neeta Raj Sharma is an internationally acclaimed academic leader, researcher, and editor with nearly three decades of experience in biotechnology, environmental sustainability, and AI-integrated Smart at Lovely Professional University, India, where she leads multidisciplinary research focused on AI-enabled Nature-based solutions, sustainable urban ecosystems, decentralized wastewater treatment, and circular green innovations. Dr Sharma has successfully led and collaborated on multi-million INR national and international research projects funded by prestigious agencies including the British Council (UK), ICSSR (India), Indian Space Research Organization (ISRO), European Commission (Erasmus+), IC-IMPACTS (Indo-Canada), MITACS Canada, and leading UK and Canadian universities. She is presently leading the UK–India collaborative project "SSG-Bridge: Enhancing Urban Resilience and Sustainability," focusing on AI-integrated sustainable city solutions. Her global academic collaborations include Birmingham City University (UK), Cardiff Metropolitan University (UK), McGill University (Canada), University of British Columbia (Canada), and Sunway University (Malaysia), where she also serves as Adjunct Professor and Visiting Professor. Dr Sharma has published over 100 peer-reviewed research papers, holds 40 patents (filed, published, and granted), and 11 copyrights. She is a Fellow of the Association of Biotechnology and Pharmacy and an experienced Book Series Editor with leading international publisher: Taylor & Francis along with a strong track record of collaboration with top-tier international publishers namely Springer Nature, Wiley, IGI, Bentham and CRC Press. Her strong editorial experience, global collaborations, funded research leadership, and expertise in AI-driven sustainability and SSG technologies make her exceptionally well suited to serve as editor for this proposed volume.

Amit Verma has been a faculty, career counselor, and career counselor for Twenty-One Years. He has a Ph.D. from Ranchi University, a Master's in Business Administration from MDU University in Rohtak, and a Bachelor of Technology in Computer Science from IET in Lucknow. Additionally, he holds a Ph.D. in Management from Panama's Swahili University. He most recently served as the ICSSR Project Director for the collaborative short-term empirical research project

"Digital Transformation of Health Services: A Comparative Study of Bhopal and Lucknow." The Communication Institute of Greece, located in Athens, Greece, has nominated him as an Academic Member. Additionally, he received an invitation to speak as the chief speaker at the Media Management Seminar from Saint Augustine University of Tanzania (SAUT), located in Mwanza, Tanzania, East Africa. Dr. Verma has held positions as a Lecturer, Senior Lecturer, Assistant Professor, and Associate Professor at prestigious institutions in Delhi NCR, Uttar Pradesh, and Uttarakhand. He has been accepted as a faculty member for the M. Com. and B. Com. programs offered by Lucknow University and is currently an Assistant Professor at Kalicharan PG College, Lucknow, Uttar Pradesh, India. He teaches management information systems, advertising management, and marketing management. He also gets ready for his career and takes part in seminar activities. Dr. Verma have values based on the philosophy of Swami Vivekananda, and he attempts to impart his principles of good character and ethics during his teachings. He is the author of 48 Research Papers in peer-reviewed journals, including 10 in ABDC, SSCI, and SCOPUS-indexed journals. He has also reviewed a large number of research papers. His publications encompass edited volumes that bridge the gap between academics and industry. His publications include edited books that cater to both academic and industrial audiences.

Navita Vijay is the Assistant professor at the School of Commerce and Management Studies, Dayananda Sagar University, Bengaluru, with over 16 years of teaching and industry experience. Her area of teaching includes courses from finance, Investment Banking, Fintech, Statistics, and Strategic Management. Her research interest is in the technology adoption in general, Technology in BFSI, Fintech, AI in finance, etc.

Ulugbek Vosiqov is an academic associated with Kimyo International University in Tashkent, Uzbekistan. He is dedicated to higher education, student development, and the advancement of academic excellence through teaching, learning, and scholarly engagement. His work contributes to building a strong educational environment that supports knowledge development, professional growth, and critical thinking among students. He is interested in fostering innovative learning approaches and strengthening the connection between academic knowledge and practical application. Through his academic involvement, he supports institutional development and the preparation of skilled graduates capable of addressing modern professional and social challenges. His commitment to education reflects a broader interest in promoting quality teaching, intellectual progress, and continuous learning within the university community. As an educator, Ulugbek Vosiqov continues to contribute to the growth of higher education in Uzbekistan by encouraging discipline, responsibility, and academic achievement among learners.

Index

A

B

C

D

E

G

I

M

P

R

S

V

W

www.ingramcontent.com/pod-product-compliance
Lightning Source LLC
LaVergne TN
LVHW080912300926
849162LV00003B/653